An Introduction to Statistical Problem Solving in Geography

J. Chapman McGrew, Jr.
Salisbury State University
&
Charles B. Monroe
The University of Akron

WCB **Wm. C. Brown Publishers**
Dubuque, Iowa•Melbourne, Australia•Oxford, England

Book Team

Editor *Lynne M. Meyers*
Publishing Services Coordinator *Julie Avery Kennedy*

Wm. C. Brown Publishers
A Division of Wm. C. Brown Communications, Inc.

Vice President and General Manager *Beverly Kolz*
National Sales Manager *Vincent R. Di Blasi*
Director of Marketing *John W. Calhoun*
Marketing Manager *Amy Schmitz*
Advertising Manager *Amy Schmitz*
Director of Production *Colleen A. Yonda*
Manager of Visuals and Design *Faye M. Schilling*
Design Manager *Jac Tilton*
Art Manager *Janice Roerig*
Publishing Services Manager *Karen J. Slaght*
Permissions/Records Manager *Connie Allendorf*

Wm. C. Brown Communications, Inc.

President and Chief Executive Officer *G. Franklin Lewis*
Corporate Vice President, President of WCB Manufacturing *Roger Meyer*
Vice President and Chief Financial Officer *Robert Chesterman*

Production by Shepherd Inc.

Interior design by Shepherd Inc.

A Times Mirror Company

Library of Congress Catalog Card Number: 92–72531

ISBN 0–697–12682–X

Printed in the United States of America by Wm. C. Brown Communications, Inc., 2460 Kerper Boulevard, Dubuque, IA 52001

10   9   8   7   6   5   4   3   2   1

This text is dedicated to our patient and tolerant wives, Kathy and Laura. The innumerable hours spent working on this manuscript meant less time could be devoted to family activities and to our children, Courtney, Grady, Caroline, and Michael. The continuous unselfish support of this project by our families is deeply appreciated.

CONTENTS

PREFACE

All well-trained geographers need to be proficient in applying statistical techniques. Geography students need a solid introduction to the variety of ways in which statistical procedures are used to explore and to solve realistic geographic problems. This book is designed to provide a comprehensive and understandable introduction to statistical methods in a practical, problem solving framework. Our hope is that students who use this text in a spatial analysis or statistical methods course will acquire a well-grounded foundation and feel comfortable in applying statistical techniques in research problems or situations that they might encounter in their subsequent geographic education and careers.

This book is targeted for undergraduate geography majors and beginning graduate students who do not have a strong background in statistical approaches to geographic problem solving. Because this is an introductory textbook, we assume that students have not taken any courses in statistical analysis and do not have prior experience with statistical methods. However, students who do have some statistical background or experience will find the geographic problem solving emphasis in this book useful.

Real-world examples and problems drawn from both human and physical geography are fully integrated into the text. Rather than illustrating statistical techniques using hypothetical data in simplistic problems, our goal is to convey the applicability of statistical analysis to the solution of real problems from diverse areas of the discipline. Geographic problems are drawn from various spatial levels, ranging from local to international. Examples and problems are presented from such topical areas as climatology, medical geography, social and population geography, landuse analysis, transportation planning, economic and urban geography, migration, economic development, and environmental planning.

Examples and problems have been selected to illustrate the complex realities and inherent difficulties of using statistical methods in geographic research. As in the real world, the geographic problems discussed in this book are not simple and often cannot be fully solved through statistical techniques. At times there may be shortcomings with the data or difficulties with the variables being analyzed. This text includes a "summary evaluation" with many of the geographic applications of statistical methods. These summary evaluations discuss various shortcomings, difficulties, and unresolved problems associated with the research design, geographic data, or statistical procedure.

A primary feature of this text is the presentation of statistical techniques using a flexible, investigative approach to geographic problem solving. Meaningful research in geography often requires use of a statistical methodology. For these methods to be useful and valid, the goals and objectives of a particular research problem must be carefully considered. Because geographers ask a variety of questions and geographic research has many different purposes, it follows that the appropriate use of statistical procedures varies as a function of the research context for the problem. This text presents an extensive set of statistical techniques that can be used by geographers to examine a broad range of spatial questions. Our approach involves a careful blending of the following three interrelated themes in applied statistics:

1. *Use of a flexible, exploratory approach.* Application of statistical techniques should allow the researcher to learn as much as possible about the geographic problem under investigation. An

important goal is to maximize the geographic understanding that can be derived from the application of statistical procedures to spatial data. Use of a flexible, exploratory approach extends the information gained from the traditional application of statistics for solving geographic problems.

2. *Adoption of the p-value method of statistical testing.* In applied statistics, a key distinction has been made between two basic methods of hypothesis testing: the classical or traditional procedure and the *p*-value or prob-value approach. The more flexible *p*-value method provides valuable information about the exact level of conviction that can be placed on conclusions from statistical analysis. The *p*-value method allows a more flexible and exploratory application of statistics and is generally more appropriate for solving realistic geographic problems. For those who wish to use the classical or traditional approach to hypothesis testing, the appropriate statistical tables are included in the appendix.

3. *Use of a statistical software package.* Geographers can investigate problems more efficiently by using a computer to analyze spatial data. Although no specific software or system (SPSS, SAS, MINITAB, etc.) is recommended, students must work in a computing environment, even at the introductory level. In the recent past, large mainframe systems provided the only computerized approach for solving geographic problems. Now, numerous microcomputer systems and software packages allow geography students greater flexibility for data processing and statistical computation. A computer-based approach to geographic problem solving is crucial if *p*-value hypothesis testing is to be applied effectively.

This text emphasizes spatial statistics and addresses methodological issues of particular concern to geographers. While scientists share many common concerns and research procedures, certain issues and method-

ological techniques are of special importance to geographers. To meet these specialized needs, the book emphasizes such spatial issues as the nature of geographic data, the level of spatial aggregation, the effect of study area boundaries, and the impact of boundary modifications. Separate chapters are devoted to descriptive spatial statistics (which determine the central location and amount of dispersion in point patterns) and inferential spatial statistics (which analyze point and area patterns for randomness, clustering, or dispersion).

The text is organized into five major parts. Part I provides an introduction to basic statistical concepts and terminology. Chapter 1 discusses the role of geography as a spatial science, provides a brief outline of the scientific research process, and introduces some geographic problems for which statistical analysis is appropriate. The first chapter establishes the framework needed for conducting geographic research using statistical procedures. Chapter 2 explains the major characteristics of geographic data and its preparation, including measurement levels and concepts, classification methods, and graphic procedures.

Part II focuses on descriptive statistics and their use in solving geographic problems. Chapter 3 presents the array of basic descriptive statistics that provide numerical or quantitative summary measures of a data set. Issues associated with interpreting descriptive measures derived from spatial data receive special emphasis. In chapter 4, analysis of locational data with descriptive spatial statistics (geostatistics) is the focus.

The chapters in Part III make the transition from descriptive analysis to inferential problem solving. Chapter 5 discusses probability with a focus on geographic applications and examples of three important probability distributions—the binomial, Poisson, and normal. Included in this chapter is a section that discusses the characteristics and areas of application of probability maps, an important statistical tool for geographers. Sampling is discussed in chapters 6 and 7. Chapter 6 introduces the basic elements of sampling and includes an explanation of various spatial and nonspatial sampling methods. In chapter 7, the logic of statistical inference is introduced, and methodology for using sample data to estimate population characteristics is discussed.

Chapters 8 through 11 comprise Part IV of the text, a discussion of inferential statistics for geographic problem solving. Both the traditional and p-value methods of hypothesis testing are described, and the close relationship between the two is explained. Because the text stresses the importance of the p-value approach, the geographic applications for illustrating the inferential tests found in this section use the prob-value alternative. Rather than subdividing inferential statistics into the conventional parametric and non-parametric procedures, our text organizes techniques according to the type of geographic problem they can solve. Each chapter in Part IV concerns a specific type or class of difference test. Chapter 8 presents one sample difference tests for comparing population characteristics to those of a sample. Chapter 9 discusses multiple sample difference tests, where the objective shifts to comparing results from two or more samples. In chapter 10, two special types of difference tests—goodness-of-fit and categorical—are described for use with frequency data. Chapter 11 presents inferential spatial statistics, where methods for analyzing point and area patterns are examined using explicitly spatial data.

Fundamental tasks in many types of geographic problem solving are to determine the association between spatial patterns and the similarity or difference between distributions. Part V of the text examines two related techniques for exploring statistical relationships between variables. Chapter 12 concerns correlation procedures that measure the degree of association between variables, and chapter 13 covers regression, a popular technique for studying causal relationships. Spatial issues such as correlating the patterns of two maps and analyzing the spatial pattern of residuals from regression are emphasized in Part V.

Introductory quantitative methods and spatial analysis courses vary among geography programs. This text can be adapted to either a one-semester, two-quarter, or two-semester course sequence. Whatever the length of the course, however, basic instruction in the use of statistical software packages is needed.

In the shorter-length context of a single semester or two quarters, several approaches are possible. One alternative is to move quickly through the entire text, emphasizing those topics that seem most appropriate. Another alternative is to eliminate certain topics and chapters entirely. Instructors have considerable flexibility concerning the depth of treatment for the variety of topics in the text. The appropriate integration of the text into a specific spatial analysis course may best be determined from the previous statistical and computing backgrounds of the students.

In the longer-length context of a two-semester (or three-quarter) sequence, it is possible to cover the entire text in considerable detail. One option is to present Parts I through III in the first semester. Students would then be well-grounded in basic statistical concepts, descriptive statistics, probability, and sampling. Parts IV and V would then be presented in the second semester, and students could examine the variety of inferential statistics in considerable depth.

An exercise manual accompanies the textbook. This workbook contains a variety of exercises and problems keyed directly to the material presented in the text. Geographic problems and exercises are available for topics and techniques discussed in every chapter. Students are referred to one of several data sets included in the workbook to address the problems.

The data sets in the workbook are organized to provide maximum flexibility for the instructor and efficiency for the student. Instructors often disagree on the relative merits of requiring students to take samples from a population data set. Although sampling is an integral element of statistical problem solving, it can be a tedious and time-consuming task. The workbook offers the instructor several alternatives by identifying a few "pre-drawn" random samples within each data set. Different students can use different samples, thus avoiding the situation of every student analyzing the same set of values, or all students in the class can work with the same sample data. The instructor also has the option of requiring the students to draw their own sample.

ACKNOWLEDGMENTS

The groundwork for this textbook dates from our experiences together as graduate students at Pennsylvania State University. In the exciting and challenging environment of the Department of Geography, our developing interests in spatial analysis flourished. We would like to acknowledge the support provided early in our careers by the many fine geography faculty at this institution.

Completion of this text would not have been possible without the assistance of many individuals. Colleagues at both Salisbury State University and the University of Akron have provided a variety of worthwhile ideas and suggestions. Many of the geographic problems discussed in the text have been improved by comments about earlier drafts of this material. Thanks are extended to Michael E. Folkoff and Brent R. Skeeter at Salisbury State University and to Robert B. Kent at the University of Akron. Special thanks to the geography department chairs (Allen G. Noble at the University of Akron and Calvin R. Thomas at Salisbury State) for providing encouragement and support to complete this project. In addition, several faculty in the Department of Mathematical Sciences at Salisbury State (Homer W. Austin, Philip E. Luft, and Robert M. Tardiff) willingly extended advice on several statistical and organizational issues. However, the authors accept sole responsibility for any errors that may remain in the text.

We would like to thank the many students at both Salisbury State and Akron who where involved with this project. Spatial analysis classes at both institutions have "field tested" various incomplete and draft versions of the text. Their patience is appreciated. Particular thanks are extended to those students who contributed in some way to data collection or statistical analysis of a geographic data set. At Salisbury State, the assistance of the following students is especially appreciated: Dan Carnack, Russ Davis, George Draper, Steve Jones, Sharon Riley, Jessica Spring, Cheryl Strafella, and Lynn Thomas. At Akron, we gratefully acknowledge the work of Chusheng Lin, Maurice Carney, Robert Sanderson, Rebecca Weisenthal, Heather Beach, John Mikstay, and Daniel Trudel.

The figures and graphic material in the text were produced by The Laboratory for Cartographic and Spatial Analysis at the University of Akron. In particular, we acknowledge the patience and dedication of its supervisor, Joseph Stoll, and the hard work of his student assistants.

Special thanks go to Laura Monroe for her excellent editorial comments throughout the manuscript. She was able to bring a "nongeographer's" perspective to our writing and helped keep the narrative from becoming overloaded with technical jargon. The result, we believe, is a more readable product.

We wish to recognize the following reviewers who made suggestions and helpful comments on our manuscript:

William Dakan
University of Louisville

Adrian A. Seaborne
University of Regina, Canada

Nancy R. Bain
Ohio University

Wolf Roder
University of Cincinnati

David Hodge
University of Washington

Their constructive criticism contributed greatly to a stronger final product, and we appreciate their efforts.

A note of thanks is reserved for the excellent editorial and production staff at Wm. C. Brown. The professional expertise, courtesy, and general helpfulness of the entire team deserves public recognition. Particular thanks go to Geography Editor Lynne M. Meyers, who shepherded our text through the production process during an especially busy time. Finally, we would like to recognize the organizational talents and high quality of professionalism exhibited by Ed Jaffe, former Assistant Vice President and Editor-in-Chief of Wm. C. Brown. His untimely passing has saddened all who knew him.

J. Chapman McGrew, Jr.
Charles B. Monroe

PART I

BASIC STATISTICAL CONCEPTS IN GEOGRAPHY

CHAPTER 1

INTRODUCTION: THE CONTEXT OF STATISTICAL TECHNIQUES

Geography is a spatial science that attempts to explain and to predict the spatial distribution and variation of human activity and physical features on the earth's surface. Geographers study how and why things differ from place to place, as well as how spatial patterns change through time. People have been interested in such geographic concerns for thousands of years. Early Greek writers, such as Eratosthenes and Strabo, emphasized the earth's physical structure, its human activity patterns, and the relationship between them. These traditions remain central to the discipline of geography today. Contemporary geography continues to be an exciting discipline that attempts to solve a variety of problems and issues from a locational or spatial perspective.

The geographer starts by asking *where* questions. Where are things located on the earth's surface? How are features on the physical or cultural landscape distributed? What spatial patterns are observable, and how do phenomena vary from location to location? Historically, geographers have focused on trying to answer these questions. In fact, the popular image of the discipline of geography probably remains almost exclusively focused on where things are located.

However, professional geographers are no longer content to limit themselves to spatial or locational description. After a spatial pattern has been described and the where questions have been answered adequately, attention then shifts to *why*. Why does a particular spatial pattern exist? Why does a locational pattern vary in a specific observable way? What spatial process or processes have affected a pattern and why? As why questions are answered, or as speculation occurs about why a spatial pattern has a particular distribution, the geographer gains a better understanding of the processes that create the pattern. Sometimes different variables are found to be related spatially, thus providing insights into underlying spatial processes. In other instances, geographers try to determine if circumstances differ in various locations or regions and seek to understand why these differences exist.

Geographers are increasingly concerned with the practical application of this spatial information. Geographers ask *what-to-do* questions that involve the development of spatial policies and plans. More geographers now want to be active participants in public and private decision making. Some questions might include: What type of policy might best achieve more equal access for urban residents to city services and facilities? What sort of government policy would a geographer recommend to balance wetlands protection and economic development in a fragile environment?

Geography is now a problem-solving discipline, and geographers are concerned with applying their spatial knowledge and understanding toward the problems facing the world today. Noted geographer Risa Palm recently stated, "...[G]eography involves the study of major problems facing humankind such as environmental degradation, unequal distribution of resources and international conflicts. It prepares one to be a good citizen and educated human being." (Assoc. of American Geog., n.d.)

1.1 The Role of Statistics in Geography

What role do statistics play in helping geographers answer these where, why, and what-to-do questions? How can statistical analysis benefit a geographic investigation? **Statistics** is generally defined as the collection, classification, presentation, and analysis of numerical data. Among many applications, the use of statistics allows the geographer to:

- describe and summarize spatial data,
- make generalizations concerning complex spatial patterns,

- estimate the likelihood or probability of outcomes for an event at a given location,
- utilize limited geographic data (sample) to make inferences about a larger set of geographic data (population),
- determine if the magnitude or frequency of some phenomenon differs from one location to another, and
- learn whether an actual spatial pattern matches some expected pattern.

The use of statistics must be placed in the context of a general research process. Most geographers view statistics as a major research tool; and many, in fact, argue that a statistical approach is necessary for effective geographic research. The geographic research process and the roles of statistics in that process are summarized in a general organizational framework (figure 1.1). The left column of the figure shows the series of steps that lead to the formulation of hypotheses. These activities are done early in scientific inquiry. The right column of the figure diagrams the steps involved in scientific research after a hypothesis has been stated. Statistical procedures are involved in geographic research both before and after hypotheses are generated.

The sequence of tasks outlined in figure 1.1 is typical of the process many geographers follow when conducting research. This framework should not be viewed as a rigid series of steps, but as a general, flexible guide used in geographic research.

The research process begins by identifying a worthwhile geographic problem to investigate. To recognize a productive research problem, the geographer must have background knowledge and experience. There is simply no substitute for having a strong background in the appropriate branch of the discipline.

Hypothesis formulation is at the center of the research process (figure 1.1). A **hypothesis** is an unproven or unsubstantiated general state-

ment concerning the problem under investigation. The investigator may have sufficient background knowledge or information from previous research (such as a review of the literature), which allows hypotheses to be readily developed. Perhaps a hypothesis can be formulated using a **model**, which is defined as a simplified replication of the real world. A well-known model in geography is the spatial interaction model, which predicts the amount of movement expected between two places as a function of their populations and the distance that separates them.

In some cases, a geographer may have identified a possible research area, but is not yet ready to formulate a hypothesis. Perhaps more information may be needed about the problem, or questions concerning the problem may need to be answered. In these situations, the geographer cannot immediately move down the right column of figure 1.1, but must first address research tasks in the left column. Statistical analysis is central to this process. Geographers often gain spatial insights by collecting data and presenting this information by using graphical procedures and maps. Statistical analysis provides quantitative summaries or numerical descriptions of the data during this phase of the investigation. The information gathered may enable the geographer to draw conclusions about the research questions, to develop a model of the spatial situation, and to generate suitable hypotheses.

If the geographer has a workable hypothesis, the research process then follows the steps shown in the right column of figure 1.1. Additional data need to be collected and prepared so that the hypothesis can be tested and evaluated. These steps are the core of statistical analysis and are a primary focus of this book. In scientific research, if the hypothesis is repeatedly verified as correct under a variety of circumstances (perhaps at various locations or times), it gradually takes on the stature of a **law**. In other words,

Figure 1.1   The Role of Statistics in the Geographic Research Process

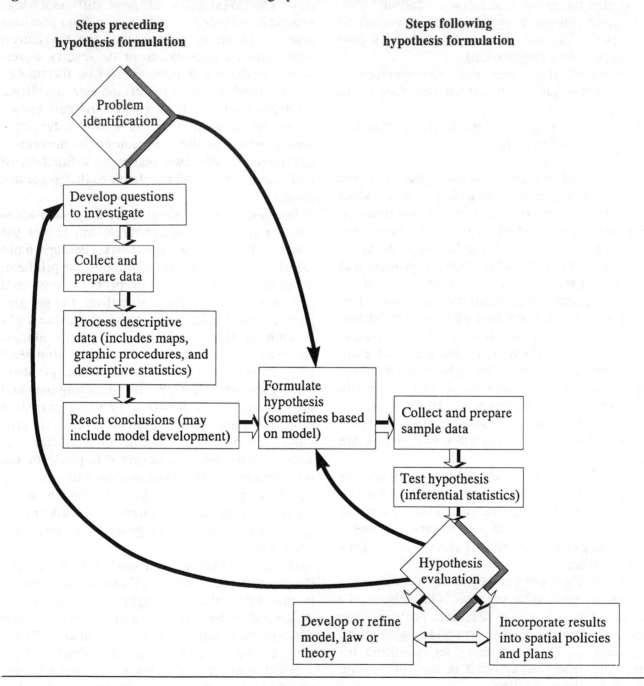

a proven hypothesis can eventually become accepted as a law. If various laws are combined, then they constitute a **theory**.

1.2 An Example of Statistical Problem Solving in Geography

How can a geographer use statistics in the geographic research process to approach and solve spatial problems? The process begins with the identification of an interesting and relevant spatial pattern or research problem. Suppose a geographer is interested in analyzing growth trends in the United States during the 1980s (figure 1.2). According to the figure, the fastest growing states are generally located in the South and West. Much of the national growth occurred in California, Florida, and Texas. Nearly all of the states losing population or experiencing slow growth are located in the Northeast and Midwest.

Given this map of population growth by state, the geographer may want to explore the spatial pattern further and ask why this spatial pattern exists. Why does the growth rate pattern vary in this way? What factors can be suggested to help explain the nature of this spatial distribution? What spatial process or processes might have been operating?

Various factors may be at work to produce the growth patterns shown on figure 1.2. Previous geographic studies help the researcher by identifying potential relationships or explanations that may be relevant. For example, many geographers feel that climatic amenities and environ-

Figure 1.2 Population Change by State, 1980-1990

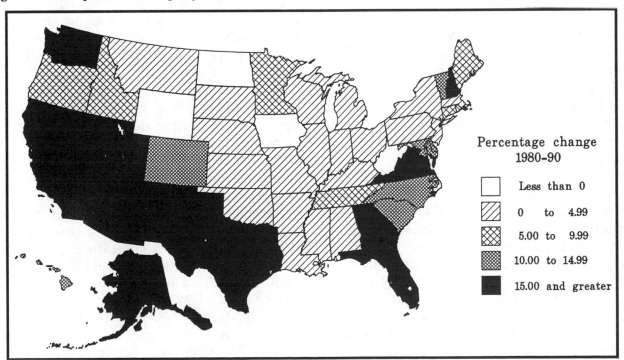

Source: Data from Bureau of the Census, U.S. Dept. of Commerce.

mental considerations influence the spatial pattern of population change in the United States. The southern and western growth regions, which include Florida, California, Arizona, and Hawaii, are known as the "sunbelt" because of their warm and sunny climates. Conversely, the northeastern and midwestern states, which are either losing population or experiencing slow growth, are called the "snowbelt" or "frostbelt." The implication is that people are fleeing the cold winters and snow for year-round warm weather and outdoor recreation opportunities.

Economic factors undoubtedly influence growth in a number of ways. In surveys examining why people change residence, respondents frequently cite job opportunities and related economic reasons. Over the last few decades, many more new jobs have been created in the southern and western states than in the Northeast and Midwest. The region containing the states of Pennsylvania, Ohio, Michigan, Indiana, and Illinois is sometimes called the "rustbelt" because of its traditional economic reliance on heavy industry and manufacturing, which are sectors of the U.S. economy that have suffered a relative decline in recent years.

Patterns of immigration also influence population growth. In the 1980s, a large number of migrants from Asia and Latin America settled in California and Florida, contributing to the high growth rates of these states. However, many immigrants also settled in Illinois and New York, states that have experienced little recent change in population size.

Geographers studying recent migration trends suggest that low-density residential areas, such as rural regions and small towns, are increasingly attractive locations for many Americans. If this trend toward nonmetropolitan growth or decentralization of the population is true, then states with smaller populations and low population densities should have high growth rates.

Other investigators cite the strong magnetic draw of the oceans for water-based recreation and economic activity as an important growth factor.

Rapidly growing coastal (peripheral) communities contrast with stagnant or declining interior (heartland) locations.

It appears that the spatial pattern of population change is likely the result of a set of complex spatial processes. A number of questions are worthy of geographic investigation: Does a relationship exist between climate and population change? Are sunbelt states in the South and West growing faster than the snowbelt states in the Northeast and Midwest? How important are economic structure and employment in "explaining" population growth? Are states with economies dependent on heavy industry and manufacturing losing population faster than other states? What is the relationship between the residential choices of recent immigrants and state population growth rates? Are coastal states growing faster than noncoastal states?

Once these questions have been posed, the geographer must collect and prepare data that can help generate answers (figure 1.1). Some of the research steps can be done easily, such as classifying each state as coastal or noncoastal, allocating states to the sunbelt or frostbelt, and collecting data on the number of immigrants settling in each state during the 1980s. Other data collection and preparation tasks may not be quite as direct. For example, some form of operational definition is needed to measure the economic structure of a state. No single clear definition of "economic structure" exists, and it is not immediately obvious how this term should be defined for this particular problem.

After all data have been collected and prepared, they are "processed" using various graphic procedures and statistical measures (figure 1.1). A series of choropleth maps could be constructed, such as the map showing state-level population change (figure 1.2). Displaying information with a histogram, ogive (cumulative frequency diagram), or scatterplot may also be useful. For example, a graph showing the relationship between 1980 to 1990 population change for each state and the number of 1980 to 1990 immigrants who chose to reside in each state would be informative. These graphic procedures are discussed in detail in section 2.5.

The use of summary measures or numerical descriptions of data may also help the geographer answer the research questions under study. Statistical procedures provide the average or typical value of a variable and determine the amount of variability within the set of values. For example, calculating and comparing the average growth rate of coastal states to noncoastal states might provide some insights about the spatial pattern. Various descriptive statistics are presented in chapter 3.

After the data have been processed, the geographer can reach some initial conclusions about the geographic research questions under study. By using graphic evidence, map summaries, and statistical summary descriptions, the research questions can be understood more fully. The initial investigation of population change in the United States could reveal general answers to some of the questions posed: Are sunbelt states growing faster than snowbelt states? Are coastal states growing faster than noncoastal states?

Some geographic studies may be complete at this point, content with a descriptive analysis of a geographic pattern and some of its relationships. However, most geographers continue the research process down the right column of figure 1.1 through an expanded inferential approach.

In addition to offering insights into the research problem, the initial investigation may allow the geographer to propose a model describing the situation and to formulate hypotheses for testing (figure 1.1). For example, analysis of the population change data at the state level may permit a researcher to develop hypotheses relating the spatial pattern to factors that produced the change. By testing these hypotheses with inferential procedures, the geographer will be able to probe the underlying spatial processes causing the pattern in figure 1.2.

However, spatially aggregate data (like the state-level data used for the choropleth map) are often not adequate for developing hypotheses in research problems. Individual and family responses to questions about their own patterns of migration and underlying reasons for their moves are needed. Since the researcher cannot possibly survey all migrants, a sample of movers must be collected. If done properly, this sample information will adequately reflect what is occurring in the overall population of movers from which the sample is drawn.

How might a geographer investigating the pattern and process of population change generate hypotheses for testing with sample data? As shown on figure 1.2, Michigan gained little population during the 1980s, whereas Florida's population grew substantially. Michigan is a noncoastal, snowbelt state with an economic and employment structure considered less favorable for growth than Florida, a coastal, sunbelt state with many new service (tertiary) and information processing (quaternary) jobs. What factors influenced population change? What reasons do former Michigan residents give when asked why they moved? What reasons do newly arrived Florida residents give when asked why they decided to relocate? A preliminary survey of out-migrants from Michigan and in-migrants to Florida would provide information to help answer these questions. Analysis of the survey data may allow the researcher to draw general conclusions and construct research hypotheses for the problem. Remember that a hypothesis is an unproven general statement regarding a research question. The following are possible hypothesis statements for this problem:

- the reasons for leaving a sunbelt state differ greatly from reasons given for leaving a frostbelt state,
- most people moving from a frostbelt state to a sunbelt state say they have done so to live in a warmer climate,
- more people who move to coastal communities from noncoastal communities say they have done so more for water-based recreation than for employment reasons,

- recent immigrants select a community that already has a large number of residents from their country,
- the majority of already-employed people who move from one state to another have a higher income in their new state of residence, and
- the majority of unemployed people who move from one state to another join the work force less than a month after moving.

After the researcher develops hypotheses concerning the problem under study, test procedures need to be established to evaluate the hypotheses (figure 1.1). Sample data are collected and prepared for analysis. The nature and exact form of this information depends on the research question and hypotheses formulated. A number of different statistical procedures are available for hypothesis testing, many of which are discussed in the following chapters. The concepts of hypothesis testing are central to statistical analysis and play an important role in geographic problem solving.

After testing a research hypothesis, the results must be evaluated and conclusions drawn. Several strategies or actions are possible (figure 1.1):

1. The research findings may be incorporated into actual or recommended spatial policies and plans. The applied geographer might be suggesting actions or addressing the what-to-do question posed at the start of the chapter. This could include recommendations on ways to handle population growth or decline. Decision makers may be given information regarding the apparent reasons for population change. Suppose the majority of sampled migrants to Florida indicate that they moved to this state because of its strict regulation of coastal zone management and environmental protection. This finding can help Florida decision makers evaluate the state's environmental statutes and recommend changes in future legislation.
2. If the hypothesis is verified as correct and valid, the results could be refined further

into a spatial model that explains the reasons for migration and population growth in different regions of the country. Repeated verification of a hypothesis or model under a variety of circumstances (perhaps at various locations or at different times) might lead to the eventual development of laws and perhaps theory concerning population growth and migration patterns.
3. If the hypothesis is tested and found to be partially or completely incorrect, the researcher may need to return to an earlier step in the geographic research process. With a partially validated hypothesis, the researcher could return to hypothesis formulation and restate or refine the original hypothesis. If a hypothesis is proven totally wrong, the researcher may want to go all the way back to question development.

The general research process outlined in figure 1.1 takes on many operational forms in geography. However, geographic research centers on the investigation of geographic patterns and processes through descriptive analysis, hypothesis generation, and inferential statistical tests, with the eventual goal of developing laws and theories. Collection, presentation, and processing of data all play central roles in the research process.

1.3 Basic Terms and Concepts in Statistics

The most basic element in statistics is **data** or numerical information. Geographers often use groups of data, which are referred to as a **data set** and presented in tabular format (table 1.1). A data set consists of observations, variables, and data values. The elements or phenomena under study, for which information (data) are obtained or assigned, are often referred to as **observations**. Observations are sometimes called "individuals" or "cases." Geographers use many types of observa-

Table 1.1 A Geographic Data Matrix

Observations	Variables			
Country	Population (millions) mid-1989	Area (1000s of sq km)	Avg. Annual Inflation Rate (%) 1980-89	Percent of Total Household Consumption to Food
Argentina	31.9	2767	334.8	35
Canada	26.2	9976	4.6	11
Ethiopia	49.5	1222	2.0	50
Italy	57.5	301	10.3	19
Thailand	55.4	513	3.2	30
United States	248.8	9373	4.0	13

Source: World Bank. 1991. *World Development Report 1991: The Challenge of Development.* New York: Oxford University Press.

tions in their research, including spatial locations, such as cities or states, and nonspatial items, such as people or households. A property or characteristic of each observation that can be measured, classified, or counted is called a **variable** because its values vary among the set of observations. The resulting measurement, code, or count of a variable for each observation is a **data value**.

The data set in table 1.1 shows a two-dimensional geographic data array with several variables chosen from a larger set of observations. In this matrix, variables are presented across the top of the table, and observations (countries) listed down the left side. A value or unit of data (inside the table) represents the magnitude of a single variable for a particular observation. Two examples of data values are Thailand's population of 55.4 million and the 50 percent of total household consumption allocated to food products in Ethiopia. Collecting data from different time periods would add a third dimension to the matrix.

In the discussion and application of the geographic research process, two forms of statistical analysis were briefly described. The early steps focused on the descriptive processing of data, and later stages involved testing of hypotheses using inferential methods. The fundamental distinction between descriptive and inferential statistics needs further explanation.

Descriptive statistics provide a concise numerical or quantitative summary of the characteristics of a variable or data set. Descriptive statistics describe—usually with a single number—some important aspect of the data, such as the "center" or the amount of "spread" or "dispersion." For most geographic problems, using such descriptive summary measures is superior to working directly with a large, cumbersome group of values. Descriptive statistics allow geographers to work efficiently and to communicate effectively.

Replacing a set of numbers with a summary measure necessarily involves the loss of information. Various descriptive statistics (with different advantages and limitations) are available. Selecting the descriptive measure whose characteristics seem appropriate to the geographic problem being analyzed is important. In many geographic situations, the objective is to minimize the loss of relevant information when moving from unsummarized data to descriptive measures. Descriptive statistics are discussed in chapters 3 and 4.

The purpose of **inferential statistics** is to make generalizations about a statistical population based on information obtained from a sample of that population. In the context of inferential statistics, a **statistical population** is the total set of information or data under investigation in a geographic study. In the state growth and migration example, the geographer might be interested in the statistical population of *all* U.S. residents and their migration patterns. Another study might focus on the reasons why *all* newly arrived Florida residents moved into the state. Another research problem could examine the statistical population of *all* former snowbelt residents who have moved to a sunbelt state.

It is not practical to question all U.S. citizens, survey all newly arrived Florida residents, or collect data from all snowbelt-to-sunbelt migrants. However, collecting information from a **sample** of people in these statistical populations may be feasible. The sampled subset must be typical or representative of an entire population. If the sample is representative (which means it is not biased), then the researcher can make inferences about the nature of the population, based on sample data. Inferential statistics use descriptive measures obtained from samples and link this descriptive information to probability theory. General statements can then be made about the nature or characteristics of the population from which the sample has been drawn. For example, from a representative (unbiased) sample of new Florida residents, general statements can be made about the beliefs and attitudes of all incoming Florida residents.

Estimation and **hypothesis testing** are the two basic types of statistical inference. In some instances, sample statistics are used to estimate the population characteristic. The sample proportion of snowbelt-to-sunbelt migrants indicating that climate was the most important reason for moving can be used to estimate the proportion of all snowbelt-to-sunbelt migrants with that belief. Estimating population characteristics from samples is discussed in chapter 7.

In inferential hypothesis testing, sample data are used to reach conclusions about population characteristics. For instance, an inferential hypothesis test could be applied to determine if the reasons for moving differ significantly between a sample of interstate migrants from the Midwest and another sample of interstate migrants from the South. If the responses from the two sample groups are significantly different, the researcher can infer that the samples were drawn from two different statistical populations. It could also be concluded that the reasons for moving are different among the populations of all midwesterners and southerners. Chapters 8 through 10 focus on inferential hypothesis testing.

Key Terms and Concepts

data (data value)	10 & 11
data set	10
descriptive statistics	11
estimation	12
geography	4
hypothesis	5
hypothesis testing	12
inferential statistics	11
law	5
model	5
observation	10
sample	12
statistical population	11
statistics	4
theory	7
variable	11

References and Additional Reading

Abler, R. F., J. S. Adams, and P. R. Gould. 1971. *Spatial Organization: The Geographer's View of the World.* Englewood Cliffs, NJ: Prentice-Hall.

Amedeo, D. and R. G. Golledge. 1975. *An Introduction to Scientific Reasoning in Geography.* New York: John Wiley and Sons, Inc.

Association of American Geographers. no date. *Geography: Today's Career for Tomorrow* (informational brochure). Washington, DC: Association of American Geographers.

Costanzo, C. M. 1983. "Statistical Inference in Geography: Modern Approaches Spell Better Times Ahead." *Professional Geographer* 35:158-64.

Gaile, G. L. and C. J. Willmott (editors). 1989. *Geography in America.* Columbus, OH: Merrill.

Haring, L. L. and J. F. Lounsbury. 1982. *Introduction to Scientific Geographic Research* (3rd edition). Dubuque, IA: Wm. C. Brown.

Harvey, D. 1969. *Explanation in Geography.* New York: St. Martin's Press.

CHAPTER 2

GEOGRAPHIC DATA: CHARACTERISTICS AND PREPARATION

Before performing statistical processing and analysis, a number of characteristics of spatial data must be known. The researcher needs to understand how variables are organized as well as how data are arranged within this organization. In this chapter, the basic concepts regarding the use of geographic data are introduced. These concepts provide the background information needed for determining the characteristics of the data before statistical techniques are applied.

Questions about data and variables arise early in the scientific research process. When identifying an appropriate geographic research problem and formulating meaningful hypotheses, decisions must be made about the sources of data available, about the method of collecting data, and about the variables to be included in the analysis. In section 2.1, the various dimensions of geographic data that help in making these decisions will be discussed.

Several measurement issues must be considered before any statistical analysis. Variables may be organized and displayed in various ways, and different levels of measurement are used in different geographic problems. The characteristics of nominal, ordinal, and interval/ratio measurement scales are reviewed in section 2.2. Section 2.3 discusses potential measurement errors and addresses the issues of precision, accuracy, validity, and reliability.

Basic methods for classifying data are discussed in section 2.4. The goals and purposes of classification are emphasized, including its importance in geographic research. The basic classification strategies of subdivision and agglomeration are reviewed, and several specific classification methods are applied to a set of spatial data.

The emphasis in section 2.5 is on graphic procedures commonly used by geographers to summarize, classify, or display spatial data, including discussions of frequency distributions, histograms, and ogives. Also, scatterplots are introduced as a mechanism to graphically display relationships between geographic variables.

2.1 Selected Dimensions of Geographic Data

In the scientific research process, questions about data arise almost immediately. In identifying appropriate geographic research problems and formulating hypotheses, questions such as these usually emerge: What sources of data are available? Which method(s) of data collection should be used? What type of data will be collected and then analyzed statistically? Various aspects of a research problem must be carefully considered, to ensure that the data collected and analyzed will answer research questions effectively and to allow meaningful conclusions to be reached. The dimensions of geographic data now discussed include sources of information, methods of data collection, and selected characteristics of data that distinguish geographic research problems.

A simple distinction can be made between primary and secondary data sources. **Primary data** are acquired directly from the original source. This information is often collected "in the field" by the geographer conducting the study. Primary data collection is often quite time-consuming and generally involves some basic decisions about sample design to acquire a set of representative data.

Secondary (or archival) data sources are provided directly for the geographer and are usually collected by an organization or government agency. Because the data have already been collected and are probably organized in an accessible, convenient form, such as a written report or census tape, secondary sources are generally less expensive and time-consuming. In addition, many of the problems associated with sampling and survey design may not be experienced with archival sources. Secondary sources are often very comprehensive, such as a census or total enumeration from a very large population. Duplicating these efforts with primary data collection would be virtually impossible for the researcher.

Potential difficulties can also occur with secondary data sources. The data may have been collected, organized, or summarized improperly. Errors may have occurred in the editing and collating of data, especially if the information was obtained from a number of different original sources. Information from secondary sources is not always totally precise, accurate, valid, and reliable, resulting in other potential problems.

Geographic data can be collected by using one of several methods. If primary data are being used in a research study, sampling will almost certainly be necessary, and survey questionnaires may need to be designed. Among the options for primary data collection are direct observation, field measurement (especially in research for physical geography), mail questionnaire, personal interview, and telephone interview.

To select the appropriate method of data collection, the nature of the research problem must be evaluated carefully. Even if a suitable method has been selected, survey design problems are likely. In collecting data through a survey, each question must be properly worded, all possible responses to a question must be considered in advance, and the sequence of questions must be determined. In fieldwork, logistic problems may occur or special arrangements may be necessary. Preliminary site reconnaissance may not reveal all the difficulties. A more detailed discussion of data collection methods is found in chapter 6.

Other dimensions or characteristics of data help to distinguish geographic research problems. Some studies can be termed **explicitly spatial** because the locations or placement of the observations or units of data are themselves directly analyzed. For example, a geographer serving as a location analyst for a retailer might calculate the "center of gravity" or average location of a set of households in an urban area to help select an accessible central site for a new store. In another application, a biogeographer might analyze the spatial pattern of a sample of diseased trees in a national forest to determine whether the diseased trees are randomly distributed throughout the area or clustered in certain portions. In both of these examples, the data are spatially explicit, because the locations of the observations or units of data are analyzed directly. An important set of **spatial statistics** can be used to investigate these problems. In chapter 4, a variety of descriptive spatial statistics are discussed, and in chapter 11, inferential spatial statistics are used to analyze the distribution of sample point and area patterns for randomness.

Other studies in geography are **implicitly spatial**. Often the observations or units of data represent locations or places, but the locations themselves are not analyzed directly. A geomorphologist might wish to determine if significant differences occur in alluvial fan development in two different basins in the basin-and-range region of the American Southwest. A random sample of alluvial fans from each of the basins could be taken, and the relevant aspects of alluvial fan development could be compared. An urban geographer may be investigating the relationship between the assessed valuation of homes and the age of those structures in a suburban neighborhood. In both of these examples, the observations (alluvial fans in a basin or homes in a suburban neighborhood) obviously have locations on the earth's surface, but the locational pattern itself is not under scrutiny. For the alluvial fan study, a two-sample difference test is appropriate (chapter 9), whereas a correlation analysis may be needed to study the suburban housing question (chapter 12).

Another important dimension of geographic research problems is the contrast between **individual-level** and **spatially aggregated** data sets. In some geographic problems, each data value represents an individual element or unit of the phenomenon under study. In other problems, each "value" entered into the statistical analysis is a summary or spatial aggregation of individual units of information for a particular place or area. The best way to see this distinction is with an example. Suppose a population geographer is researching current fertility patterns in Nigeria. One approach would be to collect a set of individual-level data, perhaps through personal interviews of a random sample of Nigerian women. Another possible approach would be to obtain birth rate estimates from officials in each of Nigeria's administrative divisions (21 states and 1 territory) and use these 22 spatially aggregated values as units of datum to estimate the nationwide fertility pattern.

Using spatially aggregated data raises special issues. Geographers must always be extremely cautious when trying to transfer results or apply conclusions "down" from larger areas to smaller areas or from smaller areas to individuals. If conclusions are derived from the analysis of data spatially aggregated for large areas, it may not be valid to reach conclusions about smaller areas or individuals. For example, in the study of Nigerian fertility patterns, valid, spatially aggregated conclusions can be drawn about the degree of acceptance or nonacceptance of family-planning programs in each of these states by using birth rate estimates from the administrative divisions. However, taking these aggregate conclusions down to the level of the individual family will probably result in deductive errors. Even in the Nigerian state with the lowest birth rate, a number of families will not be practicing any form of birth control. This invalid transfer of conclusions from spatially aggregated analysis to smaller areas or to the individual level is known as the **ecological fallacy**.

Conversely, taking individual-level data and aggregating it to larger spatial units is generally not a problem. In fact, Nigerian officials probably collected data at the individual and village level, then aggregated that information to obtain state-level estimates. The effects of level of spatial aggregation on descriptive statistics is a topic in chapter 3.

Variables in data sets can be characterized as either discrete or continuous. A **discrete variable** has some restriction placed on the values that the variable can assume. A **continuous variable** has an infinitely large number of possible values along some interval of a real number line. In general, discrete data are the result of counting, and potential values are limited to whole integers. Continuous data are the result of measurement, and values can be expressed as decimals. The following are examples of discrete variables: the number of households in a county with videocassette recorders, the number of immigrants currently living in a particular city, the number of respondents to a survey in favor of a local bond issue, and the number of active volcanoes in a chosen country. Continuous variables could include inches of precipitation during a year's time at a specific weather station, total area under irrigation in a country, distance traveled by a family on their annual vacation, and average wind speed at the summit of a mountain.

The "rounding-off" of data values must not be confused with the distinction between discrete and continuous data. For example, the elevation of a mountain is almost always expressed to the nearest foot or meter. Representation of elevation as a whole number gives the impression that this variable is discrete. However, since elevation can be measured more precisely with decimal places, it is considered a continuous variable.

In the discussion about probability distributions (chapter 5), the distinction between discrete and continuous data is important. Geographic problems with discrete variables often require the application of different probability distributions than problems having continuous variables. Several practical geographic examples of both discrete and continuous distributions are presented in chapter 5.

Variables in a set of data are either quantitative or qualitative. If a variable is **quantitative**, the observations or responses can be expressed numerically. On the other hand, each observation or response for a **qualitative** variable is placed in one of two or more categories. An agricultural geographer asked 80 farmers to identify their primary cash crop, and received the following responses: 43

corn, 28 wheat, and 9 barley. It might be tempting to conclude that these are quantitative data, since 43, 28, and 9 are clearly numerical values. However, the *variable responses* (corn, wheat, barley) are nonnumeric. The values of 43, 28, and 9 are not the raw data, but rather frequency counts of observations assigned to the nonnumerical categories, making this an example of qualitative (or **categorical**) data. Other examples of qualitative variables are land use type, sex (male or female), political party affiliation, religious preference, and climate type. Special types of statistical tests have been developed to handle qualitative (categorical) data, and these are discussed in chapter 10.

The organization of variables determines how data can be analyzed statistically. The related issue of variable measurement, discussed in the following section, is an important factor in selecting the appropriate statistical technique to solve a geographic problem.

2.2 Levels of Measurement

Nominal Scale

The simplest scale of measurement for variables is the assignment of each value or unit of data to one of at least two qualitative classes or categories. In **nominal scale** classification of variables, each category is given some name or title, but no assumptions are made about any relationships between categories—only that they are different. Values are "different" if they are assigned to different categories, or "similar" if assigned to the same category. Thus, problems using variables with nominal scales are considered qualitative.

An urban planner could assign each parcel of land in a city to one of several nominal land use categories (residential, commercial-retail, industrial, recreation-open space, etc.) without inferring that residential land use is "greater than" recreation-open space or "less than" industrial.

In fact, the only necessary conditions for a proper nominal scale classification of variables are that the categories are **exhaustive** (every value or unit of data can be assigned to a category) and **mutually exclusive** (it is not possible to assign a value to more than one category because the categories do not overlap).

Geographers create nominal variables in many ways: individuals can be classified by religious affiliation (Baptist, Catholic, Methodist, Presbyterian, etc.) or political party (Democrat, Republican, Independent); cities can be classified by primary economic function (manufacturing, retail, mining, transportation, tourism, etc.); counties can be organized by primary type of home heating fuel (fuel oil, utility gas, coal, wood, electricity, etc.); and countries can be distinguished by predominant language family (Indo-European, Afro-Asiatic, Sino-Tibetan, Ural-Altaic, etc.).

If a variable has only two categories, a special subset of nominal classification is used. This sort of dichotomous (binary) assignment is used in geographic studies when no greater degree of qualification is necessary or possible. For example, each person could be assigned a 1 if they attended a private elementary school, and a 0 if they did not. Although this type of information is clearly limited, frequency counts of the number of values or individuals in each category can be obtained, and statistical analysis on these frequency counts can be conducted. Many geographic problems have only "yes-no" or "presence-absence" data available.

When variables are organized nominally, geographers are limited in the statistical analysis that can be applied. Nevertheless, appropriate techniques are available and are illustrated in chapter 10.

Ordinal Scale

The next higher level of measurement involves placement of the values themselves in some rank order to create an **ordinal scale** variable. The relationship between observations takes on a form of "greater than" and "less than." With data in rank order, more quantitative distinctions are possible than with nominal (qualitative) scale variables.

Geographers can easily identify examples of ordinal scale variables. An important distinction needs to be made, however, between a strongly-ordered ordinal variable (which is not categorical in nature) and a weakly-ordered ordinal variable (which is categorical). When each value or unit of data is given a particular position in a rank order sequence, the variable is considered **strongly-ordered**. The city ranking schemes that occasionally appear in newspapers or magazines, such as the "ten best places to live" or "fifty best American cities," are typical examples of this sort of survey. A popular publication is the *Places-Rated Almanac,* which provides a ranking of U.S. cities, based on the aggregate compilation of many variables. Since each city is assigned its own particular rank, these "preference-rankings" are examples of strongly-ordered ordinal variables. Other examples of strongly-ordered variables include the ranking of countries by gross national product per capita or the ranking of states in terms of dollars spent per resident on higher education.

By contrast, in a **weakly-ordered** variable, the values are placed in categories and the categories are then rank ordered. Suppose the task is to construct a choropleth map that shows the percent of population change in each county of the United States from 1980 to 1990. To depict the population change cartographically, six ordinal categories are selected (greater than 10 percent increase, 5 to 9.9 percent increase, and so on), and each of the more than 3,000 counties nationwide must be assigned to one of these six categories. When frequency counts of counties are made in each category, the variable is weakly- rather than strongly-ordered. It is "weak" or "incomplete" in the sense that two counties assigned to the same category on the map cannot be distinguished, even though in reality the counties almost certainly have different population change values.

As with nominally-scaled variables, specific statistical tests are designed to deal with both strongly- and weakly-ordered ordinal variables. Some of these techniques will be discussed in chapters 9 and 10.

Interval and Ratio Scales

With variables measured on either an **interval or ratio scale**, the difference between values can be determined. That is, the interval between any two units of data can be measured on the scale. This means that not only is the relative position of each value known (a value is above or below another), but also how different each unit of data is from all other values on the measurement scale.

Interval and ratio measurement scales can be distinguished by the way in which the origin or zero starting point is determined. With interval scale measurement, the origin or zero starting point is assigned arbitrarily. The Fahrenheit and Celsius scales used in the measurement of temperature are two widely known interval scales in geography. The placement of the zero degree point on both of these measurement scales is arbitrary. With the Fahrenheit scale, zero degrees is the lowest temperature attained with a mixture of ice, water, and common salt, whereas with the Celsius scale zero degrees corresponds to the melting point of ice.

In ratio scale measurement, by contrast, a natural or nonarbitrary zero is used, making it possible to determine the ratio between values. If Montreal, Canada receives 40 inches of annual precipitation and Chihuahua, Mexico receives only 10 inches, the ratio between these two measures is easily calculated (40/10 = 4). Furthermore, it is correct to conclude that Montreal has received four times as much precipitation as Chihuahua. Zero inches of precipitation is a natural or nonarbitrary zero. Such "ratio-type" statements cannot be made with interval-scaled variables. For example, since zero degrees is an arbitrary value on the Fahrenheit scale, 60 degrees Fahrenheit cannot be considered twice as warm as 30 degrees. Many other variables of interest to geographers are measured on a ratio scale, including distance, area, and a number of demographic and socio-economic variables.

Values or units of data from the same variable can be expressed at different measurement scales, depending on how they are collected, organized, and displayed. A state resource planner interested in the type of energy used in homes across the state could use data collected and measured in several different ways. Data could be collected at the individual household level and organized nominally by type of energy used (number of homes in the state using coal, utility gas, fuel oil, etc.). Alternatively, data might be available as county-level summaries, and the counties in the state could be arranged in a strongly-ordered sequence by the percent of households in each county using coal (with the county having the highest percentage of households using coal assigned rank one, and so on). A choropleth map of the county values could also be created, displaying a weakly-ordered graphic representation of the percentage of households using coal. As yet another alternative, data showing the number of households per county using coal could be organized and displayed on a ratio scale.

2.3 Measurement Concepts

As someone becomes more proficient in working with a variety of descriptive and inferential statistics, it becomes tempting to believe that the analysis is truly error-free and that the geographic research problem has been solved. Results from statistical analysis often seem very exact and definitive. If the data are analyzed on the computer, the results are nicely displayed, and the same result is obtained if the data are resubmitted. It may seem appropriate to accept the answers automatically as error-free. However, an error-free result cannot be guaranteed just because a set of data is submitted correctly into some statistical analysis. In fact, several inter-related sources of measurement error can operate separately or in combination to produce problems for the geographer.

Precision

Precision refers to the level of exactness associated with measurement. Precision is often associated with the calibration of a measuring instrument, such as a rain gauge. As a frontal system moves through an area, the amount of rainfall is recorded by two standard rain gauges—each with a different calibration system. On the coarsely calibrated gauge, the amount of rainfall might be estimated as somewhere between 1.2 and 1.3 inches. However, the more finely calibrated gauge provides a more precise estimate of between 1.26 and 1.27 inches.

In many geographic problems, the issue of **spurious precision** must be considered. The computer (or calculator) output will often provide statistics with six or more decimal places, even when the data are in integer form. Reporting seemingly precise statistics based on less precise input is a deceptive but relatively commonplace occurrence. Unless confidence in such a level of measurement precision is warranted, this should be avoided.

Accuracy

The concept of **accuracy** refers to the extent of systemwide bias in the measurement process. It is quite possible for measurement to be very precise, yet inaccurate. Return to the rain gauge example for a moment. Suppose another finely calibrated (more precise) rain gauge is used, but the gauge is improperly calibrated. The person reading the gauge estimates the amount of rainfall to be 1.19 inches rather than the actual rainfall total of about 1.26 or 1.27 inches. The result is an inaccurate reading. Unfortunately, discovering systematic bias in a measurement instrument that is providing inaccurate readings is often quite difficult.

The "target analogy" is frequently used to illustrate the relationships between precision and accuracy (figure 2.1). In case 1, the five bullet holes are closely clustered (precise) and centered on the middle of the target (accurate), making this the best of the four alternatives. Cases 2 and 3 both have problems: the inaccuracy of the bullet holes in case 2 and the imprecision of the holes in case 3 result in different types of errors. Case 4 appears to have the severest problems, with an inaccurate systematic bias toward the upper-left corner of the target as well as considerable scatter of bullet holes that make the results imprecise as well.

Validity

In many geographic problems, the spatial distribution or locational pattern being analyzed is the result of complex processes. If a geo-graphic concept is complex, then expressing the "true" or "appropriate" meaning of that concept through the measurement of any simple variable or set of variables is understandably difficult. **Validity** addresses the measurement issues regarding the nature, meaning, or definition of a concept or variable. Geography is literally saturated with multifaceted, complex variables, such as "level of poverty," "environmental quality," "economic well-being," "level of pollution," "quality of education," and "quality of life." To express the true meaning of such variables or concepts is often not possible, so geographers find it necessary to create **operational definitions** that serve as indirect or surrogate measures. The question then becomes whether the operational definition is valid. A geographer studying the spatial pattern of the "quality of education" in a metropolitan area might evaluate elementary schools on the basis of "average student score on the California Achievement Test" and evaluate high schools by "percent of graduates who subsequently go to college." Clearly, the concept "quality of education" involves much more than what is reflected by these operational definitions, and their validity must be questioned. It should be asked, for example, whether "average CAT score" is a fully valid, somewhat valid, or invalid measure of "quality of education."

Admittedly, the degree of validity in a geographic problem may be difficult to evaluate. Consequently, the question is often ignored, and problems with validity are sometimes "assumed away" as being inconsequential. A good geographic study involving complex variables will discuss the degree of validity of any operational definitions used in the analysis.

Figure 2.1 The Measurement Concepts of Precision and Accuracy: The Target Analogy

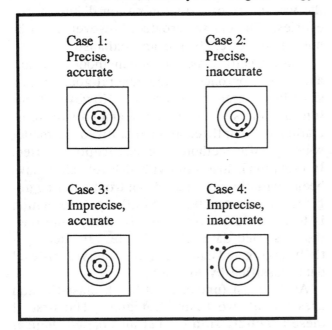

Case 1:
Precise,
accurate

Case 2:
Precise,
inaccurate

Case 3:
Imprecise,
accurate

Case 4:
Imprecise,
inaccurate

Reliability

A final measurement concept often of concern in geographic problems is **reliability**. When changes in spatial patterns are analyzed over time, the geographer must ask questions about the consistency and stability of the data. For example, a geographer could examine whether the spatial pattern of poverty across the United States is more variable today than it was 50 or 100 years ago. A reliable and consistent measure of poverty is needed to answer this question.

Reliability problems often occur when using international data. Fully comparable and totally consistent methods of data collection rarely exist from country to country. A developing country has fewer resources (personnel, money) than a more developed country, and sources of measurement error inevitably affect the data collection process. International comparative statistics therefore become unreliable. Even within the same country or region, locational variations in data collection and processing methods can render the data unreliable. Data collection procedures can change drastically from one time period to the next, again resulting in unreliable data.

One way to assess the degree of reliability of a measurement instrument is to compare at least two applications of the data collection method used at different times. This "test-retest" procedure is used, for example, to evaluate the reliability of IQ test scores, medical diagnoses, and SAT scores. Reliability checks need to be used more frequently in behavioral geography, particularly when analyzing the results from a survey or questionnaire containing individual attitudes and opinions, which are sometimes uncertain and often subject to rapid change.

2.4 Basic Classification Methods

Although categories and classification have been discussed, methods describing *how* to classify data into categories and reasons *why* classification is necessary have not been covered. In this section, the purposes and importance of classification in geographic research are explained. In addition, basic classification methods are reviewed.

Geographers face the problem of deciding how to classify or group spatial data on a fairly routine basis. Classification is used for several important reasons. Classification schemes organize, simplify, and generalize large amounts of information into effective or meaningful categories, bringing relative order and simplicity to complexity. As a result, effective communication is enhanced, detailed spatial information is better understood, and complex spatial patterns are represented more clearly. In many geographic problems, cartographic communication is the goal, and properly classified data result in a map pattern readily understandable to the map reader. Classification is also an integral part of the scientific research process, helping in the formation of hypotheses and guiding further investigations.

Simply stated, an effective classification procedure organizes values according to their similarity. That is, similar values should be placed in the same category, whereas dissimilar values should be placed in different categories. The result is a grouping of data that generally minimizes the amount of fluctuation or variability of values within the same category and maximizes the variability of values between different categories. The classes created, however, must be both mutually exclusive and exhaustive.

Many specific classification methods are available, each approaching the general goals from a slightly different perspective. However, some information is lost when large amounts of information are simplified and generalized, no matter what specific method of classification is used. Information is lost if individual-level values have been spatially aggregated, and only the aggregated data are available. Similarly, information is lost if values have been classified, and only the classified data are available. In fact, spatially aggregated data are simply individual-level data classified by location.

At the most fundamental level, classification uses one of two basic strategies. The first of these is **subdivision** (sometimes called **logical** subdivision). In subdivision, all units of data in a population are grouped together at the start. Then, through a series of steps or iterations, individual values are allocated to the appropriate subdivision using carefully defined criteria. This strategy "works down" or disaggregates all values into logically subdivided classes. Most practical geographic examples of subdivision are hierarchical, with multiple levels of subdivision, depending on the problem or situation. A clear and consistent set of rules is always needed to assign values to the proper category at each stage of the subdivision procedure. The characteristics or values associated with each category are generally defined before the classification procedure begins.

The subdivision strategy of classification is illustrated with two examples. The Soil Conservation Service has developed a basic system of soil classification, which is used for making and interpreting soil surveys. Soils are first subdivided into orders, then further divided into suborders, great groups, subgroups, families, and specific soil series. This subdivision system is a hierarchical taxonomy, concerned with relationships among many different soil properties, the assemblage of soil horizons, and soil-forming processes. The sorting process in this classification is complex: Properties relevant to the sorting in one soil order may have little meaning in another.

Ten soil orders have been determined for the United States, with 47 suborders, more than 200 great groups, and (at the lowest level of the hierarchy) about 9,500 soil series. A small portion of the soil classification system is shown in table 2.1, including examples of the descriptive nomenclature used to identify different soil classes.

The standard industrial classification (SIC) system used to categorize manufacturing products is another illustration of subdivision. In the United States, data on manufacturing activity by product grouping is reported in the *Census of Manufactures,* which is published every five years by the Bureau of the Census. The SIC classification scheme subdivides all industrial activity according to numerical code, ranging from a 1-digit general product grouping at the top level of the hierarchy to a more specific 4-digit breakdown of activity at the lowest level. A small portion of the SIC classification is shown in table 2.2.

Table 2.1   A Portion of the Soil Classification System Developed by the Soil Conservation Service

Order	Suborder	Great Group	Subgroup	Family	Series
Alfisols					
Aridisols					
Entisols	Aquents				
Arents					
	Fluvents	Cryofluvents	Typic Cryofuvents	Coarse-loamy, mixed, acid	Susitna
		Torrifluvents	Typic Torrifluvents	Fine-loamy, mixed (calcareous), mesic	Jocity and Youngston
			Vertic Torrifluvents	Clayey over loamy, mixed, (calcareous), hyperthermic	Glamis
	Orthents	Cryorthents	Typic Cryorthents	Loamy-skeletal, carbonatic	Swift Creek
			Pergelic Cryorthents	Loamy-skeletal, mixed (calcareous)	Durelle
Histosols					

Source: Soil Conservation Service, U.S. Dept of Agriculture.

The second basic classification strategy is **agglomeration**. With this general approach, each unit of data or value in a population or data set is separate and distinct from other individuals at the start of the classification process. The agglomeration procedure then "works up" by allocating values into classes according to well-defined grouping criteria. Agglomeration is accomplished when similar values are combined into the same category, while dissimilar values are placed in different categories. Agglomeration groups similar values in a method that is opposite to subdivision. The agglomeration strategy of classification is very important in the geographic research process. Data for many geographic problems are summarized numerically or graphically using agglomeration.

The appropriate classification strategy is determined only by examining the nature of the geographic research problem and the goals of classification that are specific to that problem. However, some scientists argue that the agglom-

Table 2.2   A Portion of the Standard Industrial Classification (SIC) System

1 Digit	2 Digit	3 Digit	4 Digit
0 Agriculture, forestry, fisheries			
1 Mining and construction			
2 Manufacturing (nondurable)			
3 Manufacturing (durable) -------	30 Rubber and plastics		
	31 Leather		
	32 Stone, clay, glass, concrete		
	33 Primary metals		
	34 Fabricated metals		
	35 Machinery, except electrical		
	36 Electrical and electronic machinery		
	37 Transportation equipment -------	371 Motor vehicles ----- and equipment	3711 Motor vehicles and passenger car bodies
	38 Instruments, photographic goods, optical goods, watches, and clocks		3713 Truck and bus bodies
	39 Miscellaneous		3714 Motor vehicle parts and accessories
			3715 Truck Trailers
4 Transportation, communication, utilities (electric, gas, sanitary services)		372 Aircraft and parts	
5 Wholesale trade		373 Ship and boat building	
6 Finance, insurance, and real estate		374 Railroad equipment	
7 Services (personal, business)		375 Motorcycles, bicycles, and parts	
8 Services (professional, educational)		376 Guided missiles, space vehicles and parts	
9 Public administration (federal, state, and local government)		379 Miscellaneous transportation equipment	

Source: Bureau of the Census, U.S. Dept. of Commerce.

eration strategy is superior both operationally and practically to the subdivision method. Usually faster and easier, agglomeration has the advantage of having at least one value or unit of data in each category. With subdivision, classes can be defined that have no members.

A variety of complex classification methods and powerful clustering procedures that classify multiple variables simultaneously are available to geographers. However, these methods are beyond the scope of this introductory review. Several references that discuss cluster analysis and other multivariate classification procedures in more detail are cited at the end of the chapter. The remainder of this section focuses on four basic methods of classification often used in geographic applications (table 2.3).

1. **Equal intervals based on range.** The **range** is simply the difference in magnitude between the largest and smallest values in an interval-ratio set of data. To determine class breaks (the values that separate one class from another), the range is divided into the desired number of equal-width class intervals. The procedure is easy to apply and results in class intervals of equal width, which may be an advantage. However, because the class breaks are derived from the two units of data having the most extreme values in a data set, results can sometimes be misleading. To maintain the equal width of each class, the class breaks are usually not rounded-off numbers. The number of values in each category may also vary considerably. This may be an advantage or a disadvantage, depending on the purposes of the classification and the goals of the analysis.

2. **Equal intervals not based on range.** This method of classification also designates class breaks to create equal interval classes, but the exact range is not used to select the class breaks. Instead, a convenient or practical interval-width is selected arbitrarily, based on rounded-off class break values. Units of data are then assigned to the categories. This method of classification is preferred for constructing a frequency distribution, histogram, or ogive to represent the data graphically (see section 2.5). Many institutions and government agencies use this method to map complex spatial patterns. The convenient class break values generally result in maps that are easier to understand and interpret. The number of values in each category will vary, which may be either an

Table 2.3 Summary of Basic Classification Methods

Classification Method	Brief Description
Equal intervals based on range	class breaks determined by dividing range (difference between the lowest- and highest-valued units of data) into desired number of equal-width class intervals
Equal intervals not based on range	based on convenience or practical considerations; "rounded-off" class breaks and class interval widths arbitrarily selected
Quantile breaks	equally divide the total number of values into the desired number of classes; two commonly-used divisions are quartiles (4 categories) and quintiles (5 categories)
Natural breaks	place units of data in rank order, identify "natural breaks" or separations between adjacent ranked values, and locate class breaks in the largest of these natural breaks. Iterative process, with largest natural break selected as first class break location, next largest natural break selected second, and so on until desired number of classes created

advantage or disadvantage, depending on the goals of the classification.

3. **Quantile breaks.** This method approaches classification from a somewhat different perspective. The total number of values is divided as equally as possible into the desired number of classes. Two frequently used alternatives are the division of data into quartiles (four categories) or quintiles (five categories). The allocation of an equal number of values to each category is often an advantage in choropleth mapping, particularly if an approximately equal area on the map is desired for each category. However, the possible disadvantages with quantile breaks classification should be evaluated before deciding to use this method. Class breaks are frequently not convenient rounded-off values, and class interval widths are usually not equal for different categories. If a large number of data units are clustered relatively close together (a frequent occurrence with geographic data), these similarly sized values are likely to be "split" unnaturally by a class break to keep an equal number of values in each category.

4. **Natural breaks.** The logic in this classification method is to identify "natural breaks" in the data and separate values into different classes based on these natural breaks. The process is done iteratively, with the largest natural break between adjacent values on a number line selected as the first class break location. The next largest natural break is selected second, and so on until the desired number of classes has been created. If important objectives in the classification process are to ensure that dissimilar values are separated into different categories and that "gaps" in the data are incorporated into the process, then natural breaks may be a good choice. This method often highlights extreme values, placing unusual outliers of data into their own unique categories. Depending on the research problem, highlighting extreme

values may (or may not) be a primary goal of the classification. Another common consequence of natural breaks is the clustering of large numbers of values into one or two categories. Again, it must be decided whether this is an advantage or disadvantage.

To illustrate how each of these methods works, 1989 unemployment rates by state and the District of Columbia are classified. The set of 51 unemployment rates (table 2.4) is too detailed

Table 2.4 Rate of Unemployment by State (in percent), 1989

State	Labor Force Unemployed	State	Labor Force Unemployed
Alabama	7.0	Missouri	5.5
Alaska	6.7	Montana	5.9
Arizona	5.2	Nebraska	3.1
Arkansas	7.2	Nevada	5.0
California	5.1	New Hampshire	3.5
Colorado	5.8	New Jersey	4.1
Connecticut	3.7	New Mexico	6.7
Delaware	3.5	New York	5.1
District of Columbia	5.0	North Carolina	3.5
		North Dakota	4.3
Florida	5.6	Ohio	5.5
Georgia	5.5	Oklahoma	5.6
Hawaii	2.6	Oregon	5.7
Idaho	5.1	Pennsylvania	4.5
Illinois	6.0	Rhode Island	4.1
Indiana	4.7	South Carolina	4.7
Iowa	4.3	South Dakota	4.2
Kansas	4.0	Tennessee	5.1
Kentucky	6.2	Texas	6.7
Louisiana	7.9	Utah	4.6
Maine	4.1	Vermont	3.7
Maryland	3.7	Virginia	3.9
Massachusetts	4.0	Washington	6.2
Michigan	7.1	West Virginia	8.6
Minnesota	4.3	Wisconsin	4.4
Mississippi	7.8	Wyoming	6.3

Source: Bureau of Labor Statistics, U.S. Dept. of Labor.

and cumbersome for convenient study and interpretation. This is particularly true if one of the goals in the analysis is to illustrate the spatial pattern of unemployment in a choropleth map, with each state allocated to one of several unemployment classes.

Deciding the number of classes to be included is very important in choropleth mapping. If the data are allocated into too few categories, key details in the spatial pattern are likely to be lost. If, on the other hand, too many categories are used, the map reader becomes overwhelmed with detail and misses certain important generalizations of the spatial pattern. The decision on the number of classes to use in a particular problem always involves the trade-off between effective

generalization and the communication of sufficient detail. Five classes are used here, a number many cartographers consider reasonable for choropleth mapping of geographic patterns.

Each method of classification is applied to the state-level unemployment rates, and exactly five classes and four class breaks are used in each case. The results are shown in two ways, with number lines (figure 2.2) and with choropleth maps (figure 2.3, cases 1-4). On the number lines, a short vertical bar indicates each class break. On the choropleth maps, the class intervals derived from application of each classification scheme are shown in the map legend.

The results of the classification produce different class breaks along the number lines as

Figure 2.2   Number Lines and Class Break Values for Each Method of Classification: Rate of Unemployment by State, 1989

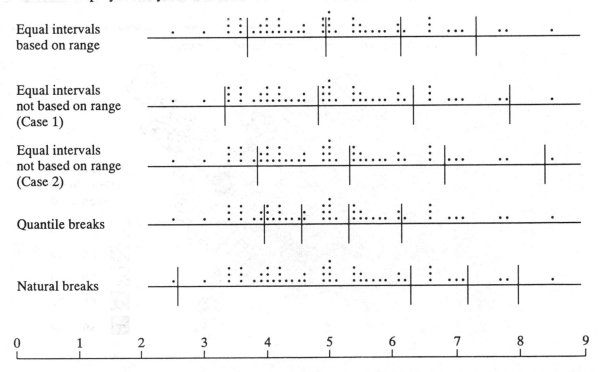

Rate of unemployment by state (in percent)
Source: Data from Bureau of Labor Statistics, U.S. Dept. of Labor.

Figure 2.3   Choropleth Maps for Each Method of Classification: Rate of Unemployment by State, 1989

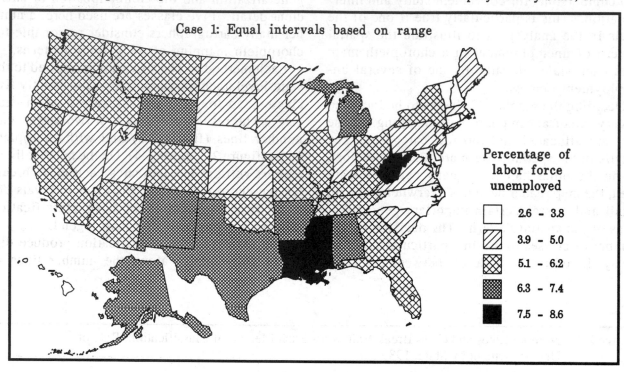

Case 1: Equal intervals based on range

Percentage of labor force unemployed

	2.6 - 3.8
	3.9 - 5.0
	5.1 - 6.2
	6.3 - 7.4
	7.5 - 8.6

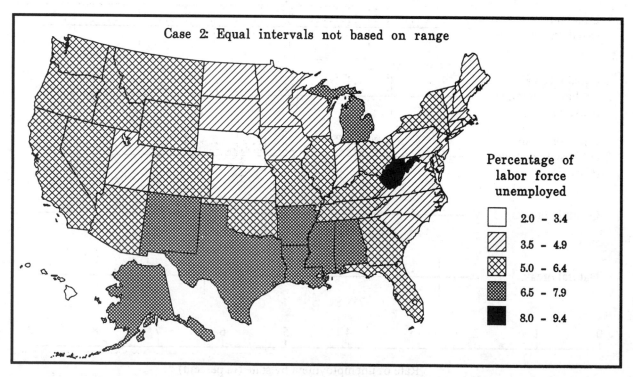

Case 2: Equal intervals not based on range

Percentage of labor force unemployed

	2.0 - 3.4
	3.5 - 4.9
	5.0 - 6.4
	6.5 - 7.9
	8.0 - 9.4

Figure 2.3 Continued

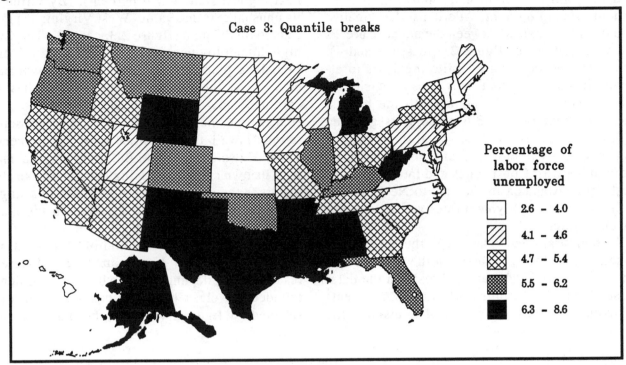

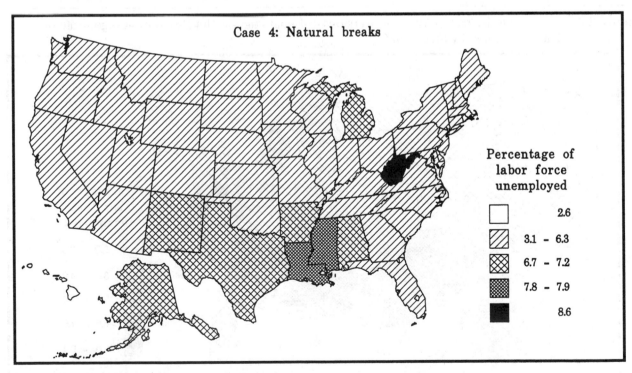

Source: Data from Bureau of Labor Statistics, U.S. Dept. of Labor

well as visibly different maps showing the pattern of unemployment. Perhaps the greatest visual contrast exists between the quantile breaks and natural breaks (figure 2.3, cases 3 and 4). The objective in using quantile breaks is to allocate an equal number of values to each category. Thus, West Virginia (with the highest unemployment rate of 8.6 percent) is grouped with nine other states. The visual impression is that unemployment rates are generally high through the Gulf Coast states and into the Southwest: from Alabama in the east to New Mexico in the west, unemployment rates are in the highest class interval.

With natural breaks, however, the objective is to separate values at places on the number line where there are large natural breaks in the data. West Virginia's unemployment rate (8.6 percent) differs substantially from that of Louisiana (the next highest state at 7.9 percent). By virtue of its unusual extreme value, West Virginia is highlighted on the map (figure 2.3, case 4). The very high unemployment rates of Mississippi and Louisiana place these states in their own category. At the other extreme, Hawaii has an extremely low unemployment rate of 2.6 percent, with a "natural break" of 0.5 percent separating Hawaii from Nebraska (at 3.1 percent). Thus, Hawaii is also highlighted on the map with its own category. Using natural breaks also results in the clustering of most states in a single category: 40 states group together in the class interval that ranges from 3.1 to 6.3.

Even with the same method of classification, equal intervals not based on range, different results occur, depending on which "convenient" rounded-off class break values are arbitrarily selected. In figure 2.4, case 1, the class intervals

Figure 2.4 Choropleth Maps for Alternate Versions of Equal Interval Not Based on Range: Rate of Unemployment by State, 1989

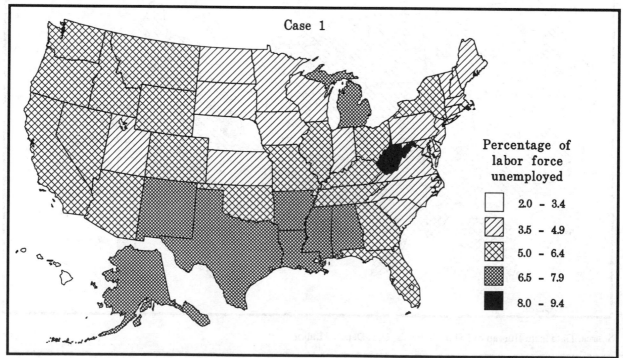

are equal, and the lower bound of each class interval is rounded-off (2.0, 3.5, 5.0, 6.5, and 8.0). Note that the map in figure 2.4, case 1 is the same as figure 2.3, case 2. However, compare this result with figure 2.4, case 2, which also has rounded-off lower bounds on each class interval (2.5, 4.0, 5.5, 7.0, and 8.5). The choropleth map patterns are clearly different, but neither map can be designated as a "better" representation of the data.

What can be concluded about these disparities among classification methods? Depending on the method of classification used, the outcomes can be quite different, even though the same data set is used, and the same number of classes are created. The different class break values in figure 2.2 and visually distinctive choropleth maps of the cases in figures 2.3 and 2.4 illustrate the important effects of classification on spatial patterning and possible conclusions that could be drawn about these spatial patterns.

2.5 Graphic Procedures

When geographers classify data, they are attempting to simplify and generalize information to understand it more fully. Representing data graphically is another way to simplify complex data and increase understanding. In addition to working with a data set as a group of numerical values, geographers can also describe or summarize data graphically to show the same information visually. In section 2.4, a number line was used to display the distribution of state-level data on the percent unemployment data graphically.

Geographers often work with the frequencies for different values or classes of values in a data

---

Figure 2.4   Continued

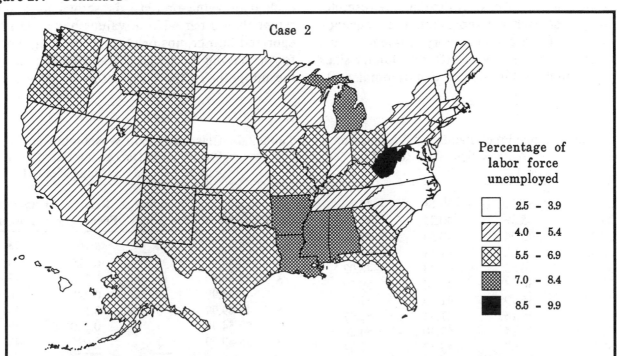

Source: Data from Bureau of Labor Statistics, U.S. Dept. of Labor.

set. For example, in the previous section the frequency of states grouped into unemployment categories by each classification method can easily be determined and presented in a table. However, the same results can often be shown more effectively with a frequency diagram. Figures can be constructed using several different formats, but they all share some common characteristics. Usually the frequency of values is shown on the vertical axis and the range of data (the minimum value to the maximum value) is presented on the horizontal axis. Information can be shown as absolute frequencies (actual frequency counts) or as relative frequencies (percentages or probabilities). Annual precipitation data from Washington, D.C. collected over a 40-year period is used to illustrate several different types of frequency diagrams (table 2.5).

The frequency of different values in a distribution can be shown with a histogram or a polygon: the two most common frequency diagrams. In a **histogram**, the frequency of values is shown as a series of vertical bars, one for each value or class of values. If the data are discrete, the height of each bar represents the frequency of values in a particular category or integer value. To show the frequency of different family sizes for a sample of families, the horizontal scale

could be divided into integer values starting at one and continuing to six or more. The height of each bar would represent the number of families having the specified number of persons.

However, if data have continuous values or the range of data is large, the histogram is usually constructed to show the frequency of each class or group derived from an appropriate classification system (table 2.6). When using categories instead of actual values along the horizontal scale of a histogram, the classification by "equal intervals not based on range" is usually the best technique. With this procedure, the class breaks occur at convenient, "rounded-off" positions, and class interval widths are uniform. Since the precipitation data for Washington, D.C. are continuous and extend from a low of 26.87 inches to a high of 57.54 inches, the histogram of the data is constructed using a series of five-inch intervals from 25 inches to 60 inches (figure 2.5).

A **frequency polygon** is very similar to a histogram, except that the vertical position of each data value or class is shown as a point rather than a bar. If the values have been categorized into groups or classes, the single point for displaying the frequency is usually placed at the midpoint of the class interval. The set of

Table 2.5 Annual Precipitation for Washington, D.C: A Ranked 40-year Record (in inches)

26.87	35.20	39.86	45.62
26.94	35.38	40.21	46.02
28.28	35.96	40.54	47.73
29.48	36.02	41.11	47.90
31.56	36.65	41.34	48.02
32.78	36.83	41.44	50.50
33.07	36.99	41.46	51.17
33.62	38.15	41.94	51.97
34.98	39.34	43.30	54.29
35.09	39.62	43.53	57.54

Source: National Climatic Data Center, U.S. Dept. of Commerce.

Table 2.6 Classification of Washington, D.C. Precipitation Data Using Equal Intervals Not Based on Range Method

Class	Interval	Absolute Frequency	Relative Frequency	Cumulative Frequency
1	25-29.99	4	0.100	4
2	30-34.99	5	0.125	9
3	35-39.99	12	0.300	21
4	40-44.99	9	0.225	30
5	45-49.99	5	0.125	35
6	50-54.99	4	0.100	39
7	55-59.99	1	0.025	40
		40	1.000	

points are then connected by straight lines to produce the frequency polygon. In figure 2.5, the frequency polygon for the Washington, D.C. precipitation example is superimposed on the histogram.

Instead of displaying the absolute frequency count on the vertical axis, a histogram or frequency polygon can easily be converted to show relative frequency. For example, by dividing the individual frequency values by the sum of all frequencies for the data set, the diagrams would display the frequency percentages for each value or class. This change would not affect the general shape of the graphic, only the scale of values along the vertical axis. Histograms and frequency polygons are used in various applications throughout the rest of the book. These graphic tools are especially useful for displaying descriptive statistics and probability.

Another useful method for displaying data in a relative frequency format is a **cumulative frequency diagram**, also known as an **ogive**. Instead of showing actual frequencies for each value or class, this graphic is designed to aggregate frequencies from value to value or class to class

and display the cumulative frequencies at each position. By starting at the lowest value and cumulating higher values, this technique is equivalent to presenting the number of values that are "equal to or less than" each value or class along the horizontal axis. When the cumulative frequency values from table 2.6 are plotted on the vertical axis for each data category or class on the horizontal axis, the result produces a typical pattern for the ogive that has an "s" shape (figure 2.6). Such a diagram is useful for comparing the data value for a given observation with all values for the distribution. The researcher can tell how many values were "less than or equal to" a particular data value. In addition, instead of displaying absolute cumulative frequencies, the data can easily be converted to relative cumulative frequencies (or cumulative percentages). Following the procedure discussed earlier, the absolute cumulative frequencies can be divided by the sum of all frequencies to obtain relative cumulative values or percentages. Cumulative relative frequencies are used to test distributions for normality in section 10.1.

The graphic procedures discussed up to now focus on ways of displaying the distribution or frequency

Figure 2.5   Histogram and Frequency Polygon for 40-year Annual Precipitation Data in Washington, D.C.

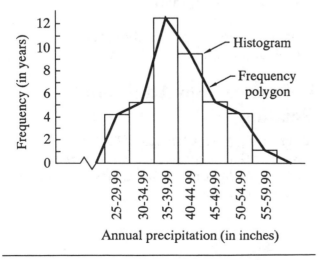

Figure 2.6   Cumulative Frequency Polygon (Ogive) for 40-year Annual Precipitation Data in Washington, D.C.

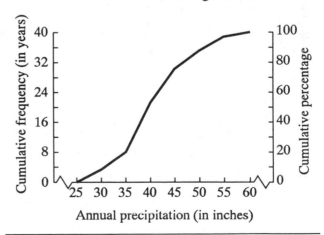

for one variable. Another type of graph, termed a **scatterplot** or scattergram, shows the pattern of association or relationship between two variables. A scatterplot requires that the two variables be measured on the interval/ratio scale for a common set of observations. When one variable is represented on the vertical or Y axis and the other on the horizontal or X axis, each observation can be shown as a point within the graph. Analysis of the scatter of points suggests the amount and nature of association or relationship that exists between the two variables graphed. For example, the 40 years of precipitation data in Washington, D.C. can be used with the corresponding yearly data for mean annual temperature to construct a scatterplot of the two variables (figure 2.7). The arrangement of points within the graph is generally random, suggesting the lack of meaningful association between precipitation and temperature. In chapters 12 and 13, scatterplots are used with correlation and regression analyses to investigate bivariate relationships.

Figure 2.7 Scatterplot Showing Association between Precipitation and Temperature for 40-year Period in Washington, D.C.

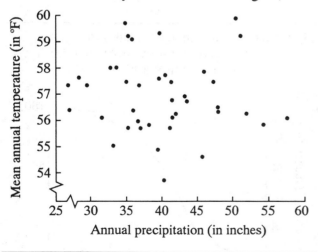

Key Terms and Concepts

References and Additional Reading

Abler, R. F., J. S. Adams, and P. R. Gould. 1971. *Spatial Organization: The Geographer's View of the World.* Englewood Cliffs, NJ: Prentice-Hall.

Blalock, H. M. Jr. 1982. *Conceptualization and Measurement in the Social Sciences.* Beverly Hills, CA: Sage Publications.

Jenks, G. F. and R. C. Coulson. 1963. "Class Intervals for Statistical Maps." *International Yearbook of Cartography* 3:119-34.

Monmonier, M. S. 1975. *Maps, Distortion, and Meaning* (Resource Paper No. 75-4). Washington, DC: Assoc. of Amer. Geographers.

Robinson, A., R. Sale, and J. Morrison. 1984. *Elements of Cartography* (5th edition). New York: John Wiley and Sons, Inc.

Sokal, R. R. 1966. "Numerical Taxonomy." *Scientific American* 215, No. 6:106-16.

PART II

DESCRIPTIVE PROBLEM SOLVING IN GEOGRAPHY

CHAPTER 3

DESCRIPTIVE STATISTICS

In chapter 1, a basic distinction is made between descriptive and inferential statistics. Recall that the overall goal of descriptive statistics is to provide a concise, easily understood summary of the characteristics of a data set. For most geographic problems, such quantitative or numerical summary measures are clearly superior to working with unsummarized raw data. With these easily understood and widely used descriptive statistics, geographers can communicate effectively.

Some of the advantages of summarizing spatial information have already been demonstrated in the discussion of basic classification methods (section 2.4) and through the use of such visual procedures as histograms and ogives (section 2.5), which provide *graphic* summary descriptions of a data set. Here, through the use of basic descriptive statistics, complementary *numerical* or *quantitative* summary measures "describe" a data set. The descriptive statistics discussed in this chapter are illustrated with the same set of precipitation data used to construct the graphics in section 2.5.

A data set can be summarized in several different ways:

1. *Measures of central tendency*—numbers that represent the center or typical value of a frequency distribution, such as mode, median, and mean (section 3.1)
2. *Measures of dispersion*—numbers that depict the amount of spread or variability in a data set, such as range, interquartile range, standard deviation, variance, and coefficient of variation (section 3.2)
3. *Measures of shape or relative position*—numbers that further describe the nature or shape of a frequency distribution, such as skewness (which indicates the amount of symmetry of a distribution) or kurtosis (which describes the degree of flatness or peakedness in a distribution). (See section 3.3.)

The choice of descriptive statistics for a particular geographic problem depends partly on the level of measurement; namely, on nominal, ordinal, or interval-ratio. In addition, different calculation procedures are applied if the data are grouped (weighted) or ungrouped (unweighted).

Geographers use descriptive statistics in a variety of ways. Spatial insights are gained through the innovative, comparative application of these measures, and answers to sophisticated spatial questions are often derived through the informed, knowledgeable use of these simple measures. In this chapter, examples of such spatial questions are explored through descriptive statistics.

Geographers must be cautious when applying descriptive statistics to spatial or locational data. The way in which a geographic problem is structured can affect the resultant descriptive statistics. The following issues are discussed in section 3.4: (1) the effects of boundary line delineation and study area location on descriptive measures; (2) the effect of altering internal subarea boundaries within the same overall study area—the so-called modifiable areal units problem; and (3) the impact of using different levels of spatial aggregation or different scales on descriptive statistics.

3.1 Measures of Central Tendency

The central or typical value of a set of data can be described numerically in several different ways. Each of these measures of central tendency has advantages and disadvantages, and the logic underlying the calculation procedure for each measure is different. To select the most appropriate measure in a particular geographic situation requires an understanding of each measure and its characteristics. The discussion in this section will be limited to three common measures of central tendency: mode, median, and arithmetic mean.

Mode

The **mode** is simply the value that occurs most frequently in a set of ungrouped data values. When nominal data are used, the mode would be the category containing the largest number of observations. With ordinal or interval-ratio data grouped into classes, the category with the largest number of observations is defined as the **modal class**. The midpoint of the modal class interval is the **crude mode**. A mode can be calculated for data at all levels of measurement, but with nominal data, the mode is the only available descriptive measure of central tendency.

Although a useful measure of central tendency in many data sets, the mode may not always provide a practical result. For example, the mode is not appropriate when applied to the annual precipitation data from Washington, D.C. (table 2.5). In this particular data set, no annual precipitation figure occurs more than once over the 40-year sample period, making the mode ineffective. In fact, a large number of tied values is unlikely in geographic situations where data are interval or ratio scale. However, if the precipitation data are grouped (table 2.6), the modal class is 35-39.99 (with 12 values), and the crude mode of 37.5 is the midpoint of this modal class interval.

Median

The **median** is the middle value from a set of observations that has been ranked. If values from a set of data are placed in ranked numerical order, the median is the value with an equal number of data units both above it and below it. With an odd number of observations, the middle value is unique and defines the median. With an even number of observations, the median is defined as the midpoint of the values of the two middle ranks. For the Washington, D.C. data, there are 40 values, so the two "middle" values are rank 20 (39.62 inches of precipitation) and rank 21 (39.86 inches), and the median is their midpoint (39.74 inches).

Mean

The **arithmetic mean** (also called the mean or average) is undoubtedly the most widely used measure of central tendency. It is usually the most appropriate measure when using interval or ratio data. The arithmetic mean ($\overline{X}$) is simply the sum of a set of values divided by the number of observations in the set. In standard statistic notation, the mean is defined as follows:

$$\overline{X} = \frac{\sum_{i=1}^{n} X_i}{n} = \frac{X_1 + X_2 + \ldots + X_n}{n} \qquad (3.1)$$

where: $\overline{X}$ = mean of variable X

   X_i = value of observation i

   Σ = summation symbol (uppercase sigma)

   n = number of observations

Because it is generally understood that summation is over all n observations, the symbols above and below sigma are usually omitted:

$$\overline{X} = \frac{\Sigma X_i}{n} \qquad (3.2)$$

The calculation of mean annual precipitation for the Washington, D.C. data is shown in table 3.1.

In many geographic situations it is imperative to differentiate a sample mean from a population mean. Following conventional statistical notation, lowercase n refers to sample size and uppercase N to population size. Population characteristics are customarily defined using Greek letters, while sample measures are not. The formula for a population mean (μ = the Greek letter mu) with N values is:

$$\mu = \frac{\Sigma X_i}{N} \qquad (3.3)$$

Because it is possible to draw many samples of size n from a large population of size N, a different-magnitude sample mean may result from each sample drawn. However, the population mean is a fixed value. In chapter 7, when applying sample information to estimate population characteristics, the distinction between samples and populations is explored further.

The mean can also be calculated for grouped data. To permit this calculation, two related assumptions must be made about the distribution of values within each category or class interval: (1) the data are assumed to be evenly distributed within each class interval, and (2) the values in each interval are best represented by the midpoint. The formula for the weighted mean ($\overline{X}_w$) is:

$$\overline{X}_w = \frac{\displaystyle\sum_{j=1}^{k} X_j f_j}{n} \qquad (3.4)$$

where: $\overline{X}_w$ = weighted mean

X_j = midpoint of class interval j

f_j = frequency of class interval j

k = number of class intervals

n = total number of values = $\displaystyle\sum_{j=1}^{k} f_j$

The weighted mean for the Washington, D.C. precipitation data (40.25 inches) is slightly larger than the unweighted mean (39.95 inches), but the difference is not substantial (table 3.2). In general, weighted and unweighted means should be expected to differ somewhat, depending on

Table 3.1 Work Table for Calculating Arithmetic Mean of Washington, D.C. Precipitation Data

Observation (i)	Precipitation (X_i)
1	41.11
2	54.29
3	35.09
...	...
38	34.98
39	35.96
40	50.50
Sum	1598.00

$$\overline{X} = \frac{\Sigma X_i}{n} = \frac{41.11 + 54.29 + \ldots + 50.50}{40} = \frac{1598.00}{40} = 39.95$$

Table 3.2 Work Table for Calculating Weighted Mean of Washington, D.C. Precipitation Data

Class Interval j	Class Midpoint X_j	Class Frequency f_j	$X_j f_j$
25-29.99	27.5	4	110.0
30-34.99	32.5	5	162.5
35-39.99	37.5	12	450.0
40-44.99	42.5	9	382.5
45-49.99	47.5	5	237.5
50-54.99	52.5	4	210.0
55-59.99	57.5	1	57.5
Sum		40	1610.0

$$\overline{X}_w = \frac{\sum X_j f_j}{n} = \frac{1610.0}{40} = 40.25$$

the overall effect of using class midpoints as representative summary values for each category. For example, a weighted mean will be higher than the corresponding unweighted mean, if the majority of original values in most of the categories happen to fall below the class midpoints.

Selecting the Proper Measure of Central Tendency

Which measure of central tendency to use depends both on the geographic application and certain key characteristics of the data set. On the surface, the mean seems to have certain advantages. It is the most widely applied and generally understood measure of central tendency. In addition, it is always affected to some extent by changes or modifications of any value in a data set. This is not always true of either the mode or the median. The mean is also an element in many tests of statistical inference when estimates of population parameters are made from sample information. When working with interval-ratio data, the mean is the measure of choice.

Despite these inherent advantages, sometimes the mean is not the best measure of central tendency:

1. If a frequency distribution is **skewed** (not symmetric in nature), the mean will be the measure of central tendency most dramatically displaced in the direction of the skew—whether **positive or negative.** Figure 3.1 illustrates the location of the mean, median, and mode under different frequency distributions. In a symmetric, evenly balanced distribution (case 1), all three "centers" are similarly located: no advantage accrues to any statistic. With slight positive skew (case 2), the median is pulled slightly to the right, toward the positive tail of the distribution. The location of the mean is even more significantly affected, positioned even further out the positive tail than the median. In a frequency distribution with a large negative skew (case 3), the magnitude of displace-

ment of the median and, especially the mean, is even more pronounced.

2. If a frequency distribution is **bimodal** (having two distinct modes) or **multimodal** (with more than two modes), the mean may be located on the distribution at a place which would not be considered "typical." In figure 3.2, neither the mean nor median depict useful central locations; the values of mode A and mode B are more informative.

3. If a frequency distribution contains a small number of extreme or atypical values, the mean will be heavily influenced by these values, and its effectiveness as a measure of

Figure 3.1 Measures of Central Tendency Placed on Symmetric and Skewed Frequency Distributions

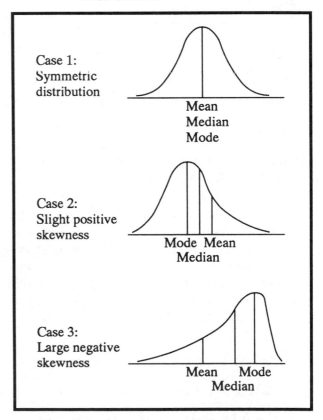

Figure 3.2   Measures of Central Tendency Placed on a Bimodal Frequency Distribution

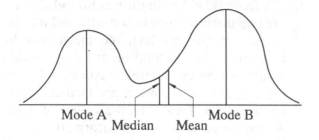

centrality may be reduced. The existence of an extreme value or values is actually a form of skewness. A simple example using the incomes of seven families illustrates the sensitivity of the mean to a single extreme value:

$21,000
 21,000   mode = $21,000
 22,000
 26,000   median = $26,000
 27,500
 32,500
349,000   mean = $\dfrac{500,000}{7}$ = $71,428.57

500,000

The mean is ineffective as a measure of central tendency in this example because it is greatly affected by the income of a single atypical wealthy family. The mode is too low an income to represent a useful "central" value from this frequency distribution. The median provides the most suitable measure in this situation. In fact, with variables such as income, which often contain a high amount of skewness, the median is generally the descriptive statistic of choice.

3.2 Measures of Dispersion and Variability

Simple Measures of Dispersion

The three measures of central tendency—mode, median, and mean—specify the center of a distribution, but provide no indication of the amount of spread or variability in a set of values. Dispersion can be calculated in several different ways. The level of measurement and nature of the frequency distribution determine which dispersion statistic is most appropriate.

The simplest measure of variability is the **range**, defined as the difference between the largest and smallest values in an interval-ratio set of data. Because it is derived solely from the two most extreme or atypical values and ignores all other values, the range can be a misleading measure. The range measured for the 40 Washington, D.C. precipitation values is (57.54 − 26.87) = 30.67.

If data are grouped, the range is defined as the difference between the upper value in the highest numbered class interval and the lower value in the lowest numbered class interval. With the grouped precipitation data, these values are (59.99 − 25) = 34.99. As with the ungrouped range, the grouped range can be a deceptive measure of dispersion. Given only the range, the degree of clustering or dispersion of the values between these two extremes is unknown.

To provide more information, certain intervals, portions, or percentiles within the frequency distribution can be examined. The data can be divided into equal portions or percentiles (**quantiles**), and the range of values within any quantile can be calculated and graphed. Although any logical subdivision is possible, data are often classified into quartiles (fourths), quintiles (fifths) or deciles (tenths). The median, which is the 50th percentile, may also be one of these subdivisions.

When data are divided into four portions or quartiles, the **interquartile range** is defined as the difference between the 25th percentile value (the lower quartile) and the 75th percentile value (the upper quartile). The interquartile range thus encompasses the "middle half" of the data. In the Washington, D.C. precipitation example, the interquartile range is (43.53 − 35.20) = 8.33.

A **dispersion diagram** can be constructed to display graphically the entire frequency distribution within lines that box in the interquartile range and show the upper quartile, median, and lower quartile. Figure 3.3 shows dispersion diagrams of annual precipitation for three American cities: San Diego, California; St. Louis, Missouri; and Buffalo, New York. This graphic representation of data offers useful information for comparing the variability of several distributions.

Other measures of dispersion are based on an examination of individual deviations from the mean value. The difference between each value and the mean is calculated, and these deviations are used as the building blocks to measure dispersion. An individual deviation (d_i) is calculated as follows:

$$d_i = \left(X_i - \overline{X} \right) \qquad (3.5)$$

where:

d_i = deviation of value i from the mean

X_i = value of observation i

$\overline{X}$ = mean of variable X

The **average deviation** or **mean deviation** is based on the mean of the set of individual deviations. That is, the mean deviation (m) for a sample of values is:

$$m = \frac{\Sigma \left| X_i - \overline{X} \right|}{n} \qquad (3.6)$$

where: $\left| X_i - \overline{X} \right|$ = the absolute value of the difference between X_i and $\overline{X}$

Some units of data are larger than the mean, making the difference ($X_i - \overline{X}$) a positive number. Conversely, values below the mean make ($X_i - \overline{X}$) a negative number. However, the sum of the deviations about a mean is always zero. To avoid this problem of negative differences offsetting positive differences, the absolute value of each individual deviation is calculated. With absolute values, negative deviations are converted to positive deviations. For example: $|-2| = |2| = 2$.

Standard Deviation and Variance

The most common measure of variability is **standard deviation**. Given a sample of values, the standard deviation (s) is defined as:

$$s = \sqrt{\frac{\Sigma \left(X_i - \overline{X} \right)^2}{n-1}} \qquad (3.7)$$

In equation 3.7, the deviation of each value from the mean is squared, then all of these squared deviations are summed. This squaring process removes the problem of negative deviations offsetting positive deviations more effectively than absolute values. The sum of the squared deviations is then divided by ($n - 1$). Finally, to reverse the effect of squaring, the square root of this quotient is taken.

Figure 3.3 Dispersion Diagrams of Annual Precipitation for Three Cities

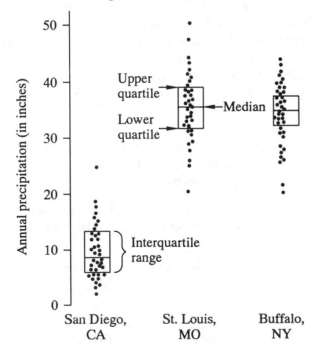

Source: National Climatic Data Center, U.S. Dept. of Commerce.

For a population, the standard deviation formula is written:

$$\sigma = \sqrt{\frac{\Sigma(X_i - \mu)^2}{N}} \qquad (3.8)$$

where: σ = lowercase Greek sigma

Note the different denominators of the two standard deviation formulas. For a sample, the denominator is $n - 1$, but is N for a population. If the standard deviation is being calculated for a large sample, $(n > 30)$, the difference between n and $(n - 1)$ is small, and the difference between s and σ will also be small. However, with a small sample $(n < 30)$, the true population standard deviation (σ) is underestimated if division is by n. When dividing by $(n - 1)$, s becomes larger, correcting the underestimation problem. The incorporation of this sample size correction provides a better estimate of the true population standard deviation for smaller samples.

The **variance** of a data set, defined as the square of the standard deviation, provides a measure of the average squared deviation of a set of values around the mean. Variance is seldom used directly as a summary descriptive statistic, since the calculated variance is generally a very large number and is difficult to interpret. Nevertheless, variance is an important descriptive statistic, which has many applications in inferential procedures. The formulas for sample variance (s^2) and population variance (σ^2) are as follows:

$$s^2 = \frac{\Sigma(X_i - \overline{X})^2}{n - 1} \qquad (3.9)$$

$$\sigma^2 = \frac{\Sigma(X_i - \mu)^2}{N} \qquad (3.10)$$

Both standard deviation and variance are fundamental building blocks in statistical analysis.

In chapter 7, which discusses sampling procedures, these descriptive measures are used as the basis for making probabilistic statements. Confidence intervals will be placed around estimates derived from samples to predict population parameters. These statistics are also integral components in inferential hypothesis testing (see chapter 9). For example, variance is used in the statistical technique of analysis of variance (ANOVA), which attempts to determine whether multiple samples differ significantly from one another.

A final issue regarding standard deviation and variance requires clarification: the procedural question of using more convenient *computational* formulas rather than the more cumbersome and time-consuming *definitional* formulas. The formulas based on the definition of standard deviation are valuable because they show how variability is a direct function of individual deviations $(X_i - \overline{X})$ or $(X_i - \mu)$. However, if it is necessary to calculate standard deviation manually, without computer assistance, more efficient computational formulas are available. These alternative formulas, based on algebraic manipulations of the definitional formulas, are summarized in table 3.3. Since variance is the square of the standard deviation, it is calculated by simply removing the square root symbol from each formula in table 3.3.

Table 3.3 Definitional and Computational Formulas for Standard Deviation of Sample and Population

	Definitional Formula	Computational Formula
Sample (s)	$\sqrt{\dfrac{\Sigma(X_i - \overline{X})^2}{n - 1}}$	$\sqrt{\dfrac{\Sigma X_i^2 - n\overline{X}^2}{n - 1}}$
Population (σ)	$\sqrt{\dfrac{\Sigma(X_i - \mu)^2}{N}}$	$\sqrt{\dfrac{\Sigma X_i^2}{N} - \mu^2}$

The calculation procedures for both the definitional and computational formulas of sample standard deviation are demonstrated in table 3.4 using the Washington, D.C. precipitation data. The statistical parity of the definitional and computational formulas can be easily seen.

For calculating standard deviation and variance from grouped data, the same assumptions are made as when calculating a weighted mean. That is, the values are assumed to be evenly distributed within each class interval, making the midpoint the best representation of the middle of that class interval. As with the weighted average, the weighted standard deviation (s_w) and variance (s_w^2) formulas utilize these class midpoints (X_j) and class frequency counts (f_j). Although a

definitional formula for weighted standard deviation exists, table 3.5 uses the simpler computational formula for calculating this index with the Washington precipitation data.

The weighted standard deviation $(s_w = 7.59)$ is similar in magnitude to the unweighted standard deviation $(s = 7.50)$. A slight difference between the two descriptive statistics is expected, given the contrasting assumptions which underlie the calculation procedures.

Coefficient of Variation

For many types of geographic research, it is extremely valuable to compare the amount of variability in different spatial patterns directly in order to examine which has the greatest spatial variation. In other geographic problems, comparing variability in some phenomenon as it changes over time may be useful. For example, a climatologist might want to compare the

Table 3.4 Work Table for Calculating Sample Standard Deviation of Washington, D.C. Precipitation Data

Observation (i)	X_i	$X_i - \overline{X}$	$(X_i - \overline{X})^2$	X_i^2
1	41.11	1.16	1.35	1690.03
2	54.29	14.34	205.64	2947.40
3	35.09	-4.86	23.62	1231.31
...	...	...	...	...
38	34.98	-4.97	24.70	1223.60
39	35.96	-3.99	15.92	1293.12
40	50.50	10.55	111.30	2550.25
Sum	1598.00		2192.76	66032.86

$n = 40$
$n - 1 = 39$

$$\overline{X} = \frac{\Sigma X_i}{n} = \frac{1598.00}{40} = 39.95$$

Using the definitional formula for sample standard deviation:

$$s = \sqrt{\frac{\Sigma(X_i - \overline{X})^2}{n-1}} = \sqrt{\frac{2192.76}{39}} = 7.50$$

Using the computational formula for sample standard deviation:

$$s = \sqrt{\frac{\Sigma X_i^2 - n\overline{X}^2}{n-1}} = \sqrt{\frac{66032.86 - 40(39.95)^2}{39}} = 7.50$$

Table 3.5 Work Table for Calculating Weighted Standard Deviation of Washington, D.C. Precipitation Data

Class Interval j	Class Midpoint X_j	Class Frequency f_j	$X_j f_j$	$(X_j)^2 f_j$
25-29.99	27.5	4	110.0	3025.00
30-34.99	32.5	5	162.5	5281.25
35-39.99	37.5	12	450.0	16875.00
40-44.99	42.5	9	382.5	16256.25
45-49.99	47.5	5	237.5	11281.25
50-54.99	52.5	4	210.0	11025.00
55-59.99	57.5	1	57.5	3306.25
Sum		40	1610.0	67050.00

Using the computational formula for sample standard deviation:

$$s_w = \sqrt{\frac{\Sigma(X_j)^2 f_j - (\Sigma X_j f_j)^2 / n}{n-1}}$$

$$= \sqrt{\frac{67050 - (1610)^2 / 40}{39}} = 7.59$$

variability in annual rainfall at different meteorological stations to learn which locations have the greatest variation from year to year. An economic development planner might be interested in comparing family income within several different counties to learn which region has the greatest internal variation. Asking these investigative, comparative questions allows geographers to explore practical problems in a highly productive fashion.

Using standard deviation or variance to make direct comparison between locations or regions is inappropriate, because they are both **absolute measures**. Their values depend on the size or magnitude of the units from which they are calculated. A data set with large numbers (i.e., in the thousands) will be described with a large average, standard deviation, and variance. Conversely, analysis of a data set containing single digit numbers will result in small absolute descriptive measures. Clearly, direct comparison of averages, standard deviations, and variances has limited utility.

To resolve this problem and compare the spatial variation of two or more geographic patterns, a **relative measure** of dispersion has been developed. Called the **coefficient of variation (CV)** or **coefficient of variability**, this index is simply the standard deviation expressed relative to the magnitude of the mean:

sample $\quad CV = \dfrac{s}{\overline{X}} \text{ or } CV = \dfrac{s}{\overline{X}} \cdot 100 \quad\quad (3.11)$

The coefficient of variation is usually expressed as a proportion or percentage of the mean and may be used with either sample or population data. Dividing the standard deviation by the mean removes the influence of the magnitude of the data and allows direct comparison of the relative variability in different data sets.

In the following example, variability for locational data whose magnitude varies from place to place is directly compared using the coefficient of variation. In studying annual precipitation data over a 40-year period for three sample locations—Buffalo, St. Louis, and San Diego—sharp contrasts are evident (table 3.6).

Although it appears that Buffalo and St. Louis have similar average amounts of precipitation (35.47 versus 35.56), St. Louis has a large absolute amount of variability in precipitation from year to year ($s = 6.62$ versus $s = 4.70$). This result can be explained in terms of location. St. Louis is found in the middle of the continent, while Buffalo is in close proximity to a consistent source of moisture from Lake Erie. But of the three cities, San Diego has by far the greatest degree of relative variability in precipitation from year to year. Although San Diego's standard deviation is only 4.42 (the smallest of the three values), it is a relatively large amount of variability when compared to San Diego's rather low average precipitation of 9.62. The higher coefficient of variation for San Diego clearly demonstrates the relatively large fluctuations in precipitation from year to year in a semiarid climate. This type of investigative comparison of coefficient of variation values could be taken further and has been used to explore practical problems dealing with spatial patterns of climatic variability.

Only ratio-scale data should be used when calculating the coefficient of variation. Data measured at the interval scale are not appropriate because the interval metric has an arbitrary zero. As a result, numerical manipulations involving multiplication or division are meaningless. Since the coefficient of variation is a ratio

Table 3.6 Descriptive Statistics Using 40 Years of Annual Precipitation Data for Three Cities

City	Mean	Standard Deviation (Absolute)	Coefficient of Variation (Relative)
Buffalo, NY	35.47	4.70	13.25
St. Louis, MO	35.56	6.62	18.62
San Diego, CA	9.62	4.42	45.95

of the standard deviation to the mean, it follows that a coefficient of variation value has no meaning when calculated using interval scale data.

The coefficient of variation could be used more frequently by geographers in their research. This relative index of dispersion is easily calculated from simple descriptive statistics and has many potential areas of application. As a numerical measure of relative variability, geographers can use the coefficient of variation to summarize maps and other spatial patterns quantitatively. Given multiple sets of locational or explicitly spatial data, geographers can use the coefficient of variation to measure and to compare the relative variability in the data. As another application, changes in a regional or areal pattern over time can be numerically summarized, and regional trends toward dispersal or clustering observed. With general access to computers and computer data bases, such direct comparisons between locations or regions can be made with relative ease.

3.3 Measures of Shape or Relative Position

Skewness and Kurtosis

In addition to the coefficient of variation, other relative measures are available to describe the nature or character of frequency distributions. Two of the more useful measures are skewness and kurtosis. **Skewness** measures the degree of symmetry in a frequency distribution by determining the extent to which the values are evenly or unevenly distributed on either side of the mean. **Kurtosis** measures the flatness or peakedness of a data set. Like the coefficient of variation, these indices are underutilized by geographers, yet they provide important descriptive insights about a frequency distribution and have considerable potential in spatial research.

Introducing the concept of moments of a distribution about the mean allows better understanding of skewness and kurtosis. The first moment is the sum of individual deviations about the mean and must equal zero:

$$\text{First moment} = \Sigma d_i = \Sigma \left(X_i - \overline{X} \right)^1 = 0 \qquad (3.12)$$

The second moment of a frequency distribution is the numerator in the expression which defines variance:

$$\text{Second moment} = \Sigma \left(X_i - \overline{X} \right)^2 \qquad (3.13)$$

Skewness involves use of the third moment of a frequency distribution. One commonly used measure of relative skewness contains the third moment in the numerator:

$$\text{Skewness} = \frac{\Sigma \left(X_i - \overline{X} \right)^3}{ns^3} \qquad (3.14)$$

The denominator of this expression contains the cubed standard deviation, which effectively standardizes the third moment. This allows geographers to compare directly the amount of relative skewness in different frequency distributions.

If a frequency distribution is **symmetric**, with an equal number of values on either side of the mean, the distribution has little or no skewness. If a value in a distribution is greater than the mean, its cubed deviation will be positive. However, if a value is less than the mean, it will produce a negative cubed deviation. In a symmetric distribution, these positive and negative cubed deviations will counterbalance each other, and the sum (the third moment) will be zero. In a distribution having a tail to the left, large negative cubed deviations will cause the sum of all deviations (the third moment) to be negative. The resultant distribution is said to be **negatively skewed**. On the other hand, in a distribution with a tail to the right, large positive cubed deviations will dominate the sum, and a **positively-skewed** distribution will result.

Another measure of skewness, Pearson's coefficient of skewness, is based on a comparison of the mean and median:

$$\text{Pearson's Skewness} = \frac{3(\overline{X} - \text{Median})}{s} \quad (3.15)$$

With a symmetric distribution, the mean and median have the same value, and the skewness coefficient is zero (figure 3.1, case 1). When the mean is greater than the median (as in case 2), positive skewness results, and when the mean is less than the median (as in case 3), the skewness measure is negative. Division by the standard deviation provides a standardization of the Pearson's skewness values, allowing direct comparisons.

Kurtosis measures the fourth moment of a frequency distribution using one of the following formulas:

$$\text{Kurtosis} = \frac{\Sigma(X_i - \overline{X})^4}{ns^4} \quad (3.16)$$

$$\text{Kurtosis} = \frac{\Sigma(X_i - \overline{X})^4}{ns^4} - 3 \quad (3.17)$$

If a large proportion of all values is clustered in one part of the distribution, it will have a "pointed" or "peaked" appearance, a high level of kurtosis, and be considered **leptokurtic**. In a data set having low kurtosis (a "flat" or **platykurtic** distribution), values are dispersed more evenly over many different portions of the distribution.

The interpretation of kurtosis is enhanced by comparing the peakedness of a distribution to that of a normal probability distribution. Although the importance of the normal curve is discussed in more detail in chapter 5, it is relevant here because kurtosis formulas assign characteristic values to a normal distribution. In equation 3.16, a normal distribution has a kurtosis of 3.0,

a leptokurtic distribution has a kurtosis greater than 3.0, and a platykurtic distribution has a value less than 3.0. Many computer packages use equation 3.17, which subtracts 3.0 from the quotient, so that a normal distribution has zero kurtosis. In this way, platykurtic distributions have negative values, and leptokurtic distributions are positive.

Exploratory Comparison of Skewness and Kurtosis Values

Comparative examination of skewness and kurtosis values can provide geographers with useful insights about the nature of spatial patterns. Geographers may want to compare descriptive statistics for a particular variable at different locations during a single time period. They may ask questions such as: During this period of time, which locations had the greatest relative variability? The greatest amount of skewness? The most leptokurtic distribution? After these questions have been examined and the appropriate data analyzed, geographers can continue the investigation by exploring the reasons *why* the observed skewness or kurtosis values occurred. By using these descriptive statistics more fully, geographers gain a better understanding of their data.

In the previous section, annual precipitation data from Buffalo, St. Louis, and San Diego were compared over the same 40-year period using means, standard deviation, and coefficient of variation. In table 3.7, intercity comparisons are made for skewness and kurtosis. Note both the high positive level of skewness and large positive (leptokurtic) kurtosis value for San Diego relative to Buffalo and St. Louis. Because skewness and kurtosis are standardized third and fourth moments, respectively, they are quite sensitive to values that have a large relative difference from the mean. This sensitivity is particularly high with kurtosis, where the deviations between individual values and the mean are taken to the fourth power.

The precipitation data for San Diego clearly demonstrate the sensitivity of skewness and kurtosis to extreme values. With a low mean annual precipitation of 9.62 inches and a semi-arid climate, San Diego's 40-year rainfall record shows several very wet years. For example, the 40 years of data include annual totals of 24.93 inches, 17.74 inches, and 19.03 inches, all of which are several standard deviations above average ($s = 4.42$). These unusually high annual precipitation values are shown on the dispersion diagram (figure 3.3), and their existence dominates the subsequent descriptive statistics. The result is a large positive skewness value, with a few very wet years extending the tail of the distribution to the right.

San Diego also has a peaked kurtosis value, with all but a few atypically wet years clustered together at the lower end of the frequency distribution. In fact, 24 of the 40 values are grouped in the range between 6 inches and 12 inches. In contrast, Buffalo has a slightly negative skewness value, with more of a tail to the left (toward relatively dry years). In Buffalo, the highest precipitation total over the 40-year record is 44.78 inches, only about 9 inches over the 40-year mean. Although Buffalo receives a considerable amount of precipitation each year, the total seldom deviates very much from the mean, especially in the "wetter" direction.

Table 3.7 Additional Descriptive Statistics Using 40 Years of Annual Precipitation Data for Three Cities

City	Mean	Standard Deviation	Skewness	Kurtosis
Buffalo, NY	35.47	4.70	− 0.389	0.672
St. Louis, MO	35.56	6.62	0.080	− 0.096
San Diego, CA	9.62	4.42	1.393	2.684

3.4 Spatial Data and Descriptive Statistics

Potential difficulties associated with the analysis of spatial or location-based data must be recognized by geographers. The problems addressed here include: boundary delineation, modifiable areal units, and level of spatial aggregation or scale. Although these issues are rarely discussed outside the discipline, an understanding of the nature of these problems is essential in the conduct of geographic research.

Impact of Boundary Delineation

The location of an external study area boundary and the consequent positioning of internal subarea boundaries affects various descriptive statistics. Consider first the possible impact of study area size on absolute measures, such as the mean or standard deviation. Suppose a social geographer is conducting a study on the number and spatial pattern of families below the poverty level in census tracts of a large urban area. If the geographer chooses a smaller study area boundary, corresponding to older inner-city tracts (boundary A in figure 3.4),

Figure 3.4 Alternative Study Area Size

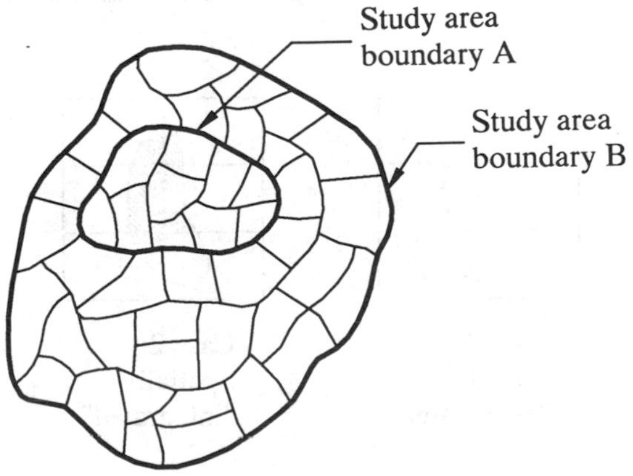

Study area boundary A

Study area boundary B

the average number of families below the poverty level will likely be higher than if the average is calculated using a larger study area boundary which includes suburban tracts (boundary B). Other absolute measures such as standard deviation or variance, whose values are partially a function of the mean, will likewise be affected. It is clear, therefore, that absolute descriptive statistics should be comparatively evaluated only with regard to a particular study area.

Boundary location and the placement of internal subarea boundaries may also influence statistical analysis. Suppose the shaded area in figure 3.5 represents the location of a high concentration of a particular demographic group. Notice how the positioning of the study area and subarea boundaries seem to "determine" whether the group is segregated (case 1) or integrated (case 2) within the large region. In each case, the location of the demographic group is the same; only the location of the study area and subareas have been changed. Again, it is advisable to conclude that summary descriptive statistics can have valid interpretations only for the

area and subarea configuration over which they were calculated.

Modifiable Area Units

Descriptive statistics can also be influenced significantly by using alternative subdivision or regionalization schemes within the same overall study area. In many cases, geographers are given little or no choice in the way data are subdivided. For example, certain demographic data may be available only at the county level within a state. Therefore, county-level areal units and boundaries must be used for any substate analysis. However, for studies where alternative subdivision or regionalization schemes are possible, geographers should realize that each subdivision scheme may produce a different result.

Consider the impact of the alternative areal subdivision schemes shown in figure 3.6. The descriptive measures calculated from each scheme convey completely different statistical summary impressions, even though the same 12 data points are used in each case. Regionalization scheme A suggests spatial separation of different magnitude values: low values (and a low mean) in Region A_1, intermediate-sized values in Region A_2, and high values in Region A_3. Only slight intraregional variation occurs with scheme A, as indicated by the low standard deviation values. The means in regionalization scheme B, however, show little difference in the magnitude. However, considerable intraregional variation occurs in scheme B, as denoted by the relatively large standard deviation values. The inclusion of large, intermediate, and small values within each subdivision in scheme B results in larger coefficient of variation figures than those derived from scheme A.

The effects of subarea boundary modification are often not as predictable as in this example. When constructing a regionalization scheme, the geographer must realize that the resultant summary statistics are a direct function of the sub-

---

Figure 3.5 Effect of Alternative Study Area Boundaries on Subarea Representation of a Demographic Group

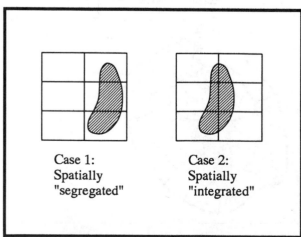

Case 1:
Spatially
"segregated"

Case 2:
Spatially
"integrated"

Figure 3.6 Impact of Alternative Subdivision Schemes on Descriptive Statistics

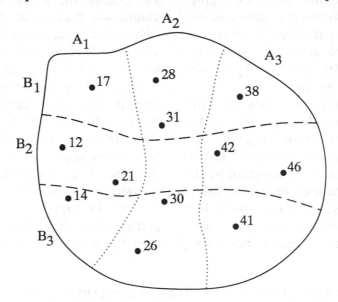

| ·········· | Subdivision scheme A |
| — — — | Subdivision scheme B |

Region	Mean	Standard deviation	Coefficient of variation
A_1	16.00	3.9	0.244
A_2	28.75	2.2	0.077
A_3	41.75	3.3	0.079
B_1	28.50	8.7	0.305
B_2	30.25	16.4	0.542
B_3	27.75	11.1	0.400

area boundary configuration. Summary statistics can be obtained to convey different impressions (e.g., slight internal variation in one configuration and considerable internal variation in another). One must ask whether a particular regionalization scheme contains too much "spatial bias." All subarea boundary schemes have bias, however, since the resultant statistics are affected by whatever configuration is chosen. Descriptive statistics should be interpreted and evaluated carefully, keeping in mind the particular boundary scheme used in the study.

Spatial Aggregation or Scale Problem

In data analysis, geographers often can choose different levels of spatial aggregation or scale. Socioeconomic variables, for example, are available at the block, census tract, enumeration district, election district, county, planning region, state, and other levels. When the same data are aggregated at different spatial levels or scales, the resultant descriptive statistics will vary, sometimes in a systematic, predictable fashion and sometimes in an uncertain way.

The following example uses data on farm population from the last six decades, analyzed at three different levels of spatial aggregation: census region, census division, and state. The 50 states of the United States are aggregated by the Bureau of the Census into nine census divisions and four census regions, allowing convenient comparison of descriptive statistics across several spatial scales at different times (table 3.8).

For each of the time periods from 1930 to 1980, both the mean and standard deviation values are highest at the census region level (where the greatest level of aggregation exists) and smallest at the state level (where the lowest level of aggregation exists). This result is expected because absolute descriptive measures are directly influenced by the magnitude of the data from which they are calculated. The absolute measures decrease in size at a particular level of aggregation over time. For example, the mean farm population by state steadily decreased from over 600,000 in 1930 to barely 100,000 in 1980. This trend reflects the dramatic reduction in farm population nationally over the last half century.

Table 3.8   Descriptive Statistics of Farm Population Data at Various Dates and Levels of Aggregation

| | Descriptive Statistics | | | | |
| | Absolute Measures (in 1000s) | | Relative Measures | | |
Date and Level of Aggregation	Mean	Standard Deviation	Coefficient of Variation	Skewness	Kurtosis
1930 Census Region	7632.25	6760.35	88.68	0.795	–1.602
Census Division	3392.11	2180.21	64.27	–0.233	–2.260
State	636.02	521.02	81.92	0.876	0.861
1940 Census Region	7636.75	6698.12	87.71	0.867	–1.227
Census Division	3394.22	2141.05	63.08	–0.202	–2.161
State	636.40	512.85	80.59	0.712	0.024
1950 Census Region	5762.25	4860.69	84.35	0.676	–2.226
Census Division	2560.89	1610.64	62.89	–0.168	–2.025
State	480.13	376.98	78.52	0.482	–0.804
1960 Census Region	3361.00	2662.28	79.21	0.022	–5.740
Census Division	1493.78	989.46	66.24	0.046	–1.878
State	268.90	224.74	83.58	0.525	–0.894
1970 Census Region	2073.04	1677.04	80.90	0.287	–4.165
Census Division	921.35	691.59	75.06	0.778	–0.457
State	165.84	145.80	87.92	0.724	–0.633
1980 Census Region	1404.48	1161.32	82.69	0.679	–1.867
Census Division	624.21	499.57	80.03	1.113	0.286
State	112.36	103.63	92.23	0.895	–0.157

Source: Calculated from data published by Economic Research Service, U.S. Dept. of Agriculture.

In comparison to the absolute measures, the columns of table 3.8 showing relative measures are more difficult and geographically challenging to interpret. Figure 3.7 summarizes graphically the differences in coefficient of variation, skewness, and kurtosis values at various levels of spatial aggregation. Each of the graphs shows substantial differences by spatial scale over all time periods. In figure 3.7 (case 1), coefficient of variation values at all spatial levels have increased since 1960 and have been climbing at the census division and state levels since 1950. This increase in relative dispersion indicates a spatial concentration of farm population in a few places. The concentration is especially noticeable at the state level, where coefficient of variation values are highest. The five states with the largest farm populations in 1950 (North Carolina, Texas, Mississippi, Tennessee, and Kentucky) accounted for less than 25 percent of the total farm population nationally. By 1980, the five states with the largest farm populations (Iowa, Minnesota, Illinois, Wisconsin, and Missouri) contained over 28 percent of the nation's total farm population.

Skewness values also differ significantly at various levels of spatial aggregation for most time periods (figure 3.7, case 2). By 1980, all skewness figures are strongly positive, with the greatest skewness in both 1970 and 1980 at the census division level. The recent positive skewness results from the concentration of farm population in the 2 midwestern census divisions: East North Central and West North Central. In fact, the 5 states with the largest farm populations in 1980 are all located in these two census divisions. The spatial concentration in the Midwest also accounts for the increasing positive skewness at the census region level.

Kurtosis values differ by spatial level and vary over time (figure 3.7, case 3). The kurtosis statistic is unstable for data sets having a small number of values. This instability is particularly evident in the fluctuation of the kurtosis values over time at the census region level. In contrast, the variation in kurtosis values over time is less dra-

Figure 3.7 Relative Descriptive Statistics of Farm Population Data at Different Spatial Levels

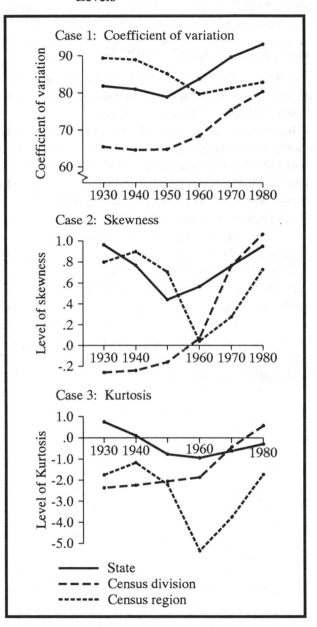

matic at the state level. Generally, kurtosis levels over the last half century have been platykurtic, reflecting frequency distributions flatter than a normal distribution. Since 1960, however, kurtosis values have become less platykurtic, as farm population becomes increasingly concentrated in the Midwest. If this concentration intensifies in the future, increasingly leptokurtic distributions should be expected.

Key Terms and Concepts

References and Additional Reading

Barber, G. M. 1988. *Elementary Statistics for Geographers.* New York: Guilford Press.

Matthews, J. A. 1981. *Quantitative and Statistical Approaches to Geography: A Practical Manual.* Oxford: Pergamon Press.

Silk, J. 1979. *Statistical Concepts in Geography.* London: George Allen and Unwin.

Taylor, P. J. 1977. *Quantitative Methods in Geography: An Introduction to Spatial Analysis.* Boston: Houghton Mifflin.

CHAPTER 4

DESCRIPTIVE SPATIAL STATISTICS

In the preceding chapter, a variety of basic descriptive statistics were examined, including the mean, standard deviation, and coefficient of variation. To summarize point patterns, a set of descriptive spatial statistics has been developed that are areal or locational equivalents to the nonspatial measures (table 4.1). Since geographers are particularly concerned with the analysis of locational data, these descriptive spatial statistics, appropriately referred to as **geostatistics**, are often used to summarize point patterns and to describe the degree of spatial variability of some phenomenon. Geostatistics are also useful to summarize an areal pattern on a choropleth map, if each area can be operationally represented by a point.

In section 4.1, spatial measures of central tendency, such as the mean center and median center, are examined. Each of these measures has characteristic properties and a set of practical geographic applications. The most important absolute measure of spatial dispersion is standard distance, the spatial equivalent to standard deviation. Standard distance will be discussed in section 4.2, along with relative distance, a measure of relative spatial dispersion. In section 4.3, attention will be given to selected locational issues that relate to the use of descriptive spatial statistics.

4.1 Spatial Measures of Central Tendency

Mean Center

In chapter 3, the mean was discussed as an important measure of central tendency for a set of data. If this concept of central tendency is extended to locational point data in two dimensions (X and Y coordinates), the average location, called the **mean center**, can be determined.

Consider the spatial distribution of points shown in figure 4.1. These points might represent any spatial distribution of interest to geog-

raphers—the only stipulation is that the phenomenon can be displayed graphically as a set of points in a two-dimensional coordinate system. The directional orientation of the coordinate axes and location of the origin are both arbitrary.

Once a coordinate system has been established and the coordinates of each point determined, the mean center can be calculated by separately averaging the X and Y coordinates, as follows:

$$\overline{X}_c = \frac{\Sigma X_i}{n} \text{ and } \overline{Y}_c = \frac{\Sigma Y_i}{n}$$

Table 4.1   Nonspatial and Spatial Descriptive Statistics

Statistic	Central Tendency	Absolute Dispersion	Relative Dispersion
Nonspatial	Mean	Standard Deviation	Coefficient of Variation
Spatial	Mean Center or Median Center or Euclidean Median	Standard Distance	Relative Distance

Figure 4.1   Graph of Locational Coordinates and Mean Center

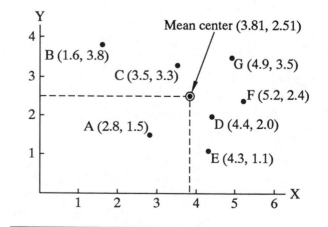

where: $\overline{X}_c$ = mean center of X

  $\overline{Y}_c$ = mean center of Y

  X_i = X coordinate of point i

  Y_i = Y coordinate of point i

  n = number of points in the distribution

For the point pattern shown in figure 4.1, the mean center coordinates are $\overline{X}_c$ = 3.81 and $\overline{Y}_c$ = 2.51 (table 4.2).

The mean center is strongly affected by points at atypical or extreme coordinate locations. This potential weakness was discussed as an attribute of the nonspatial mean as well (section 3.1). Suppose, for example, that one additional point with coordinates (15, 13) is included in the previous example. The mean center location would dramatically shift from (3.81, 2.51) to (5.21, 3.82), a relocation to a coordinate position with larger X and Y coordinates than any of the other 7 points. Thus, while the mean center represents an average location, it may not represent a "typical" location.

Table 4.2   Work Table for Calculating Mean Center

Point	Locational Coordinates*	
	X_i	Y_i
A	2.8	1.5
B	1.6	3.8
C	3.5	3.3
D	4.4	2.0
E	4.3	1.1
F	5.2	2.4
G	4.9	3.5
$n = 7$	$\Sigma X_i = 26.7$	$\Sigma Y_i = 17.6$

$$\overline{X}_c = \frac{\Sigma X_i}{n} = \frac{26.7}{7} = 3.81 \qquad \overline{Y}_c = \frac{\Sigma Y_i}{n} = \frac{17.6}{7} = 2.51$$

Mean center coordinates: (3.81, 2.51)*

*See figure 4.1 for graph of locational coordinates and mean center.

The mean center may be considered the **center of gravity** of a point pattern or spatial distribution. Perhaps the most widely known application of the mean center is the decennial calculation of the **geographic "center of population"** by the U.S. Bureau of the Census. This is the point where a rigid map of the country would balance if equal weights (each representing the location of one person) were situated on it. Over the last 2 centuries, the westward movement of the U.S. population has continued without significant interruption, which is reflected in the concomitant westward shift of the center of population (figure 4.2).

In the seven-point example and work table, each point is given an equal weight in the mean center calculation; that is, each point is equally important statistically. With many geographic applications, however, it is appropriate to assign differential weights to points in a spatial distribution. The weights are analogous to frequencies in the analysis of grouped data (e.g., weighted mean). The points might represent cities, and the frequencies the number of people; or the points could be retail store locations, and the frequencies could be volume of sales per store. The **weighted mean center** is defined as follows:

$$\overline{X}_{wc} = \frac{\Sigma f_i X_i}{\Sigma f_i} \text{ and } \overline{Y}_{wc} = \frac{\Sigma f_i Y_i}{\Sigma f_i} \qquad (4.2)$$

where: $\overline{X}_{wc}$ = weighted mean center of X

  $\overline{Y}_{wc}$ = weighted mean center of Y

  f_i = frequency (weight) of point i

The set of points in figure 4.1 is expanded to include weights at each point (figure 4.3), and weighted mean center coordinates are calculated (table 4.3). The locational coordinates of the weighted mean center are somewhat different from the coordinates of the comparable unweighted mean center. The weighted mean center is heavily affected by the large frequency

Figure 4.2   Geographic Center of U.S. Population, 1790-1990

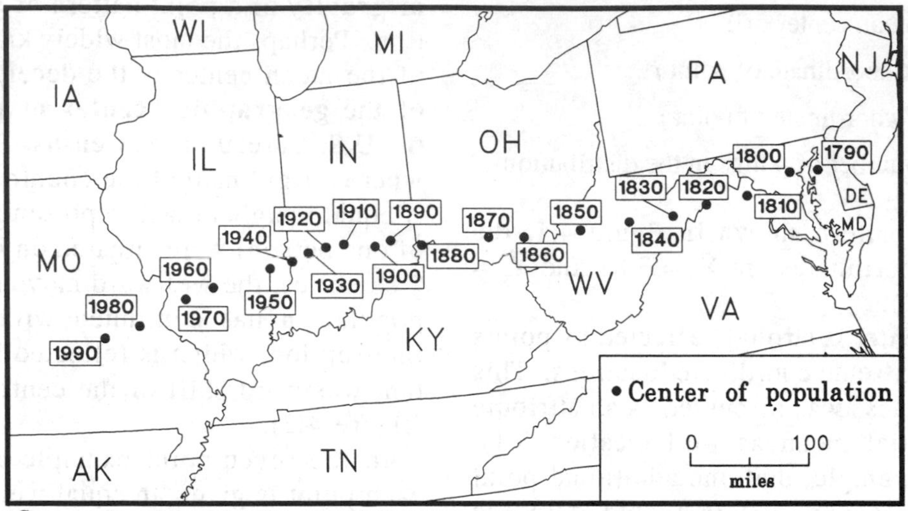

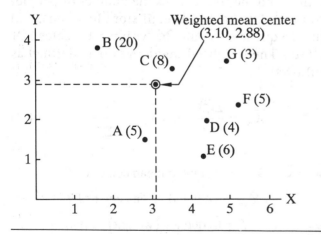

Source: Bureau of the Census, U.S. Dept. of Commerce.

Figure 4.3   Graph of Point Locations, Frequencies (in Parentheses) and Weighted Mean Center

Table 4.3   Work Table for Calculating Weighted Mean Center

Point	Locational Coordinates* X_i	Y_i	Weight f_i	Weighted Coordinates f_iX_i	f_iY_i
A	2.8	1.5	5	14.0	7.5
B	1.6	3.8	20	32.0	76.0
C	3.5	3.3	8	28.0	26.4
D	4.4	2.0	4	17.6	8.0
E	4.3	1.1	6	25.8	6.6
F	5.2	2.4	5	26.0	12.0
G	4.9	3.5	3	14.7	10.5
$n = 7$			$\Sigma f_i = 51$	$\Sigma f_iX_i = 158.1$	$\Sigma f_iY_i = 147.0$

$$\overline{X}_{wc} = \frac{\Sigma f_i X_i}{\Sigma f_i} = \frac{158.1}{51} = 3.10$$

$$\overline{Y}_{wc} = \frac{\Sigma f_i Y_i}{\Sigma f_i} = \frac{147.0}{51} = 2.88$$

Weighted mean center coordinates: (3.10, 2.88)*

*See figure 4.3 for graph of point locations, frequencies, and weighted mean center.

associated with point B. Gravitation of the center toward a point with an unusually heavy weight must occur even if that point is in a peripheral location within the spatial distribution.

The mean and mean center share an important characteristic that has locational ramifications. Recall from the discussion on mean deviation (section 3.2) that the sum of the deviations of all observations about a mean is zero. In addition, the sum of squared deviations of all observations about a mean is the minimum sum possible. That is, the sum of squared deviations about a mean is less than the sum of squared deviations about any other number. This important attribute is called the least squares property of the mean.

The mean center serves as a spatial analogue to the mean, in that it is the location that minimizes the sum of squared deviations of a set of points. Thus, the mean center has the same least squares property as the mean. The mean center $(\overline{X}_c, \overline{Y}_c)$ minimizes:

$$\Sigma\left(X_i - \overline{X}_c\right)^2 + \left(Y_i - \overline{Y}_c\right)^2 \qquad (4.3)$$

In a location coordinate system, deviations such as $(X_i - \overline{X}_c)$ and $(Y_i - \overline{Y}_c)$ are, in fact, *distances* between points. One standard procedure for measuring distances is based on straightline or **Euclidean distance.** The Euclidean distance (d_i) separating point i (X_i, Y_i) from the mean center $(\overline{X}_c, \overline{Y}_c)$ is illustrated in figure 4.4 and defined by the Pythagorean theorem as follows:

$$d_i = \sqrt{\left(X_i - \overline{X}_c\right)^2 + \left(Y_i - \overline{Y}_c\right)^2} \qquad (4.4)$$

Thus, the mean center is the location that minimizes the sum of *squared distances* to all points. This characteristic makes the mean center an appropriate center of gravity for a two-dimensional point pattern, just as the mean is the center of gravity along a one-dimensional number line.

Figure 4.4 Calculation of Euclidean Distance (d_i) from Mean Center

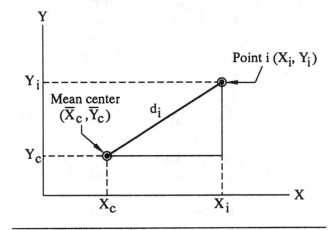

Euclidean Median

For many geographic applications, another measure of "center" is more useful. Often, it is more practical to determine the central location that minimizes the sum of *unsquared,* rather than squared, distances. This location, which minimizes the sum of Euclidean distances from all other points in a spatial distribution to that central location, is called the **Euclidean median** (X_e, Y_e) or **median center**. Mathematically, this location minimizes the sum:

$$\Sigma\sqrt{\left(X_i - X_e\right)^2 + \left(Y_i - Y_e\right)^2} \qquad (4.5)$$

Unfortunately, determining coordinates of the Euclidean median is complex methodologically. Computer-based iterative algorithms (step-by-step procedures) must be used to derive a solution. These algorithms evaluate a sequence of possible coordinates and gradually converge on the best location for the Euclidean median. (See the references at the end of the chapter for more information and examples of computer algorithms for the Euclidean median).

A weighted Euclidean median is a logical extension of the simple (unweighted) Euclidean median. The same types of algorithmic procedures are used to locate the weighted Euclidean median. The coordinates of the weighted Euclidean median (X_{we}, Y_{we}) will minimize the expression:

$$\Sigma f_i \sqrt{\left(X_i - X_{we}\right)^2 + \left(Y_i - Y_{we}\right)^2} \qquad (4.6)$$

The weights or frequencies may represent population, sales volume, or any other feature appropriate to the spatial problem.

The location of the weighted Euclidean median is important to geographers in several practical contexts. For example, a classical problem in economic geography is the so-called Weber problem, which seeks to determine the "best" location for an industry. The optimal location minimizes the total cost of transporting the raw material to the factory and the finished product to the market. The weighted Euclidean median is the location that minimizes these transportation costs.

Perhaps the most extensively developed applications for the Euclidean median in geography are public and private facility location. Often an important goal in facility location is minimizing the average distance traveled per person to reach a designated or assigned facility. This efficiency- based objective is equivalent to minimizing the aggregate or total distance people must travel to utilize the service systemwide. The Euclidean median achieves this goal.

Consider, for example, the problem of locating the site for an urban fire station based on a predicted pattern of fires for the region. Using the past or present pattern of fires as a reasonable estimate of future fires, the optimal central location for the station could be defined as the site that minimizes the total (and hence, average) distance traveled by the fire equipment to reach fires. That location is determined by the Euclidean median.

In another application, suppose location analysts for an exclusive women's apparel chain wish to select an accessible site for a new store. Suppose that market analysis indicates that the demographic group most likely to shop in the store is women aged 35-60 who are members of households with incomes greater than $80,000. From the compilation of census tract information in the designated trade area, each tract could be weighted by the number of women falling into this target population. The weighted Euclidean median will designate the location that minimizes the total (and average) distance traveled by these women to reach the potential store site.

Extending this procedure to the simultaneous location of multiple facilities within a spatial pattern of demand is known as the "location-allocation" problem or the "multiple facility location" problem. Suppose, for example, city health care planners wish to locate a set of neighborhood medical centers to provide selected types of remedial health care. Not only must a set of medical centers be located, but the potential clientele must be allocated to an appropriate facility, creating "catchment districts" or zones for each center. Problems such as these are extremely complex and challenging, and both theoretical issues and practical applications receive considerable attention from geographers.

4.2 Spatial Measures of Dispersion

Standard Distance

Just as the mean center serves as a locational analogue to the mean, **standard distance** is the spatial equivalent of standard deviation (table 4.1). Standard distance measures the amount of **absolute dispersion** in a point pattern. After the locational coordinates of the mean center have been determined, the standard distance statistic incorporates the straight-line or Euclidean distance of each point from the mean center. In its most basic form, standard distance (S_D) is written as follows:

$$S_D = \sqrt{\frac{\Sigma\left(X_i - \overline{X}_c\right)^2 + \Sigma\left(Y_i - \overline{Y}_c\right)^2}{n}} \qquad (4.7)$$

Equation 4.7 can be modified algebraically to reduce the number of required computations considerably:

$$S_D = \sqrt{\left(\frac{\Sigma X_i^2}{n} - \overline{X}_c^2\right) + \left(\frac{\Sigma Y_i^2}{n} - \overline{Y}_c^2\right)} \qquad (4.8)$$

Using the same point pattern as in the earlier example (figure 4.1), the standard distance is calculated (table 4.4) and shown as the radius of a circle whose center is the mean center (figure 4.5).

Like standard deviation, standard distance is strongly influenced by extreme or peripheral locations. Because distances about the mean center are squared, "uncentered" or atypical points have a dominating impact on the magnitude of the standard distance.

Weighted standard distance is appropriate for those geographic applications requiring a weighted mean center. The definitional formula for weighted standard distance (S_{WD}) is:

$$S_{WD} = \sqrt{\frac{\Sigma f_i\left(X_i - \overline{X}_c\right)^2 + \Sigma f_i\left(Y_i - \overline{Y}_c\right)^2}{n}} \qquad (4.9)$$

which may be rewritten in computationally-simpler form as:

$$(4.10)$$

$$S_{WD} = \sqrt{\left(\frac{\Sigma f_i\left(X_i^2\right)}{\Sigma f_i} - \overline{X}_c^2\right) + \left(\frac{\Sigma f_i\left(Y_i^2\right)}{\Sigma f_i} - \overline{Y}_c^2\right)}$$

Table 4.4 Work Table for Calculating Standard Distance

Point	\multicolumn{4}{c}{Locational Coordinates*}			
	X_i	Y_i	X_i^2	Y_i^2
A	2.8	1.5	7.84	2.25
B	1.6	3.8	2.56	14.44
C	3.5	3.3	12.25	10.89
D	4.4	2.0	19.36	4.00
E	4.3	1.1	18.49	1.21
F	5.2	2.4	27.04	5.76
G	4.9	3.5	24.01	12.25

From earlier calculation of mean center:

$\overline{X}_c = 3.81$ and $\overline{Y}_c = 2.51$

$\overline{X}_c^2 = 14.52$ and $\overline{Y}_c^2 = 6.30$

$n = 7 \qquad \Sigma X_i^2 = 111.5 \qquad \Sigma Y_i^2 = 50.80$

$$S_D = \sqrt{\left(\frac{\Sigma X_i^2}{n} - \overline{X}_c^2\right) + \left(\frac{\Sigma Y_i^2}{n} - \overline{Y}_c^2\right)}$$

$$= \sqrt{\left(\frac{111.55}{7} - 14.52\right) + \left(\frac{50.80}{7} - 6.30\right)}$$

$$= 1.54$$

*See figure 4.5 for graph of locational coordinates, mean center, and standard distance.

A weighted standard distance can be computed using the same point pattern as before (table 4.5). A moderate disparity exists between the relative magnitudes of the unweighted and weighted standard distances (1.54 vs. 1.70). This difference can be explained by point B. Because this point is distant from the mean center and exerts a proportionally greater influence on the standard distance measure with its larger weight, point B causes the weighted standard distance to be larger than the unweighted standard distance.

Figure 4.5 Graph of Point Locations, Mean Center, and Standard Distance

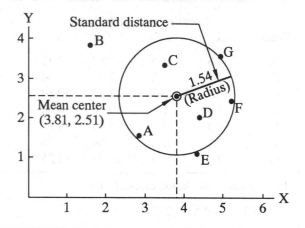

Relative Distance

The coefficient of variation (standard deviation divided by the mean) is the nonspatial measure of **relative dispersion** (table 4.1). Unfortunately, a perfect spatial analogue to the coefficient of variation does not exist for measuring relative dispersion. Although it seems logical to divide the standard distance by the mean center to produce a relative dispersion index, this procedure will not provide meaningful results.

Consider a situation in which the spatial statistics for the same point pattern are calculated twice using different positions for the coordinate system (figure 4.6). In case 1, the X coordinates

Table 4.5 Work Table for Calculating Weighted Standard Distance*

Point	f_i	X_i	X_i^2	$f_i(X_i)^2$	Y_i	Y_i^2	$f_i(Y_i)^2$
A	5	2.8	7.84	39.20	1.5	2.25	11.25
B	20	1.6	2.56	51.20	3.8	14.44	288.80
C	8	3.5	12.25	98.00	3.3	10.89	87.12
D	4	4.4	19.36	77.44	2.0	4.00	16.00
E	6	4.3	18.49	110.94	1.1	1.21	7.26
F	5	5.2	27.04	135.20	2.4	5.76	28.80
G	3	4.9	24.01	72.03	3.5	12.25	36.75

From earlier calculation of weighted mean center:

$\overline{X}_c = 3.10$ and $\overline{Y}_c = 2.88$

$\overline{X}_c^2 = 9.61$ and $\overline{Y}_c^2 = 8.29$

$\Sigma f_i = 51 \qquad \Sigma f_i(X_i)^2 = 584.01 \qquad \Sigma f_i(Y_i)^2 = 475.98$

$$S_{WD} = \sqrt{\left(\frac{\Sigma f_i(X_i)^2}{\Sigma f_i} - \overline{X}_c^2\right) + \left(\frac{\Sigma f_i(Y_i)^2}{\Sigma f_i} - \overline{Y}_c^2\right)}$$

$$= \sqrt{\left(\frac{584.01}{51} - 9.61\right) + \left(\frac{475.98}{51} - 8.29\right)}$$

$$= 1.70$$

*See figure 4.3 for graph of point locations and weighted mean center

Figure 4.6 Arbitrary Placement of Coordinate Axes and Resultant Descriptive Spatial Statistics

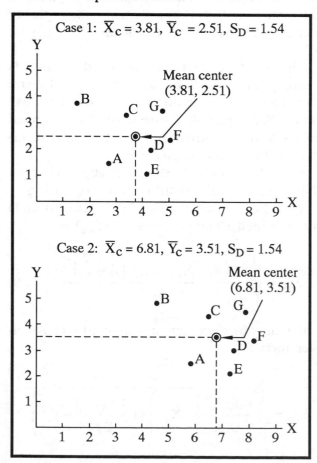

Case 1: $\overline{X}_C = 3.81$, $\overline{Y}_C = 2.51$, $S_D = 1.54$

Mean center (3.81, 2.51)

Case 2: $\overline{X}_C = 6.81$, $\overline{Y}_C = 3.51$, $S_D = 1.54$

Mean center (6.81, 3.51)

of each point are three units lower and the Y coordinates one unit lower than in case 2. Notice, however, that the coordinate system shift has no effect on standard distance. As a measure of absolute dispersion, standard distance remains unchanged. Conversely, the coordinates of the mean center will change whenever the coordinate system is shifted. Thus, because coordinate system location affects the mean center, but not standard distance, a relative dispersion metric based on the ratio of these measures will be meaningless.

Despite these difficulties, some measure of relative dispersion is necessary. Consider the three point patterns in regions A, B, and C (figure 4.7, case 1). The distribution of points in each region has the same amount of absolute dispersion and the same standard distance. However, in small region A, the points have a high degree of relative dispersal, whereas in region C, they have a low relative dispersal because the region is larger. The point patterns in regions D and E (figure 4.7, case 2) appear to have the same amount of relative dispersion. However, the point pattern in region D has a larger standard distance (absolute dispersion) than region E because of its larger size.

To derive a descriptive measure of relative spatial dispersion, the standard distance of a point pattern must be divided by some measure of regional magnitude. As discussed earlier, this divisor cannot be the mean center. One possible divisor is the radius (r_A) of a circle with the same area as the region being analyzed. A useful measure of relative dispersion, called **relative distance** (R_D), can now be defined:

$$R_D = \frac{S_D}{r_A} \qquad (4.11)$$

This relative distance measure allows direct comparison of the dispersion of different point patterns from different areas, even if the areas are of varying sizes.

Care must be taken when using this measure of relative distance. Geographers often have no

Figure 4.7   Comparisons of Absolute and Relative Point Pattern Dispersion

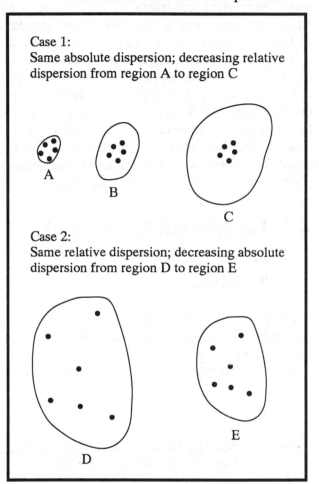

Case 1:
Same absolute dispersion; decreasing relative dispersion from region A to region C

Case 2:
Same relative dispersion; decreasing absolute dispersion from region D to region E

control over the boundary and size of the study area. Study areas are often defined by political boundaries that have no logical relationship to the spatial pattern investigated. Thus, a measure of relative dispersion based on the area of the study region will be influenced by the location of the boundary.

In addition, using a circle to represent the region's area may not be fully valid, particularly if the region has a highly irregular shape. An elongated region, for example, can have greater distances within its borders than a circular region of the same total area. Thus, a region's shape can also affect the relative dispersion of a pattern.

When absolute and relative dispersion of population for selected countries are compared, some interesting contrasts emerge (table 4.6). Although Japan has a somewhat low standard distance because of its small areal extent, the country has a rather high relative distance, indicating a dispersed population. However, this is partly a function of Japan's elongated shape. On the other hand, given its large area, China has a clustered population, reflected by a low relative distance measure. Other things being equal, a country with a compact shape, such as China, should show less relative dispersion.

4.3 Locational Issues and Descriptive Spatial Statistics

Care must always be taken when interpreting geostatistics. Because they are summary measures representing complex spatial patterns, the results may sometimes be misleading or illogi-

Table 4.6   Standard Distance and Relative Distance Measures of Population for Selected Countries

Country	Standard Distance (km)	Relative Distance (km)
Australia	615	0.63
Brazil	697	0.68
China	579	0.52
India	538	0.85
Japan	256	1.20
United Kingdom	134	0.77
United States	839	0.86

Source: Neft, David S. 1966. *Statistical Analysis for Areal Distributions* (Monograph Series No. 2). Philadelphia: Regional Science Research Institute.

cal. For example, the mean center or Euclidean median of high-income population in a metropolitan area could be located in a low-income central city neighborhood, which contains few high-income families.

In addition, the magnitude or location of each measure is influenced by all points in the distribution. A single anomalous point can severely modify the resultant descriptive spatial measures, making interpretation difficult. Geographers should view geostatistics as general locational indicators, rather than as precise measurement instruments to be used in isolation.

The descriptive analysis of point patterns could benefit from consideration of other possible pattern characteristics. In fact, a spatial or locational measure analogous to each and every nonspatial descriptive statistic could be created. Spatial statistics equivalent to skewness and kurtosis have potential application in geography. For example, when comparing point patterns in geographic research, knowing which patterns are more symmetric or skewed might offer spatial insights. Geographers could also find it valuable to compare directly the degree of clustering and dispersal in different point patterns through measuring spatial kurtosis levels.

In previous discussions about mean center, Euclidean median, and standard distance, reference was made only to straight-line or Euclidean distance, based on the Pythagorean theorem. In certain spatial contexts, this may not be the most appropriate way to proceed. For example, in an urban geography study of various commercial activities (e.g., grocery stores, drug stores, etc.), neither the mean center nor the Euclidean median may be the most suitable measure of "center." In those uncommon situations where travel is confined to a grid or rectilinear street pattern, a straight-line or Euclidean *distance metric* underrepresents the actual travel distances, and could result in a misplaced center and inaccurately low measure of absolute dispersion.

A general distance formula is available to handle various geographic situations where "friction of distance" or other influences preclude simple straight-line measure. The formula measuring the Euclidean distance (d_{ij}) of point i (X_i, Y_i) from point j (X_j, Y_j) is:

$$d_{ij} = \sqrt{\left(X_i - X_j\right)^2 + \left(Y_i - Y_j\right)^2} \qquad (4.12)$$

If the Euclidean distance metric (equation 4.12) is "generalized" to allow non-Euclidean distance measurement, the result is:

$$d_{ij} = \left(\left(X_i - X_j\right)^k + \left(Y_i - Y_j\right)^k\right)^{1/k} \qquad (4.13)$$

When k is two, the formula is conventional Euclidean distance. When k is one, however, the formula approximates distances when movement is restricted to a rectangular or grid system. The term **Manhattan distance** describes the restrictive movement typical of travel in the New York City borough of Manhattan. Measuring the distance between points i and j in Manhattan space, where k equals one, gives:

$$d_{ij} = \left|X_i - X_j\right| + \left|Y_i - Y_j\right| \qquad (4.14)$$

The "center" point in Manhattan space is the **Manhattan median**, which minimizes the sum of absolute deviations between itself and all other points in the pattern. Absolute values are used to ensure that all distances are positive. Formally, the Manhattan median (X_m, Y_m) is that location which minimizes the sum:

$$\Sigma \left|X_i - X_m\right| + \left|Y_i - Y_m\right| \qquad (4.15)$$

Unfortunately, a single unique location for the Manhattan median cannot be found in a spatial pattern having an even number of points. That is, no single "middle" point exists when the number of points is even. However, a unique set of coordinates can always be determined for the Manhattan median with an odd number of points. This situation parallels the calculation of the "nonspatial" median (section 3.1), since both statistical procedures differ in problems having an even or odd number of points. A further difficulty with the Manhattan median is that its location will change if the coordinate axes are shifted.

Depending on the geographic context, the distance metric k could assume many values (figure 4.8). In case 1, a straight line is traveled, and the Euclidean distance metric ($k = 2$) is used. Case 2 illustrates Manhattan space ($k = 1$), where distances are measured along a grid-like street pattern. In other geographic situations, distances might best be measured with a metric value somewhere between Euclidean and Manhattan. For example, in case 3, an intermediate-value distance metric ($k = 1.5$) best estimates the distance separating points i and j. In case 4, travel from point i to j must be around the intervening lake. Since a circuitous route must be traveled, a k value less than one is needed ($k = 0.6$).

Key Terms and Concepts

Figure 4.8   Distance Metrics Under Various Spatial Conditions

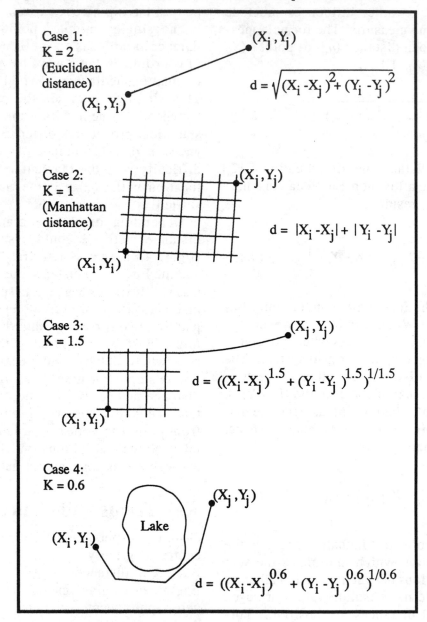

References and Additional Reading

Barber, G. M. 1988. *Elementary Statistics for Geographers*. New York: Guilford Press.

Ebdon, D. 1985. *Statistics in Geography* (2nd edition). Oxford: Basil Blackwell.

Kuhn, H. W. and R. E. Kuenne. 1962. "An Efficient Algorithm for the Numerical Solution of the Generalized Weber Problem in Spatial Economics." *Journal of Regional Science* 4:21-33.

Neft, D. S. 1966. *Statistical Analysis for Areal Distributions* (Monograph Series No. 2). Philadelphia: Regional Science Research Institute.

Rushton, G. 1979. *Optimal Location of Facilities*. Wentworth, NH: COMPress, Inc.

Rushton, G., M. F. Goodchild and L. M. Ostresh, Jr. (editors). 1973. *Computer Programs for Location-Allocation Problems* (Monograph No. 6). Iowa City: Department of Geography, University of Iowa.

Scott, A. J. 1971. *An Introduction to Spatial Allocation Analysis* (Resource Paper No. 9). Washington, DC: Assoc. of Amer. Geographers.

Taylor, P. J. 1977. *Quantitative Methods in Geography: An Introduction to Spatial Analysis*. Boston: Houghton Mifflin.

PART III

THE TRANSITION TO INFERENTIAL PROBLEM SOLVING

CHAPTER 5

PROBABILITY

Studying spatial patterns found on the physical and cultural landscape is a central concern of geographers. They seek to develop descriptions and explanations of existing *patterns* and to understand the *processes* that create these distributions. In some cases they attempt to predict future occurrences of geographic patterns. In short, the core of geographic problem solving is the description, explanation, and prediction of geographic patterns and processes.

In earlier chapters, the focus was on ways to describe or summarize spatial data. Much of the remainder of the book involves methods used for exploring relationships between spatial patterns and for understanding the nature or characteristics of processes that create these patterns. The concept of probability occupies a central position in the chapters that follow.

Chapter 5 provides an introduction to probability and the important role it plays in geographic problem solving. The first section discusses the nature of geographic patterns and the processes that produce them. Included here are examples of deterministic and probabilistic processes in geography. Section 5.2 introduces the concept of probability, including definitions and rules for using probability in geographic analysis. Sections 5.3 through 5.5 cover specific probability distributions that are important in geographic problem solving. The binomial distribution, discussed in section 5.3, concerns probability of events having only two possible outcomes and is particularly useful for studying multiple events or trials. Section 5.4 discusses the Poisson distribution, which is used to analyze patterns that are random over time or space. The most important probability distribution to geographers is the normal distribution, introduced in section 5.5. The concluding section of chapter 5 presents a probability mapping technique, which extends the procedure of normal probabilities into a spatial context.

5.1 Deterministic and Probabilistic Processes in Geography

Real-world processes that produce physical or cultural patterns on the landscape are often complex and not usually totally identifiable. The nature of these processes can be described by two general categories: deterministic and probabilistic. **Deterministic processes** create patterns with total certainty, since the outcome can be exactly specified with 100 percent likelihood. Because of uncertainty in human behavior and decision making, virtually no cultural processes are completely deterministic, but some physical processes established through scientific principles fall into this category. For example, the length of time solar insolation strikes a point on the earth's surface is determined by both latitude and day of the year. Given these two components, geographers can determine the exact hours of daylight and darkness at any location.

The second category, **probabilistic processes**, concerns all situations that cannot be determined with complete certainty. Given the complex character of physical and cultural patterns and processes, most geographic situations fall into this category. For example, although the number of hours of sunlight is considered deterministic, the amount of sunlight reaching the ground is probabilistic, since cloud cover and particulate matter absorb and reflect solar energy as it passes through the atmosphere.

Probabilistic processes can be subdivided into two useful categories. The likelihood of outcomes for some processes can be specified as a set of probabilities. Such processes result in intermediate situations between total certainty and total uncertainty, which can be labeled **stochastic**. In other situations, probabilities cannot be assigned to possible outcomes. The process in this extreme situation of total uncertainty is often termed **random**.

Forecasting future use for a currently undeveloped parcel of land is one way to illustrate the nature of stochastic processes in geography. Land use selection entails numerous complex factors, such as the monetary value of the parcel, its accessibility to regional activities, government restrictions (like zoning), and the land's physical characteristics (e.g., slope, drainage, and soil type). Knowledge of these factors can be used to estimate the probability that a land parcel will have a given use. For example, a particular parcel of land could be developed as an office park, an industrial park, a shopping center, a residential subdivision, or simply left undeveloped. Because the land development process is complex, it is not possible to specify future land use with certainty. The most that can be done in this situation is to specify the likelihood or probability of each potential land use.

Some spatial patterns are totally unpredictable. For example, the location of tornado "touch down" points within a region is the result of random meteorological processes. Sometimes the size of the study area affects the degree of randomness. For example, the number of tornadoes next year within Kansas can be estimated as a probability. However, if the focus is narrowed to a small area within Kansas, the location of these tornado touch down points is random.

In summary, most geographic patterns result from processes that are stochastic in nature. Few patterns studied by geographers occur as a result of either totally deterministic or totally random processes. In addition to spatial distributions, geographers also study temporal patterns that can be described as occurring from either deterministic or probabilistic processes. In such investigations, use of probability again plays an integral role, since most temporal patterns are neither completely deterministic nor totally random.

5.2 The Concept of Probability

Given that most spatial and temporal patterns are produced by processes that have some degree of uncertainty, geographers need to understand and use probability for solving problems. For example, every location on the earth's surface receives a variable amount of precipitation. These data can be recorded over time and across space, and precipitation patterns summarized with calculations such as the mean and standard deviation. However, since precipitation results from complex atmospheric processes, its prediction can only be stated in terms of probabilities, not exact certainty. Therefore, geographers make statements such as "50 percent of the time snowfall in January exceeds 20 inches" or "9 years out of 10, at least 5 inches of rain fall in June." Although probabilities can be stated, the exact amount of snowfall next January or the exact amount of rainfall next June cannot be predicted with certainty.

The study of probability focuses on the occurrence of an **event**, which can usually result in one of several possible **outcomes**. Once all possible outcomes have been considered for the event, probability represents the likelihood of a given result or the chance that any outcome actually takes place. Because of the obvious connection to chance, examples from gaming situations are often used to illustrate probability. What is the likelihood of rolling a 6 on a die? In this problem, the event studied is the roll of the die, and the possible outcomes are the 6 sides of the die. Since each outcome or side has an equal chance of occurring, the likelihood of rolling a 6 is one out of six or 0.167.

Probability can also be thought of as **relative frequency**—the ratio between the absolute frequency for an outcome and the frequency of all outcomes:

$$P_a = \frac{F_a}{F_E} \qquad (5.1)$$

where: P_a = probability of outcome a

F_a = absolute frequency of outcome a

F_E = absolute frequency of all outcomes

for event E

Probabilities can also be interpreted as percentages when the denominator of equation 5.1 is converted to 100. In the example of the die roll, the probability of a 6 (0.167) can be interpreted as an outcome that occurs 16.7 percent of the time.

Some of the basic elements of probability can be observed through a simple geographic example. Suppose a day is classified as wet (w) if measurable precipitation (0.01 inch or more) falls during the 24-hour period. The day is termed dry if measurable precipitation does not occur. By keeping a record of wet and dry days over a 100-day period, precipitation frequencies can be determined and probabilities calculated from the data (figure 5.1). In this example, 62 days

are categorized as dry and 38 as wet. The probability of a wet day occurring (P_w) is:

$$P_w = \frac{\text{number of wet days}}{\text{total days}} = \frac{38}{100} = 0.38$$

Thus, a 38 percent chance exists that a day will have measurable precipitation.

Several rules guide the use of probability. The maximum probability for any outcome is 1.0, which indicates total certainty or perfect likelihood of a particular occurrence. The lowest probability for an outcome is 0.0, which suggests no chance of this occurrence. Most outcomes have probabilities falling between the maximum (1.0) and minimum (0.0) values.

Because each event actually takes place with one outcome occurring, the sum of the probabilities for all outcomes must equal 1.0. In the previous example of precipitation, the probability of a wet day was 0.38 and the probability of a dry day was 0.62. This simple example has only two outcomes, and the probabilities clearly sum to 1.0. Many other rules of probability exist for more complex applications. Because this discussion of probability is only an introduction, readers are directed to sources at the end of the chapter for more thorough treatments of the subject.

Probabilities from Known Mathematical Distributions

The probability of outcomes in certain problems follows consistent or typical patterns. Such patterns, called probability distributions, relate closely to frequency distributions discussed in section 2.5. In a frequency distribution, the frequency of occurrences appears on the vertical axis, whereas in a probability distribution, the probability of occurrence is displayed on the vertical axis. In both cases, the horizontal axis shows the actual outcomes, occurrences, or values of the variable being studied.

Figure 5.1 Relative Frequency of Wet and Dry Days from a 100-day Period

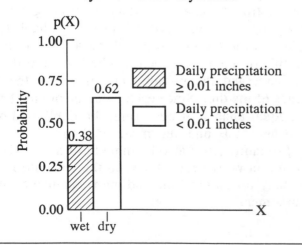

Recall that variables are termed "discrete" or "continuous" depending on whether the values occur as distinct whole numbers (discrete) or decimal values (continuous). Probability distributions for discrete outcomes are termed **discrete probability distributions,** whereas those for outcomes that can occur at an infinite number of points are termed **continuous probability distributions.** Although numerous examples of both types of distributions are used in geography, this discussion focuses on three commonly used distributions: binomial, Poisson, and normal.

5.3 The Binomial Distribution

The **binomial** is a discrete probability distribution associated with situations where only two outcomes are possible. Binary outcomes are often described as zero-one, yes-no, or presence-absence problems, and many geographic situations fit a binary framework. Consider these examples: A location has measurable precipitation over a 24-hour period or it does not; a person is either employed or unemployed; a river is either above or below flood stage; a respondent to an opinion poll either favors or opposes an issue. In each of these situations, only one of two outcomes is possible, assuming that an undecided or uncertain result is not possible.

The binomial distribution is especially useful in examining probabilities from multiple events or trials. The probability of a single event can usually be determined quite easily, perhaps using historical data. For example, the probability of a river in Bangladesh reaching flood stage during a given year may be 0.40. Thus, on average, flooding occurs 4 years out of 10. With this information, it is possible to use the binomial distribution to determine the probability of flooding over other time periods (perhaps 15 years out of 30). In this example, the assumption is that a river being in flood stage during a given year is not related to its flood stage status in other years. In other words, the outcomes from repeated events are **independent**.

When using binomial probabilities, the focus is on one of the two possible outcomes, termed the "given" outcome (X). The binomial distribution is shown mathematically as follows:

$$P(X) = \frac{n!\left(p^X\right)\left(q^{n-X}\right)}{X!(n-X)!} \qquad (5.2)$$

where: n = number of events or trials

p = probability of the "given" outcome in a single trial

$q = 1 - p$ or the probability of the "other" outcome in a single trial

X = number of times the "given" outcome occurs within the n trials

$n!$ = n factorial:

if $n > 0$, $n! = \left[n(n-1)(n-2)...(2)(1)\right]$

if $n = 0$, $n! = 1$

The best way to illustrate the characteristics and practical uses of the binomial probability equation (5.2) is with a geographic example. Suppose a vegetable grower is seeking a new location to start a business. One of the key variables in site selection is the probability of adequate precipitation to reduce the necessity of expensive irrigation. The grower has determined that at least 3 inches of precipitation are needed during the growing season to avoid irrigation, and that irrigation can be afforded only 1 year out of 5 to make a profit. Precipitation data collected for a potential site show that in 21 of the last 25 years rainfall during the growing season exceeded 3 inches. The historical record, therefore, suggests that the probability of a given year requiring supplemental irrigation is 4/25 or 0.16. What is the probability that the grower can meet the requirement of having to irrigate only 1 year in 5?

Over the specified 5-year period, the farmer faces six possible outcomes: from none of the 5 years requiring irrigation up to all 5 years requiring irrigation. However, only two of the six outcomes—no years and 1 year—would result in a profitable situation for the grower; each of the other four outcomes would be too costly. Thus, the probability that the grower will meet his profit requirement is found by summing the probabilities for the two suitable outcomes. The critical values and calculations of the binomial probabilities for the two suitable outcomes are shown in table 5.1. The binomial probabilities for the problem are listed in table 5.2 and plotted graphically in figure 5.2.

The vegetable grower would have a 0.418 probability of needing no irrigation and a 0.398 chance of needing 1 year of irrigation during the next 5 years. Adding the binomial probabilities for no years and 1 year of irrigation (0.418 + 0.398) shows that the grower has an 81.6 percent chance of meeting the profitability requirement at this potential site during the 5-year interval.

Table 5.1   Critical Values and Binomial Probabilities of Suitable Outcomes for Vegetable Grower

$$P(X) = \frac{n!(p^X)(q^{n-X})}{X!(n-X)!}$$

CRITICAL VALUES

$n = 5$ years

$p = 4/25 = 0.16$

$q = (1-p) = 0.84$

$X = 0, 1, 2, 3, 4, 5$

$n! = (5)(4)(3)(2)(1) = 120$

When $X = 0$: $P(0) = \dfrac{5!(.16^0)(.84^5)}{0!(5!)}$

$= \dfrac{120(1)(.418)}{1(120)} = 0.418$

When $X = 1$: $P(1) = \dfrac{5!(0.16^1)(0.84^4)}{1!(4!)}$

$= \dfrac{120(0.16)(0.498)}{1(24)} = 0.398$

Table 5.2   Binomial Probabilities of Vegetable Grower Needing Irrigation Over a 5-year Period

Number of Years Out of Five	Binominal Probability	Type of Outcome
0	0.418	Suitable
1	0.398	Suitable
2	0.152	Unsuitable
3	0.029	Unsuitable
4	0.003	Unsuitable
5	0.000	Unsuitable
TOTAL	1.000	

Total Probability of Suitable Outcomes   = 0.816
Total Probability of Unsuitable Outcomes = 0.184

Figure 5.2   Probability of Vegetable Grower Needing Irrigation for 0 to 5 Years

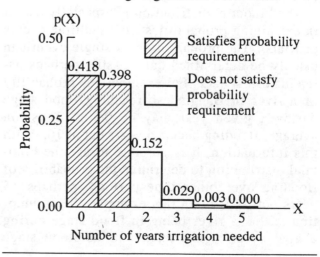

This level of risk may or may not be acceptable to the grower.

To summarize, the binomial distribution is applicable for those geographic problems in which the following conditions are met:

1. The objective is to determine the probability of multiple (n) independent events or trials,
2. Each event or trial has two possible outcomes, one termed the "given" outcome (X), with associated probability p, and the second or "other" outcome having probability $q = 1 - p$, and
3. The probabilities p and q must remain stable or consistent over the duration of the study or over successive trials.

5.4 The Poisson Distribution

Some probability problems in geography involve the study of events that occur at random over either time or space. For example, the placement of calls to an emergency response dispatcher might be considered random over a short period of time. Many weather-related phenomena, such as thunderstorms, tornadoes, and hurricanes, occur with little spatial predictability. Geographers also study various cultural entities, some of whose patterns may be the result of random processes. In instances where events are occurring at random (i.e., independent of past or future occurrences), the **Poisson probability distribution** can be used to analyze how often an outcome occurs during a certain time period or across a particular area.

For example, farmers in the Midwest may be concerned about the frequency of the devastating effects of hailstorms on crops. Because of the uncertain nature of such weather events, hailstorm occurrence varies greatly from year to year and from region to region. Farmers may want to know the probability of experiencing two hailstorms per year. Such information can be useful in deciding what types of crops to

plant or whether to invest in hail damage insurance. The occurrence of such phenomena is independent of past or future occurrences and can be considered to take place in a random manner. Therefore, the Poisson probability distribution accurately describes the expected frequency of occurrence over time or across space.

A small set of data on hailstorm occurrence over a 35-year period for a county in the Midwest illustrates a Poisson problem with a temporal occurrence (table 5.3). During these 35 years, 10 years were completely free of hailstorms, while 1 year experienced four hailstorms. The most common frequency (mode) was one hailstorm per year, which occurred 12 of the 35 years. Since a total of 43 hailstorms occurred during the period, the mean number of hailstorms per year was 1.23. The probability for any frequency of occurrence can also be calculated. For example, since the data cover 35 years and 10 of these years did not have a hailstorm, the resultant probability of no hailstorm occurring in a year is 10/35 or 0.285. Thus, given this set of county data, the area has a 28.5 percent chance of avoiding a hailstorm in any particular year.

If the process producing the pattern is truly random, the probabilities of occurrence will follow the Poisson distribution. The mean frequency

Table 5.3 Observed Hailstorm Occurrence over a 35-year Period for a County in the Midwest

Number of Hailstorms per year	Observed Frequency of Years	Total Hailstorms	Observed Probability of Occurrence
0	10	0	.285
1	12	12	.343
2	9	18	.257
3	3	9	.086
4	1	4	.029
5+	0	0	.000
TOTAL	35	43	1.000

is the average number of occurrences per time period or geographic area. Knowing only this mean value is sufficient to allow all Poisson probabilities to be calculated:

$$P(X) = \frac{e^{-z}(z^X)}{X!} = \frac{z^X}{e^z(X!)} \qquad (5.3)$$

where: X = frequency of occurrence
  z = mean frequency
  e = mathematical (exponential) constant: 2.7183
  $X!$ = X factorial

Like the binomial distribution, Poisson requires discrete or integer outcomes (X) to represent the number or frequency of occurrences. Unlike the binomial, however, the Poisson distribution is not binary. In the hailstorm example, discrete outcomes represent the different number of hailstorms occurring per year. The mean outcome or average number of hailstorms per year (z) is 1.23. Since both e and z are constants, the expression e^z in equation 5.3 is a fixed value as well ($e^z = 2.7183^{1.23} = 3.4123$). This calculation requires the exponential function on a calculator or computer. Using this constant e^z value, Poisson probabilities are calculated for a series of outcome frequencies (X = 0, 1, 2, 3, 4) and presented in table 5.4.

If hailstorms in the county occur randomly with a mean of 1.23 per year, the likelihood of no storm during a given year is 0.2923 (table 5.4). In other words, in about 29 years out of 100, no hailstorms should occur. The highest probability occurs at a frequency of one hailstorm per year (0.3605) or about 36 years out of 100 (figure 5.3).

Geographers often want to compare the frequencies obtained from a set of observed data to frequencies generated by the Poisson or random distribution. The expected frequencies are determined by multiplying the Poisson probabilities by the total frequency (n), 35 years in the

Table 5.4 Critical Values and Poisson Probabilities for Expected Hailstorm Frequency per Year

$$P(X) = \frac{z^X}{e^z(X!)}$$

CRITICAL VALUES

X = 0, 1, 2, 3, 4

N = 35 years

F = total frequency = 43 hailstorms

e = 2.71

z = mean frequency per year = 43/35 = 1.23

$e^z = 2.71^{1.23} = 3.4123$

$$P(0) = \frac{1.23^0}{3.4123(0!)} = \frac{1}{3.4123} = 0.2923$$

$$P(1) = \frac{1.23^1}{3.4123(1!)} = \frac{1.23}{3.4123} = 0.3605$$

$$P(2) = \frac{1.23^2}{3.4123(2!)} = \frac{1.5129}{6.8246} = 0.2217$$

$$P(3) = \frac{1.23^3}{3.4123(3!)} = \frac{1.8609}{20.4738} = 0.0909$$

$$P(4) = \frac{1.23^4}{3.4123(4!)} = \frac{2.2889}{81.8952} = 0.0279$$

current example. If the difference between the observed and expected frequencies is small, it is likely that a random process generated the pattern. Conversely, if the differences are large, it is less likely that the observed pattern is random. Table 5.5 shows both the actual and expected Poisson frequencies of hailstorm occurrence. By examining the observed and expected frequencies, it appears that the occurrence of hailstorms is indeed random.

Figure 5.3   Poisson (Expected) Probabilities for Number of Hailstorms per Year

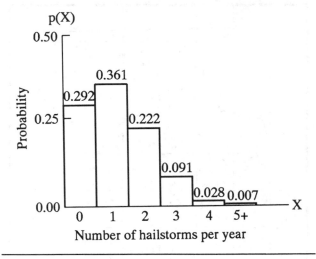

Number of hailstorms per year

Table 5.5   Observed and Expected (Poisson) Frequencies for Hailstorm Occurrence per Year

Number of Hailstorms per Year	Observed Frequencies (Years)	Poisson Probabilities	Expected Frequencies (Years)
0	10	0.2923	10.2
1	12	0.3605	12.6
2	9	0.2217	7.8
3	3	0.0909	3.2
4	1	0.0279	1.0
5+	0	0.0067	0.2
TOTAL	35		35.0

Since geographers primarily study spatial patterns rather than temporal sequences, the Poisson probability distribution is commonly used to investigate the degree of randomness in point patterns. In the temporal Poisson example, the period of study was divided into equal time intervals, such as a year. In the spatial equivalent of Poisson, the geographic region under study is divided into spatial areas, usually a series of regular sized, square cells called *quadrats*. The number of occurrences of an item or phenomenon being studied is recorded for each of the quadrats covering the study area. Given the mean number of occurrences per quadrat or cell across the region, the Poisson distribution shows the probability of a quadrat containing a certain frequency of occurrences.

The use of Poisson probabilities to examine spatial point patterns for randomness is demonstrated with an example of tornado "touch down" sites in Illinois from 1916 to 1969 (figure 5.4). Because tornadoes are produced by complex processes, their locational pattern is commonly thought to be random across small geographic regions. To calculate the expected or Poisson pattern of tornado occurrence that would be found under an assumption of randomness, a set of quadrats is first placed over the study area (figure 5.5).

Because the boundary of Illinois is irregular, some of the quadrats around the border lie only partially inside the study area. In fact, of the 85 cells covering the state, only 47 lie totally inside the boundary. Of the remaining 38 cells, 22 have less than half of their area inside the study region. To determine Poisson probabilities for the tornado pattern, the observed frequencies and mean frequency per cell are calculated using only those cells with more than half of their area in Illinois. This set of 63 quadrats covers 450 of the 480 points on the map, representing almost 94 percent of the tornadoes that occurred in Illinois over the 54-year period.

The frequency of tornadoes is determined for each cell by counting the number of points inside each quadrat. Shown in table 5.6, the observed frequencies range from a low of 1 tornado to a high of 18 tornadoes. The most frequent occurrence or mode is five points per quadrat, which is found in 10 cells. The average cell frequency (z)—total number of tornadoes (450) divided by the number of cells (63)—is 450/63 or 7.143. Thus, for this set of quadrats, an average cell contains 7.143 points (tornadoes).

Figure 5.4 Spatial Pattern of Tornado "Touch Downs" in Illinois, 1916-1969.

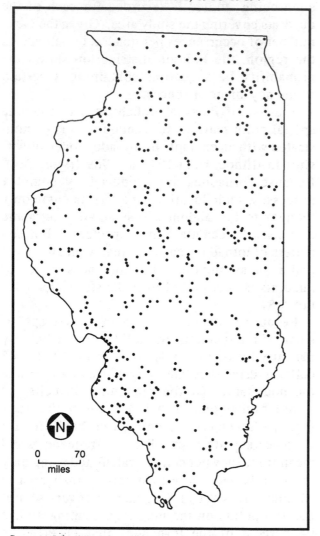

Source: Modified from Wilson, John W. and Stanley A. Changnon, Jr. 1971. *Illinois Tornadoes.* Illinois State Water Survey Circular 103, pp. 10, 24.

Figure 5.5 Illinois Tornado Pattern with Quadrats Superimposed

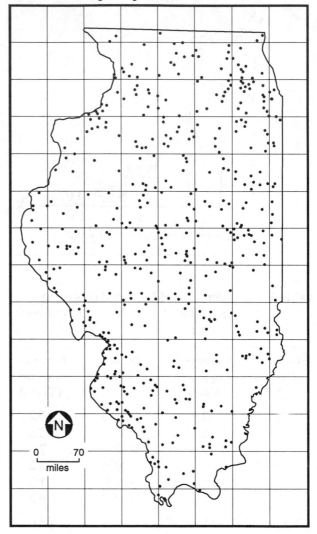

The probability of tornado occurrence under an assumption of randomness is determined from the Poisson equation (5.3) using the mean cell frequency. The calculations for zero, one, and two tornadoes per cell are shown in table 5.7, and figure 5.6 shows the Poisson probabilities graphically. To compute the expected cell frequencies, each Poisson probability is multiplied by the total number of cells. For example, since the Poisson probability of a cell having four tornadoes is 0.0857, 8.57 percent of the 63 cells or 5.40 cells should have four tornadoes (table 5.8). The largest frequency expected for a random pattern of tornadoes is seven points per cell, which should occur in 9.37 cells of the study area. This maximum expected value is consistent with the mean cell frequency of 7.143 points per quadrat.

Table 5.6 Observed Frequency of Tornado Occurrence per Cell for Illinois, 1916-1969

Number of Tornadoes per Cell	Observed Frequencies of Cells	Total Tornadoes	Observed Probability of Occurrence
0	0	0	0.000
1	1	1	0.016
2	2	4	0.033
3	7	21	0.111
4	4	16	0.063
5	10	50	0.159
6	5	30	0.079
7	8	56	0.127
8	6	48	0.095
9	8	72	0.127
10	3	30	0.048
11	3	33	0.048
12	0	0	0.000
13	0	0	0.000
14	4	56	0.063
15	1	15	0.016
16	0	0	0.000
17	0	0	0.000
18	1	18	0.016
19+	0	0	0.000
TOTAL	63	450	1.00

Figure 5.6 Poisson (Expected) Probabilities for Number of Tornadoes per Cell

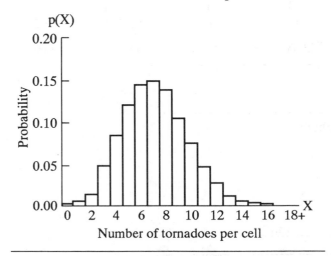

Table 5.7 Critical Values and Poisson Probabilities of Expected Tornado Frequency per Cell for Three Outcomes

$$P(X) = \frac{z^x}{e^z(X!)}$$

CRITICAL VALUES

$X = 0, 1, 2, 3, 4, \ldots, 17, 18$

$N = 63$ cells

$F = $ total frequency $= 450$ tornadoes

$e = 2.71$

$z = $ mean frequency per cell $= 450/63 = 7.14$

$e^z = 2.71^{7.143} = 1265.22$

$$P(0) = \frac{7.143^0}{1265.22(0!)} = \frac{1}{1265.22} = 0.0008$$

$$P(1) = \frac{7.143^1}{1265.22(1!)} = \frac{7.143}{1265.22} = 0.0056$$

$$P(2) = \frac{7.143^2}{1265.22(2!)} = \frac{51.02}{2530.44} = 0.0202$$

Although the observed frequencies appear to match somewhat the calculated Poisson frequencies, notable discrepancies exist between the two frequencies. Inferential statistics can be used to determine the likelihood that the expected and observed frequencies are similar. This procedure is discussed in section 11.1, which deals with inferential statistics for point patterns.

The use of quadrats to examine spatial point patterns generates several methodological decisions for the geographer. In addition to deciding how to handle quadrats that occupy border locations partially outside the study area, researchers must consider the important issue of quadrat size. How would the Poisson probabilities have differed if more quadrats of smaller size had been placed over the pattern of tornadoes? How does a decision to include quadrats only partially inside the study area affect results? These questions will be examined in section 11.1.

5.5 The Normal Distribution

The most generally applied probability distribution for geographical problems is the **normal distribution**. When a set of geographical data is normally distributed, many useful conclusions can be drawn, and various properties of the data can be assumed. The normal distribution provides the basis for sampling theory and statistical inference, both of which are discussed in later chapters. The discussion in this section concerns how probability statements can be made from a normal frequency distribution.

Although a normal distribution is fully described with a rather complex mathematical formula, it can be generally understood by a simple graph that shows the frequency of occurrence on the vertical scale for the range of values displayed on the horizontal axis (figure 5.7). Since the values on the horizontal axis are not restricted to integers, the normal curve is an example of a continuous distribution. The most striking feature of the normal curve is the symmetry or balance of frequencies between the lower (left-hand) and upper (right-hand) ends of the distribution. As discussed in section 3.3, this symmetric pattern of values in a normal distribution suggests that no skewness exists in the data. The central value of the data represents the peak or most frequently occurring value. In a normally distributed set of data, this position

corresponds to all three measures of central tendency—the mean, median, and mode.

A normal curve is also described as having a general bell-shaped pattern. This characteristic shape represents a frequency distribution with

Table 5.8 Observed and Expected (Poisson) Frequencies for Tornado Occurrence per Cell

Number of Tornadoes per Cell	Observed Frequencies (Cells)	Poisson Probabilities	Expected Frequencies (Cells)
0	0	0.0008	0.05
1	1	0.0056	0.35
2	2	0.0202	1.27
3	7	0.0480	3.02
4	4	0.0857	5.40
5	10	0.1225	7.72
6	5	0.1458	9.18
7	8	0.1488	9.37
8	6	0.1328	8.37
9	8	0.1054	6.64
10	3	0.0753	4.74
11	3	0.0489	3.08
12	0	0.0291	1.83
13	0	0.0160	1.01
14	4	0.0082	0.52
15	1	0.0039	0.25
16	0	0.0017	0.11
17	0	0.0007	0.04
18	1	0.0003	0.02
19+	0	0.0003	0.02
TOTAL	63	1.0000	63.00

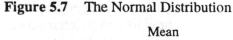

Figure 5.7 The Normal Distribution

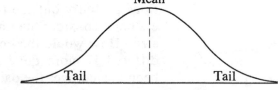

an intermediate amount of peakedness. The tails of a normal distribution are the portions of the curve farthest from the mean which have the lowest frequencies. As shown in figure 5.7, the frequency of values in a set of normally distributed data declines in a gradual manner in both directions away from the mean.

Because of its particular shape and mathematical definition, the normal distribution is very useful in making probability statements about actual outcomes in geographical problems. For example, if the amount of precipitation a location receives in a year is normally distributed over a series of years, the probability of this site receiving a given amount of precipitation can be calculated mathematically. This valuable attribute of the normal distribution will be discussed further in later chapters on sampling theory and statistical inference.

Probability estimates using the normal distribution focus on the area found under the curve. Since the distribution delimits all data values, the total area under the normal curve represents 100 percent of the outcomes. The percentage of values lying under the curve within any smaller interval along the horizontal axis can be determined. For example, due to the symmetry of the normal distribution, 50 percent of the values must lie under the curve and to the right of the central or mean value. Since the normal curve is also a probability distribution, a value taken from a normal distribution has a 0.5 probability of falling above the mean.

Although determination of the 50 percent value is obvious given the symmetric form of the normal curve, the corresponding percentages for other intervals along the horizontal axis are less apparent. Since the objective of integral calculus is to find areas under various mathematical distributions, this technique could be applied to determine areas under the normal curve. However, a simpler method uses a table of normal values that shows the proportions of total area under various parts of the normal curve and is derived mathematically from a theoretical normal distribution.

In order to use the table of normal values, the concept of **standard scores** must be introduced.

A standardized scale, which represents data as a standard score (also called a **normal deviate**), shows the number of standard deviations separating a particular value from the mean of the distribution. Standard scores can be either positive or negative. For units of data greater than the mean, the corresponding standard scores are positive; for values less than the mean, standard scores are negative. A standard score of 1 represents a value that is one standard deviation *above* the mean, whereas a score of −1 is one standard deviation *below* the mean (figure 5.8). The mean corresponds to a standard score or normal deviate of 0. The larger the standard score (in either the positive or negative direction), the farther the value lies above or below the mean.

For any given standard score or normal deviate, the table of normal values provides a probability, which can be interpreted in two ways. First, it gives the probability of a value falling between the mean and the given standard score location for a set of data that is normally distributed. Second, when the probability is multiplied by 100, it shows the percentage of all values in the normal distribution that lie between the mean and this location. Thus, the table of normal values provides information to determine the area under the normal curve for any interval.

The use of the normal curve can be explained through several examples from the table of normal values (appendix, table A). The probability value

---

Figure 5.8   Selected Areas Under
the Normal Curve

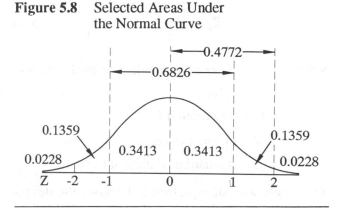

associated with a standard score of 1.0 is 0.3413 (or 34.13 percent). In a normally distributed set of data, approximately 34 percent of the values lie between the mean and one standard deviation above the mean. Since the normal curve is symmetric, 34.13 percent of the values also lie between the mean and one standard deviation *below* the mean. Therefore, by combining these two areas, a total of approximately 68 percent of the values in a normal distribution lie within one standard deviation on either side of the mean (figure 5.8).

Using the same procedure, the probability of values lying between the mean and a standard score of 2 is 0.4772. Thus, almost 48 percent of values in a normally distributed set of data lie between the mean and two standard deviations above the mean and about 95 percent of the values are within two standard deviations on either side of the mean (figure 5.8). The remaining 5 percent of the values are in the two tails of the distribution, where the standard score is either greater than 2.0 or less than −2.0.

Although the table of normal values provides probabilistic information on a standardized scale, geographic data are measured on various scales. To apply the normal probabilities to specific sets of data, values must first be converted from their original units of measurement (X) to the unitless standardized (Z) scale. In this process, data values are represented by their relative position in comparison to the mean:

$$Z_i = \frac{X_i - \overline{X}}{s} \qquad (5.4)$$

where: Z_i = Z-score or standard score for the ith value

   X_i = observation i

   $\overline{X}$ = mean of the data

   s = standard deviation of the data

The numerator of equation 5.4 shows the deviation of the ith value from the mean of the data

set. This deviation is then divided by the standard deviation for the distribution. The resulting Z-value (or standard score) can be interpreted as the number of standard deviations an observation lies above or below the mean. If the value under consideration is greater than the mean, the deviation is positive, and Z will be greater than zero. If the observation is less than the mean, the deviation and resulting Z-score will be negative. If X equals the mean, the value does not deviate from the mean, and the Z-score will be zero.

Any set of data can be converted from original measurement units into the corresponding set of standardized units. However, to determine probability estimates from information found in the table of normal values, the original data must be normally distributed. The statistical test for normality in a data set is discussed in section 10.1.

The methodology for making probability estimates is demonstrated by studying the annual precipitation data for Washington, D.C., originally presented in chapter 2. These data are assumed to be normally distributed over a 40-year period, with a mean of 39.95 inches and a standard deviation of 7.5 inches. Suppose a geographer wants to know the probability of annual precipitation in Washington, D.C. exceeding 48 inches. The table of normal values can be used to estimate probabilities of outcomes for a normal distribution with the following three-step process. First, the standard score corresponding to the precipitation level under consideration (48 inches) must be calculated. Second, using the table of normal values, the probability value is determined for this standard score. Third, this probability value is evaluated to answer the specific research question.

Step 1: Calculate the standard score:

The standard score corresponding to 48 inches is calculated, where $\overline{X}$ = 39.95 and s = 7.5.

$$Z_i = \frac{X_i - \overline{X}}{s} = \frac{48.0 - 39.95}{7.50} = \frac{8.05}{7.50} = 1.07$$

Thus, a precipitation level of 48 inches is 1.07 standard deviations *above* the mean precipitation of 39.95 inches (see figure 5.9).

Step 2: Determine the probability from the normal table:

Using the table of normal values, the *Z*-score of 1.07 corresponds to a probability level of 0.3577 (see figure 5.9). Thus, almost 36 percent of the values under the normal curve (years of precipitation in Washington, D.C.) lie between the mean and 1.07 standard deviations. In other words, about 36 years out of 100, the precipitation in Washington, D.C. should be between 39.95 and 48 inches.

Step 3: Determine position on normal curve and evaluate the probability value:

Although the table of normal values always determines probabilities for areas under the curve in relation to the mean, the actual probability being sought may represent a different part of the curve. For this problem, the answer lies in the shaded portion of the curve above (to the right of) the *Z*-score for 1.07. Since the proportion of the total area under the normal curve above the mean is .5000, the correct answer is found by subtracting the probability in Step 2 from .5000 (.5000 − .3577 = .1423). Therefore,

in 14 years out of 100, the annual precipitation in Washington, D.C. should exceed 48 inches.

Probability questions from a normally distributed data set can be stated in another way. Given the same normal frequency distribution for precipitation data in Washington, D.C., what amount of precipitation is likely to be exceeded with a probability of 0.90? That is, what amount of precipitation will be exceeded 9 years out of 10? To answer this question, the methodology is reversed.

Step 1: Determine the position on the normal curve:

As shown on figure 5.10, precipitation in Washington, D.C. will be exceeded 90 percent of the time at the position indicated by the shading. The total shaded area represents .90 of the total area under the curve, while the portion to the left of the mean is .40. Therefore, the *Z*-score corresponding to a probability of .40 must be determined.

Step 2: Determine the *Z*-score:

According to the table of normal values, a probability of 0.40 corresponds to a *Z*-score of 1.28. Since the location is below (or less than) the mean, the value is −1.28. Thus, the precipitation that occurs at least 90 percent of the time

Figure 5.9 Determining the Probability of Annual Precipitation Exceeding 48 Inches in Washington, D.C.

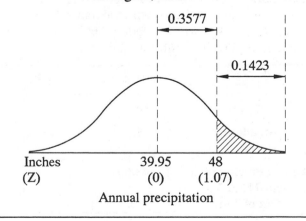

Figure 5.10 Determining the Amount of Precipitation Exceeded 9 Years Out of 10 in Washington, D.C.

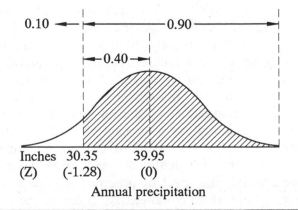

(9 years out of 10) is at a value 1.28 standard deviations below the mean.

Step 3: Calculate the precipitation value (X_i):

Using equation 5.4, the precipitation value corresponding to a Z-score of -1.28 is determined:

$$-1.28 = \frac{X_i - 39.95}{7.50}$$

$$X_i = -1.28(7.50) + 39.95 = 30.35 \text{ inches}$$

Therefore, 9 years out of 10, precipitation in Washington, D.C. will exceed 30.35 inches. Conversely, 1 year in 10, precipitation could be expected to be less than 30.35 inches.

In summary, the normal curve is a very useful probability distribution in geography. For normally distributed data, the proportions of total area under the normal curve are fixed, allowing geographers to make a variety of probability statements. Just as importantly, the characteristics of normal curves serve as the basis for sample estimation and inferential statistics, discussed in upcoming chapters.

5.6 Probability Mapping

The previous section discussed how to make probabilistic statements about an event at a single location. For example, what level of annual precipitation will be exceeded 9 years out of 10 in Washington, D.C.? This technique can be extended to allow similar statements to be made for a set of locations distributed across a region. If probability data are plotted on a map, the resulting information represents a **probability map** or "probability surface," showing spatial variation in the variable under consideration.

Suppose annual precipitation data are collected for additional cities in the eastern United States, and the mean and standard deviation of each distribution are calculated. Assuming the data are normal, the level of precipitation exceeded 9 years out of 10 could easily be determined for each location studied, using the technique dis-

cussed in the previous section. A probability map is produced by assigning each probability value to its location on a map and by connecting equal probability values with isolines. Just as contours show elevation patterns in an area, the probability surface would show the spatial pattern of precipitation probability.

How does a map of precipitation probability provide more useful information than a simple annual precipitation map? Probability maps not only consider the central tendency of the data, but also the variability as well. In fact, the key advantage of probability maps over maps of central tendency (e.g., a map of average annual precipitation) is their focus on the variation of values at each location.

The following example illustrates the importance of variability in the construction of probability maps. Consider two cities, A and B. They have equal average annual precipitation, but very different levels of variability around those averages (table 5.9). Note the contrast between the minimum annual precipitation expected 9 years out of 10 in city A versus city B. City A has little precipitation variability from year to year, and therefore, the minimum expected precipitation is only slightly lower than the mean (43.6 vs. 50). By contrast, city B has

Table 5.9 Comparison of Expected Precipitation in Two Cities with the Same Average Annual Precipitation, but Very Different Precipitation Variability

	City A	City B
Mean	50	50
Standard Deviation	5	20
Annual Precipitation Exceeded Nine Years Out of Ten	43.6	24.4

great variability in precipitation over time, making the minimum expected precipitation considerably lower than the mean (24.4 vs. 50). When mapping the minimum annual precipitation expected 9 years out of 10 for many different cities, the resulting probability map has a spatial pattern that differs greatly from that of average annual precipitation. In fact, a map of average annual precipitation is comparable to a probability map of minimum precipitation expected 5 years out of 10 or one half of the time. Since the data analyzed are assumed to be normally distributed, the mean value at each location corresponds to a 0.50 probability.

To illustrate the construction and interpretation of probability maps, consider data on cooling degree days for a set of cities across the United States. A *cooling degree day* is a surrogate measure for the amount of cooling or energy needed to produce a "comfortable" indoor climate. It is a critical index for measuring yearly energy demand in warm climates and cooler climates experiencing hot summers. A cooling degree day is the opposite of a "heating degree day," which estimates the amount of energy needed to produce indoor warmth in cold or cooler climates.

The cooling degree day statistic is calculated from the average daily temperature at a site, which is the mean of the high and low temperatures for that day. Sixty-five is then subtracted from the average daily temperature in degrees Fahrenheit. (If temperature is measured in degrees Celsius, the constant 18.3 rather than 65 is used.) If the cooling degree day statistic is negative, no cooling degree days are recorded, and the assumption is that air-conditioning will not be needed on that day. The daily values are then accumulated over a year to produce the annual number of cooling degree days needed.

One hundred-three large cities in the conterminous United States, excluding Alaska and Hawaii, have been selected as locations to create a probability map of cooling degree days. This map shows the number of cooling degree days which would be exceeded 9 years out of 10 (0.90

probability). A similar procedure could be followed for any other probability level. For example, how many cooling degree days should occur 95 years out of 100 (0.95 probability) or 8 years out of 10 (0.8 probability)?

Using the means and standard deviations for the cooling degree day data of the 103 cities, probability values are calculated according to the standard score formula (equation 5.4). The calculation for Birmingham, Alabama, indicates that a cooling degree day level of 1632.94 will be exceeded 90 percent of the time (figure 5.11).

Figure 5.11   Calculation of Cooling Degree Days Exceeded 9 Years Out of 10 for Birmingham, Alabama

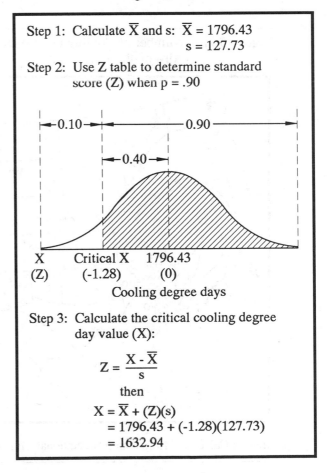

Step 1: Calculate $\overline{X}$ and s: $\overline{X}$ = 1796.43
                                              s = 127.73

Step 2: Use Z table to determine standard score (Z) when p = .90

X	Critical X	1796.43
(Z)	(-1.28)	(0)

Cooling degree days

Step 3: Calculate the critical cooling degree day value (X):

$$Z = \frac{X - \overline{X}}{s}$$

then

$$X = \overline{X} + (Z)(s)$$
$$= 1796.43 + (-1.28)(127.73)$$
$$= 1632.94$$

Following this procedure, values for all cities are placed on a map of the United States, and isolines drawn connecting locations having the same estimated number of cooling degree days (figure 5.12). Clearly, isoline mapping always requires interpolation to proceed from a finite set of point data to a continuous map surface.

The probability map of cooling degree days is roughly analogous to the spatial pattern of solar energy or heat received across the United States. Since the map portrays a situation expected to occur 9 years out of 10, it provides a fairly accurate spatial estimate of the amount of energy needed to air-condition.

The general east-west trend of isolines on the probability map reflects the important influence of latitude on the distribution of heat. This situation is clearly seen, for example, in the regular north to south increase in cooling degree days and air-conditioning need from Minnesota to Louisiana. When isolines dip to the south, the number of cooling degree days decreases, due to cooler summer temperatures in higher elevations. This pattern is especially evident in the Appalachian Mountains of West Virginia, western Virginia and North Carolina, and northern Georgia.

Isolines that curve to the north suggest warmer temperatures are found in these areas during the summer. This pattern is seen immediately west of the Rocky Mountains, for example, across Nevada and eastern Oregon. The moderating influence of the Pacific Ocean is seen along the California coast, keeping the number of cooling degree days lower than the corresponding latitudes several hundred miles inland.

Figure 5.12   Number of Cooling Degree Days Expected 9 Years Out of 10: An Estimate of Air-conditioning Need

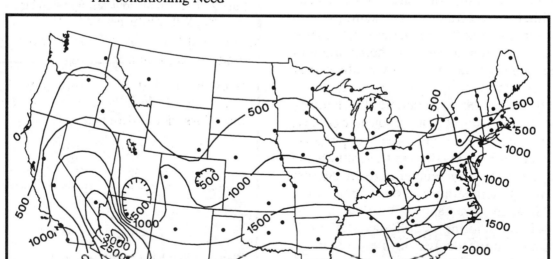

Source: Calculated from data from National Climatic Data Center, U.S. Dept. of Commerce.

A peak of cooling degree days occurs near Phoenix, Arizona, suggesting that this region will need more air-conditioning than adjacent areas. Significant elevation differences between Flagstaff and Phoenix create a very complex isoline pattern in this region. Flagstaff, with an elevation over 6,900 feet, requires virtually no air-conditioning, since the cooling degree day value is barely over 100. Phoenix, on the other hand, with a lower elevation around 1,100 feet in the Sonoran Desert, needs a great deal of air-conditioning. Interpretation of cooling degree day patterns in Arizona illustrates a common problem associated with isoline mapping. The density and placement of control points from which the isolines are drawn influence the resulting pattern. If data from additional weather stations were used, the isolines that constitute the probability surface could be located more precisely.

Probability maps can be constructed for many spatial variables. The technique seems most directly applicable for the analysis of natural phenomenon in climatology, meteorology, and environmental studies. For example, geographers could construct probability maps of atmospheric particulates, ozone levels, major winter storms, timing of killing frosts, or the acid rain level. It may be possible to extend probability mapping to selected topics in human geography, such as disease, unemployment, or poverty. Extreme care must be taken, however, because many variables in human geography are not continuously distributed over space.

Key Terms and Concepts

References and Additional Reading

Bulmer, M. G. 1967. *Principles of Statistics.* Edinburgh: Oliver and Boyd.

Gregory, S. 1978. *Statistical Methods and the Geographer.* London: Longman.

Matthews, J. 1981. *Quantitative and Statistical Approaches to Geography: A Practical Manual.* Oxford: Pergamon.

Silk, J. 1979. *Statistical Concepts in Geography.* London: George Allen and Unwin.

Taylor, P. J. 1977. *Quantitative Methods in Geography: An Introduction to Spatial Analysis.* Boston: Houghton Mifflin.

Unwin, D. 1981. *Introductory Spatial Analysis.* London: Methuen.

Winkler, R. L. and W. L. Hays. 1975. *Statistics: Probability, Inference, and Decision.* New York: Holt, Rinehart, Winston.

CHAPTER 6

BASIC ELEMENTS OF SAMPLING

Chapter 1 discussed the scientific research process and the ways in which statistical techniques are incorporated into that procedure. In most geographic research and problem solving, sampling is an essential component. In fact, whatever the area of geographic study, a basic knowledge of sampling procedures and methodology is valuable.

How does sampling fit into the overall scheme of statistical analysis and scientific research? Recall the fundamental distinction made in chapter 1 between descriptive and inferential statistics. The ultimate aim of inferential statistics is to *infer* certain characteristics about a statistical population, based on information obtained from a sample. Since the objective of sampling is to make inferences about the wider population from which the sample is drawn, sampling is at the center of the statistical inference process. The probability that sample results can be successfully generalized to the larger group or population can be determined.

Sampling is an essential skill in virtually all areas of geographic study. For example, a biogeographer interested in the spatial pattern of environmental change associated with high intensity recreation use in a national park cannot examine conditions everywhere in the park. A representative sample of study sites needs to be selected for detailed analysis of these human-environment relations. An urban geographer wishing to examine the locational variation of housing quality in a metropolitan area must select a sample of homes for detailed study. A behavioral geographer conducting natural hazards research may distribute questionnaires to a representative sample to learn public attitudes toward alternative flood-management policies along the flood plain of a river. A medical geographer concerned with neighborhood variations in the use of hospital emergency rooms as primary care centers would use sampling to select the neighborhoods and the hospitals to include in the study.

These examples illustrate the geographer's use of sampling in both spatial and nonspatial contexts. Spatial sampling takes place when the biogeographer selects *locations* to examine environmental change in the national park. Likewise, the neighborhoods chosen by the medical geographer to examine hospital use patterns also constitute a spatial sample. On the other hand, the geographer conducting the natural hazards attitude study along the river flood plain may have selected individuals from a nonspatial list of households in the area. The urban geographer conducting the housing quality study could have taken the sample of homes from either a nonspatial list (i.e., tax rolls) or from a spatial source (i.e., a map).

Whatever the topical area of geographic research, certain basic sampling concepts must be understood and proper sample design procedures followed. Section 6.1 discusses the basic concepts of sampling, including an overview of the advantages of sampling, the sources of sampling error, and the steps involved in a well-designed sampling procedure. The various types of probability sampling are reviewed in section 6.2. All probability samples contain an element of randomization, but simple random sampling is often not the best choice of sample design. Systematic, stratified, cluster, and hybrid sample designs are also examined. Section 6.3 discusses the circumstances under which spatial sampling is necessary and reviews the types of spatial sampling. Special attention is focused on point sampling designs.

6.1 Basic Concepts in Sampling

Advantages of Sampling

A variety of both practical and theoretical reasons makes sampling preferable to complete enumeration or census of an entire population. The advantages of sampling may be summarized as follows:

1. Sampling is a necessity in many geographic research problems. If the population being studied is extremely large (or even theoretically infinite), completing a total enumeration is not possible. For example, sampling is the only alternative when studying the reasons why families in the United States change residence: all those who move cannot possibly be contacted and surveyed individually. A geographer analyzing global changes in the nature and spatial extent of tropical rain forests will find it impossible to have total spatial coverage, since the number of locations in a rain forest is infinite.

2. Sampling is an efficient and cost-effective method of collecting information. An appropriate amount of data concerning the population can be obtained and analyzed quickly with a sample. Not only is sample information collected in less time, but also sampling keeps expenditures lower and logistical problems to a minimum. The overall scale of effort (time, cost, personnel, logistics, etc.) is made practicable with sampling, as opposed to an examination of the entire population.

3. Sampling can provide highly detailed information. In geographic problems where indepth analysis is necessary, only a small number of individuals or locations can be included in the study. These few elements in the sample could then be closely scrutinized with the collection of a comprehensive set of information. A study of shopping behavior patterns, for example, might require numerous questions about the number of shopping trips, locations visited while shopping, attitudes about alternative stores, as well as demographic or socioeconomic information concerning household members. Such detailed analysis can be obtained only through sampling.

4. Sampling allows repeated collection of information quickly and inexpensively. Many geographic research problems require detailed information collected over a specific period of time or focus on spatial changes that occur rather quickly. For example, a geographer studying the attitudes of citizens living in a barrier island community toward alternative coastal zone management strategies may want to follow a sample of individuals through time as pertinent legislation moves through the political process. Their attitudes may change over time, especially if a hurricane hits the island during the study period. An urban geographer analyzing the spatial pattern of growth may wish to focus on the views of residents in a neighborhood before, during, and after the construction of a nearby shopping mall. With dynamic situations such as these, sampling is required: information from all individuals in the population could not possibly be collected without unreasonable effort and cost.

5. Sampling often provides a high degree of accuracy. With sampling, an acceptable level of quality control can be assured. Complete and accurate questionnaire returns can be obtained if a small number of well-trained personnel conduct all of the interviews. This procedure may provide more accurate results than a complete census requiring a larger number of personnel, some of whom may not be fully trained. The 1990 U.S. Census of Population illustrates the difficulties of acquiring accurate information from everyone. Considerable controversy has been generated because many Americans were literally not counted. The Census Bureau has used statistical modeling procedures to estimate actual population counts from samples. In other words, the only way Bureau personnel could ensure unbiased, accurate counts was through taking samples.

Sources of Sampling Error

Certain fundamental criteria or conditions help guide any proper sampling procedure. Clearly, a central goal of sampling is to derive a truly *representative* set of values from a population. A representative sample will reflect accurately the actual characteristics of the population without bias. To ensure an unbiased representative sample, an element of **randomness** must be incorporated into the sample design procedure. However, just having randomness built into a sampling plan does not guarantee an unbiased sample, for many other **sources of sampling error** are possible.

Unfortunately, it is not usually possible to know with absolute certainty whether a sample is totally representative. The very nature of sampling means that everything cannot be known about the population containing the sample. Because only sample data are available, some uncertainty will always be associated with sample estimates, and some sampling error will occur. The geographer must try to minimize that sampling error given various practical constraints such as cost or time.

The measurement concepts of precision and accuracy (discussed in chapter 2) help categorize the many sources of sampling error. The results from a small sample may not be very exact or precise. Imprecision can be reduced by increasing the sample size, thereby deriving a more exact estimate (figure 6.1, line A). Larger samples, however, are invariably more costly and time-consuming to obtain (figure 6.1, line B). Satisfactory resolution of this difficult trade-off between sample precision and the effort required to sample is important in most real-world sampling situations.

Sample inaccuracy is a far more complex issue. Systematic bias can enter a sampling situation in many different ways. Some inaccuracies are the result of problems in the sampling pro-

Figure 6.1 The Influence of Sample Size on Amount of Imprecision and Effort Required to Sample

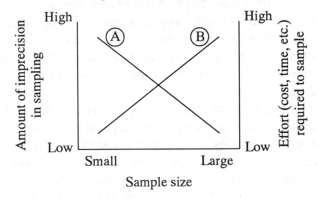

cedure itself. For example, the elements or individuals in a population could be "mismatched" with the set of elements or individuals from which a sample is taken. An urban geographer studying new home construction patterns might use a building permit list as a data source. If this list contains home renovation projects as well as new home construction, some of the renovations might be erroneously included in the new home construction sample.

Selecting an improper sampling design is another potential source of inaccuracy in sampling. The complexity of a geographic problem might call for a more sophisticated approach, rather than a simple random design. In other geographic sampling problems, an inappropriate method of data collection may be used. For example, a geographer studying population mobility might use a mail questionnaire (to keep costs down) when telephone interviews would provide more accurate results.

Other types of inaccuracy may result from some operational, logistic, or personnel problem not directly tied to the actual sampling procedure. Inconsistencies in field data collection or interviewing could adversely affect the sample results. Errors could be made in the editing, recording, or tabulating of information. Even forces of nature beyond the researcher's control could bias results. For example, a geographer examining land use patterns along the flood plain of a large river could find the entire project in jeopardy if severe flooding occurs during the study period.

Simply stated, the quality of statistical problem solving in geography depends heavily on correct samples. The geographer must carefully develop a sampling procedure that reduces sampling error to a tolerable or acceptable level. Most possible sources of error can be reduced substantially (and maybe even avoided or eliminated) if all steps of the sampling procedure are carefully planned and evaluated *before* the full set of sample data is collected and analyzed.

Steps in the Sampling Procedure

Sampling was briefly introduced in the opening chapter as an important data preparation task in the scientific research process. Recall that this process in geography begins with the identification of an appropriate research problem and leads to the formulation of precisely defined research hypotheses. Careful consideration of these general issues will ensure that sampling is placed in a productive, well-designed research context.

If faced with a geographic research problem in which the collection of sample information is necessary, a number of steps need to be followed (see table 6.1 and figure 6.2).

Collecting sample data is not the first action taken. The researcher must first anticipate and resolve various problems that might cause

Table 6.1   Steps in the Sampling Procedure

Step 1:   Conceptually Define Target Population and Target Area
    —**target population**: the complete set of individuals from which information is to be collected
    —**target area**: the entire region or set of locations from which information is to be collected
    —all samples have *both* a target population and target area

Step 2:   Designate Sampled Population and Sampled Area from Sampling Frame
    —**sampled population**: the set of individuals from which the sample is actually drawn
    —**sampled area**: the region or set of locations where the sampled elements are found
    —operationally, the sampled population and sampled area are taken from a **sampling frame** (a nonspatial sampling frame might be a list of all individuals in a population, while a spatial sampling frame might be a map with a coordinate system for locational identification of sample units)

Step 3:   Select Sampling Design
    —probability sampling preferred over nonprobability sampling
    —important types of probability samples include:

random, systematic, stratified, cluster, and hybrid designs
    —there are spatial and nonspatial variations in sampling design

Step 4:   Design Research Instrument and Operational Plan
    —methods of data collection include: direct observation, field measurement, mail questionnaire, personal interview, and telephone interview
    —establish protocols for handling all problems or situations in the sampling procedure which can be anticipated
    —complete miscellaneous logistic and procedural tasks in the preparation of sample taking

Step 5:   Conduct Pretest
    —trial run or pilot survey of sample data collection method
    —correct all discovered problems which could lead to sampling error
    —pretest results may be effectively used to determine sample size

Step 6:   Collect Sample Data
    —consistency in collection methods and procedures is essential
    —assure overall high level of quality control

Figure 6.2   Steps in the Sampling Procedure

Step 1: Conceptually define target population and target area

Conceptually define target population		Conceptually define target area

Step 2: Designate sampled population and sampled area from sampling frame

Operationally define a nonspatial sampling frame (e.g., list) which depicts the target population as closely as possible	Operationally define a nonspatial sampling frame (e.g., map) which depicts the target areas as closely as possible
The set of individuals or elements contained in the nonspatial sampling frame constitute the sampled population	The set of coordinate locations which can be identified in the spatial sampling frame constitute the sampled area

Step 3: Select sampling design

Step 4: Design research instrument and operational plan

Step 5: Conduct pretest

Step 6: Collect sample data

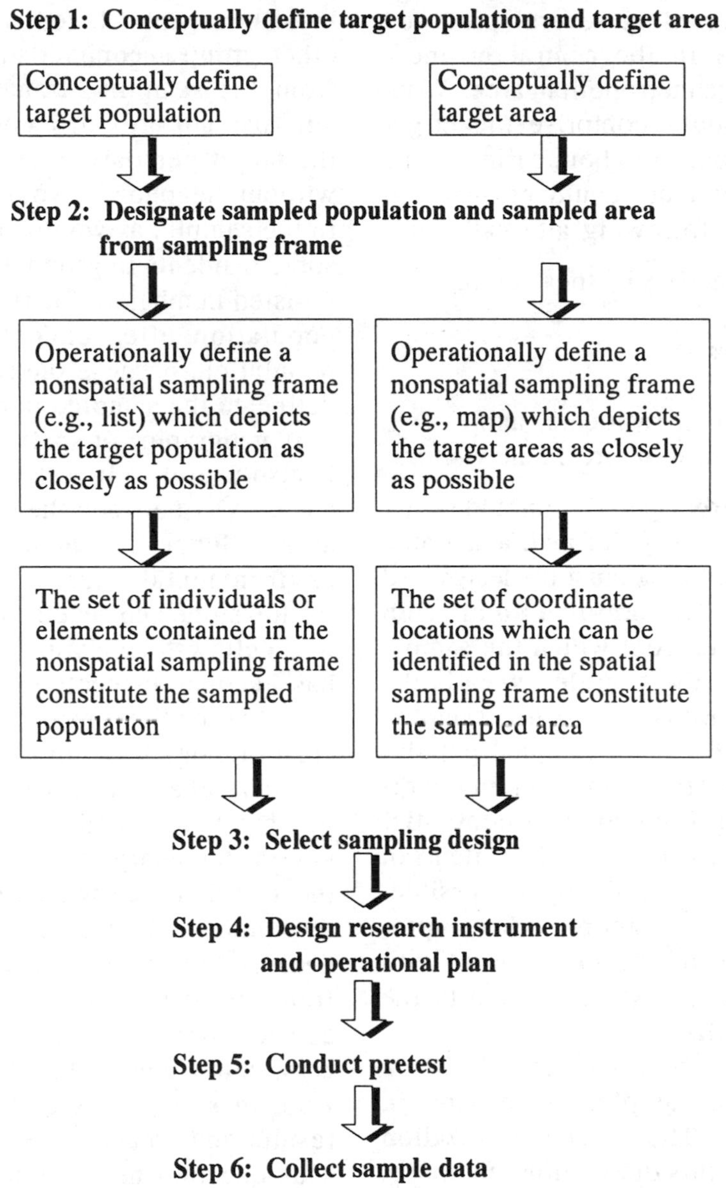

sampling error or other difficulties. Numerous safeguards or checks should be incorporated into the sampling procedure, whenever necessary. The characteristics and relevant issues at each step of a well-designed sampling procedure are now summarized.

Step 1: As hypotheses are formulated in scientific research, variables in the problem must

be conceptually and operationally defined. In sampling, the population and area "targeted" for study must be defined. The **target population** is the complete set of individuals from which information is collected, whereas the **target area** is the entire region or set of locations from which information is gathered. Precise delineation of the target population and target area is not a simple task. Suppose, for example, an urban geographer wishes to study the spatial variation of public attitudes regarding several proposed revitalization projects in the central business district (CBD) of a large metropolitan area. What set of individuals should comprise the target population? What should the boundaries of the target area be? Arguments could be made in support of any of the following alternatives:

- all city residents active in local civic groups,
- all city landowners,
- all city residents,
- all city and suburban residents, and
- all area residents plus nonresident visitors.

Step 2: Once the target population and target area have been conceptually defined, a sampled population and sampled area must be designated from a sample frame. The **sampled population** is the set of individuals from which the sample is actually drawn, and the **sampled area** is the region or set of locations where the sampled elements are found. Both the sampled population and sampled area are taken from an operationally practical **sampling frame**. A nonspatial sampling frame might list individuals or items in the target population as completely as possible. A spatial sampling frame might provide comprehensive map coverage of the target area, with some type of coordinate system to identify the location of sample units.

Why is it important to distinguish between target population and sampled population in nonspatial sampling? The CBD revitalization example can illustrate this distinction. Suppose, conceptually, the population being targeted is all metropolitan area residents, both in the city and in adjacent suburbs. Operationally, however, this target population is not listed in any single comprehensive source. A sampling frame that represents the intended target population as closely as possible needs to be constructed.

Many possible sources of data could provide an operational definition for the sampling frame.

The customer lists from electric utilities would not include all households or all residents at those households listed. People in apartments or other group accommodations would be omitted from the sampling frame. Telephone listings will also not provide a complete enumeration of the target population. Lower-income residents without telephones would be underrepresented in the sample, as would students in dormitories, some residents in group quarters, and those with unlisted numbers. The result is that the sampled population often cannot duplicate the target population, because the sampling frame used to delineate the sampled population is incomplete.

In geographic or spatial sampling problems, it is also useful to investigate whether any differences exist between the target area and sampled area. Suppose a geographer wants to study environmental "threats" within the greater Yellowstone ecosystem (figure 6.3). According to a Yellowstone National Park official, the Park has become an ecological island, with boundaries that neither encompass a complete ecological unit, nor adequately protect the area's unique geothermal activity. The target area for the geographical study is the Greater Yellowstone Ecosystem boundary, as shown in figure 6.3. But the sampled area will almost certainly not correspond exactly with this designated target area, because the geographer will likely be prohibited from entering private ranches and resorts, as well as some public lands leased by lumber and energy companies. Will the inability to sample from these portions of the target area bias the results and weaken the accuracy of the study? This question must be carefully considered before proceeding further.

The geographer's task is to construct sampling frames so that the target population matches the sampled population and so that the target area matches the sampled area as closely as possible. In the CBD revitalization problem, what is the nature and probable extent of sampling error that will occur if a telephone survey (sampled population) is used to elicit opinions

Figure 6.3 The Greater Yellowstone Ecosystem and Environmental Threats

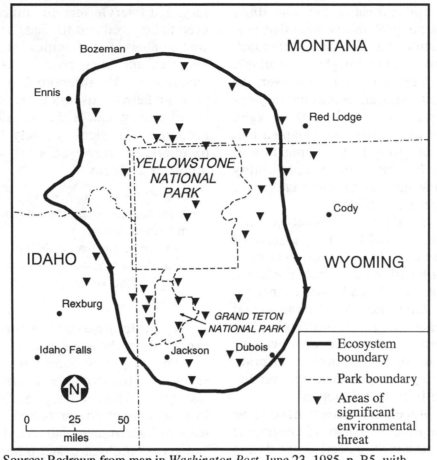

Source: Redrawn from map in *Washington Post*, June 23, 1985, p. B5, with permission.

of all metro area residents (target population)? In the Yellowstone example, how much bias will be injected into the sampling procedure if a considerable portion of private land in the region is excluded from analysis? Whenever facing mismatches like these, only qualified or conditional statements are possible, and sources of systematic bias or inaccuracy in the study must be clearly explained.

Step 3: The selection of an appropriate sample design method is crucial to the success of the entire sampling procedure. Sampling design refers to the way in which individuals or locations are selected from the sampling frame.

A fundamental distinction can be made between **nonprobability** and **probability sampling designs**. Nonprobability sampling is subjective because the selection of sampled individuals or locations is based on personal judgment. Nonprobability samples are, therefore, sometimes called "judgmental" or "purposive" samples. Criteria for sample selection might be: (1) personal experience and background knowledge; (2) convenience or better access to a nonrandom selection of individuals or locations; (3) use of only those who volunteer or respond to the survey instrument, such as a questionnaire in a newspaper; or (4) selection of a single nonrandom study area or sample region (case study) in which to conduct research.

Excellent descriptive results can sometimes be derived from a nonprobability sampling procedure. For example, the case study approach may provide considerable insight on variable interaction within the study area. However, the underlying problem with all nonprobability approaches to sample selection is that no valid inferences can be made from the sampled elements to any wider group in the population or to any other areas. In other words, nonprobability sampling limits the generalizations that can be drawn from the research findings.

By contrast, the nature of probability sampling is more objective and is closely associated with scientific research. In a probability sample, each individual or location that could be selected from the sampling frame has a known chance (or probability) of being included in the actual sample. The advantage of probability sampling is that the amount of sampling error can be estimated. Because of the importance of probability sampling in geographic research, section 6.2 is devoted entirely to this topic.

Step 4: Many procedural matters have to be resolved when sampling in real-world geographic situations. A key task is the selection and design of an appropriate measurement instrument. Optional formats for data collection include direct observation, field measurement (especially in areas of physical geography), mail questionnaire, personal interview, and telephone interview. To select the proper format, the nature of the research problem must be carefully evaluated. Even after the method of data collection is selected, instrument design problems may remain. In survey design, each question must be carefully worded, all possible responses to a question must be considered in advance, and the questions must be sequenced properly.

Various other procedural decisions and logistic arrangements may be necessary, depending on the nature of the problem under investigation. Preliminary site reconnaissance may be necessary, and interviewers and other personnel may need to be hired and trained. If not anticipated, any number of difficulties can plague the researcher, and even create bias in the sample, undermining its inferential power. It is not possible here to discuss in detail the wide variety of circumstances that might arise, as every geographic problem is likely to have its own peculiarities. However, a list of some common problems includes:

- low response to a survey or questionnaire,
- unexpected response due to ambiguity or misunderstanding,
- environmental or political change in the sampled area, and
- unqualified or incompletely trained interviewers.

Step 5: Before extensive time and money are expended on the main data collection effort, a small-scale pretest needs to be conducted to assess both the strengths and deficiencies in the research instrument design and operational plan. In a field study, the pretest should reveal problems with instrument calibration or malfunction, site access problems, and other logistical difficulties. Problems in a survey or questionnaire format include items such as nonresponse, improperly worded questions, poorly sequenced questions, and so forth. A quality pretest can identify and correct difficulties in research design and data collection.

A good pretest or pilot survey can help determine the size of the sample that needs to be taken. Too small a sample will not yield meaningful results, whereas an unnecessarily large sample will waste time, money, and effort. Pretest information can help determine the sample size needed to achieve a certain level of precision in the main sample. This topic will be discussed in section 7.3.

Step 6: The collection of sample data is a major task, in which high levels of quality control must be maintained and well-considered data management procedures carefully followed. Consistency in all aspects of data collection and processing is absolutely essential. In this step, the benefits of careful design of the research instrument and operation planning (step 4) will become evident, and the improvements made in pretesting (step 5) will expedite the major data collection effort significantly.

6.2 Types of Probability Sampling

In probability sampling, each individual or item that could be selected from the sampling frame has a known chance (or probability) of being included in the sample. This important advantage occurs because a randomization component is incorporated into the sample design in some known way. If the sample data are to be used in any inferential manner (either for estimation of population parameters or for hypothesis testing), then an element of randomness must be built into the sample design procedure. All types of probability sampling contain this characteristic of randomness and avoid the subjectivity of nonprobability sampling.

Simple Random Sampling

If randomization is fundamental to all probability sampling, then a **simple random sample** is the most basic way to generate an unbiased, representative cross-section of the population. In a simple random sample, every individual in the sampling frame has an equal chance of being included in the sample.

To select sample members, a list of random numbers is used. Computer-based random number generators can be used to obtain sets of random numbers, and such capabilities are available even on small microcomputers. For demonstration purposes, however, the traditional table of random numbers (appendix, table B) will be used here. This table is a long sequence of integers, and the selection of each integer is independent of all other selections. This usually results in a table with all ten integers (0, 1, 2, . . . , 9) present in roughly equal proportion, with no trend or pattern in their sequencing.

The task of probability sampling is to draw a set of individuals or items from the sampling frame. Each individual in the sampling frame must be identifiable, so that the randomization device (random number table or computer-based random number generator) can select units to be included in the sample. In a nonspatial sampling frame, individuals are placed in a list, with each member identified by a specific number.

If the sampled population in the sampling frame consists of fewer than 100 members, then a 2-digit number could be assigned to each individual (00 could represent the first individual in the list, 01 the second individual, and so on). With a sampled population of 100 members or more, numbers with more digits would be necessary.

Suppose in planning a new student center, a university research department wants to determine the detailed opinions of students regarding which activities to provide. Because time is limited, only a small sample of students can be surveyed. From a student body of 8,500, a simple random sample of 25 students is selected. Each student must have an equal chance of being chosen (that probability is 25/8500 = .0029—or, about 3 per 1000).

Any place in the random numbers table may be used as a starting point, but this position must itself be arbitrarily selected. Suppose the fifth column and sixth row in that column are selected as the starting position. This number is 10440 (see appendix, table B). Since the target population is 8,500, each member of that population can be assigned a 4-digit identification number. Therefore, one digit of each 5-digit sequence can be dropped. If the last digit is dropped, this would make the starting sequence 1044. Proceeding down the column, the following numbers are found:

1044	3551	6141	
0786	3454	7678	
9351*	5851	1344	
9692*	0000	8483	
9117*	7007	9273*	*(rejected—out of range)
4764	7704	3652	
1952	0079	7964	
7833	2645	5756	
1858	6983	5314	
6988	5201		(desired sample size of 25 obtained)

Note that all numbers 8500 or larger are rejected as being out of range. The first "usable" 4-digit number is 1044, the second is 0786, the third is 4764, and so on. Selection continues until the desired sample size (25) is obtained. If a 4-digit sequence repeats, that student has already been selected, and should not be chosen again. Another number, the next sampling unit in the table, would be selected instead.

Sometimes the method for operationalizing a simple random sample is awkward or inefficient. For example, the student list might be organized alphabetically or by social security number, and not numbered sequentially. To create a sequential list to use with the random numbers table, considerable effort might be required. Depending on circumstances, the geographer might want a less cumbersome and time-consuming sample design.

Systematic Sampling

Systematic sampling is a widely used design that often simplifies the selection process. A **systematic sample** makes use of a regular sampling interval (k) between individuals selected

for inclusion in the sample. A "1-in-k" systematic sample is generated by randomly choosing a starting point from among the first k individuals in the sampling frame, then selecting every k^{th} individual from that starting point. For example, if a sample of 25 ($n = 25$) is taken from a population of 500 ($N = 500$), the sampling interval (k) would be N/n ($k = 500/25 = 20$—a "1-in-20" sample). Care must be taken to ensure that no nonrandom pattern or sequencing is present in the list from which every k^{th} individual is selected. Otherwise, bias may be introduced into the sample process.

A systematic method of sampling is generally less cumbersome or awkward to operationalize than is simple random sampling. Systematic sampling provides a relatively quick way to derive a large size sample (and obtain more information) at a reasonable cost. As a result, it is used in many practical contexts. Government agencies (such as the Census Bureau) and political polling firms (such as Gallup or Harris) routinely apply systematic samples when detailed analyses or follow-up studies are needed. To estimate the quality of a product coming off an assembly line, factory management could have a more detailed inspection of every 100th item. Every 20th visitor to a park could be asked to complete a survey regarding park services, or every 50th taxpayer in a city could be asked about alternative funding or planning projects. In market analysis, the editors of a magazine could easily send a detailed survey or questionnaire concerning existing and proposed features to every 10th subscriber.

Stratified Sampling

In many geographic sampling problems, the target population or target area is separated into different identifiable subgroups or subareas, called "strata." If sample units in the different strata are expected to provide different results, such "target subdivision" is logical. With **stratified sample** design, the effect of certain pos-

Figure 6.4 Alternative Stratified Sample Designs: Student Opinion Survey

Case 1:
Stratify by place of residence

Campus dormitory	Off-campus housing
1. ___	1. ___
2. ___	2. ___
3. ___	3. ___
etc.	etc.

Case 2:
Stratify by year in school

Freshman	Sophomore	Junior	Senior
1. ___	1. ___	1. ___	1. ___
2. ___	2. ___	2. ___	2. ___
3. ___	3. ___	3. ___	3. ___
etc.	etc.	etc.	etc.

Case 3:
Stratify by place of residence and year in school

	Freshman	Sophomore	Junior	Senior
Campus dormitory	1. ___ 2. ___ 3. ___ etc.	1. ___ 2. ___ 3. ___ etc.	1. ___ 2. ___ 3. ___ etc.	1. ___ 2. ___ 3. ___ etc.
Off-campus housing	1. ___ 2. ___ 3. ___ etc.	1. ___ 2. ___ 3. ___ etc.	1. ___ 2. ___ 3. ___ etc.	1. ___ 2. ___ 3. ___ etc.

Note: Random samples of varying sizes taken from each stratum

sible influences can be controlled. Taking a simple random sample from each class or stratum makes the fullest possible use of available information and increases the precision of sample estimates.

Stratified sample designs may be either **proportional (constant-rate)** or **disproportional (variable-rate).** In a proportional stratified sample, the percentage of the total population in each stratum matches the proportion of individuals actually sampled in that stratum as closely as possible. For example, suppose 20 percent of all residents in a city are apartment dwellers. Proportional representation of the apartment stratum in a stratified sample design would require that 20 percent of the sampled individuals be apartment dwellers. Suppose further that this housing study specifically involves a resident survey on rent control. Because apartment dwellers would be particularly affected by the proposed legislation and their opinions important to decision makers, council members might want their views to be represented more heavily. A disproportional (or variable-rate) stratified sampling design would then be appropriate, and the apartment resident stratum would be "oversampled" (with a larger sample size) in this case. Another strategy is simply to weight apartment dweller responses more heavily and not increase the actual number of apartment sample units.

For the opinion survey on the new student center, a simple random sample would probably yield less precise results than would a well-designed stratified sample. A university has a diverse student population, and different student subgroups would likely have varying opinions on student center activities. Several stratified sample designs are possible (figure 6.4). For example, those living in campus dormitories may have significantly different priorities than do commuters living off-campus (figure 6.4, case 1). Also, the views of freshmen may differ from those of sophomores, juniors, and seniors (figure 6.4, case 2). Stratifying *both* by place of residence *and* year in school might prove most useful, resulting in a composite sample design structure with 8 strata (figure 6.4, case 3).

If the views of each student are considered equally important, a proportional stratification is applied. Disproportional stratification is appropriate, however, if the views of dormitory residents are considered more important. In both instances, if results are expected to vary by stratum, a stratified sample is preferable to a simple random sample of the same size. That is, strati-fication should be applied if background knowl-edge or logic suggests that it would be beneficial and practical.

Cluster Sampling

For some geographic problems, **cluster sam-pling** is most appropriate and may be more ef-ficient or cost-effective than random, systemat-ic, or stratified sampling. A cluster sample is derived by first subdividing the target popula-tion or target area into mutually exclusive and exhaustive categories (figure 6.5). An appropri-ate number of categories (clusters) is selected for detailed analysis through random sampling.

Two alternatives exist: (1) all individuals within each cluster are included in the sample, making a total enumeration or census within that cluster, or (2) a random sample of individuals is taken from each cluster. The latter option is some-times called a "two-stage" cluster sampling pro-cedure because a random process is used twice—once to select clusters and then again to select sampled individuals within each cluster. The actual approach used depends on the circum-stances and practical sampling conditions. The complete cluster sample is the composite of these selected clusters of observations.

Cluster sampling is generally an effective design option when practical or logistic problems make other choices more expensive, difficult, or

Figure 6.5   Steps in Cluster Sampling

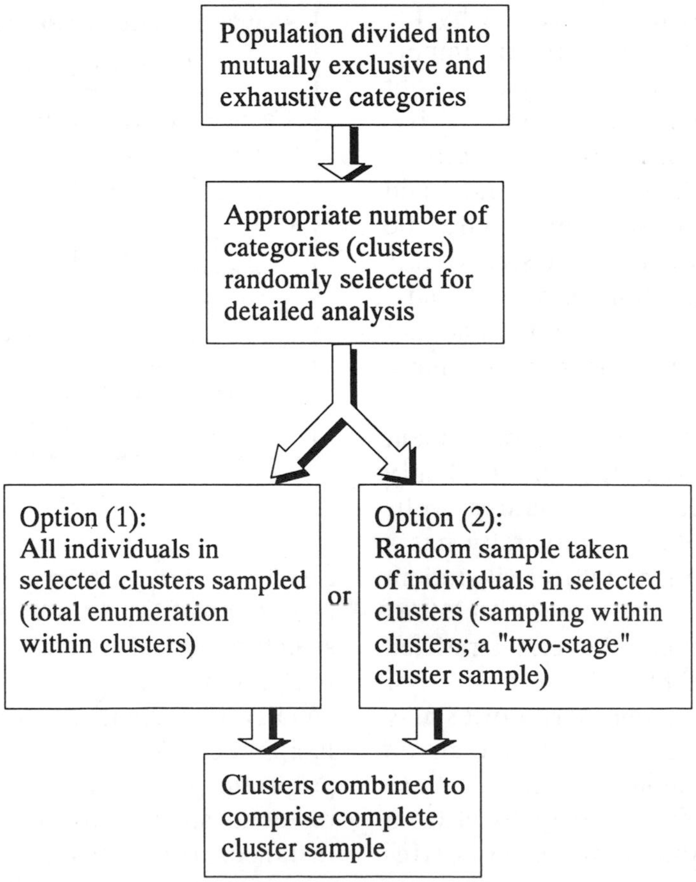

time-consuming. In some situations, a complete sampling frame or list is not available, but cluster lists can be obtained. This might occur in an urban geography study where enumerations are available for many city blocks, but no complete list of all city residents exists. A cluster could be defined as all (or many) homes on a city block. In other geographic problems, the population may be widely dispersed, resulting in high travel costs and increased logistical problems. In these circumstances, noncluster design options (e.g., simple random sample) would not be practical. The choice of a spatially contiguous or spatially concentrated cluster will keep the costs of obtaining the necessary sample to a minimum and permit an adequate sample size to be generated with reasonable effort.

A total enumeration cluster approach makes it easier to obtain large sample sizes (option 1 in figure 6.5). In the student opinion survey, interviewing all students in a randomly selected number of dormitories or sampling all students in a number of general education classes would result in an appropriately large sample size.

For cluster sampling to be most effective, however, the individuals within each cluster should be as different or heterogeneous as possible. This will make the cluster sample observations representative of the entire sampled population and avoid systematic bias. If the clusters are internally similar or homogeneous in nature, stratified sampling may be a better alternative. Therefore, the appropriateness of clusters in a geographic research problem should be weighed carefully. In this context, the total enumeration cluster sample of a dormitory or general education class may not be appropriate because these clusters are too homogeneous.

Combination or Hybrid Sampling Designs

Choosing a sampling design is seldom a simple, straightforward matter. Decision makers should consider cost, time, and convenience, as well as the various practical problems unique to the situation. Common sense and experience are also important in sample design selection.

In many cases, the simplicity of a simple random approach is an important advantage. However, practical circumstances may make the use of *any* simple type of sampling (random, systematic, stratified, or cluster) difficult or unwieldy. When practical conditions dictate, some **combination or hybrid sampling design** may be most appropriate. The following experience illustrates how practical realities influence the selection of sample design.

A geography department was asked to conduct a survey of air passengers enplaning at the nearby regional public airport (McGrew and Rosing, 1979). The survey had three general objectives: (1) to establish the airport's service area; (2) to determine selected passenger characteristics; and (3) to provide various recommendations on how to improve or expand airport services. Since a complete census of all passengers was impractical, a survey sampling procedure had to be devised. Selection of a proper sample design was critical to ensure that the passengers questioned accurately represented the entire population of passengers enplaned at the airport. A number of statistical requirements and practical realities limited the sampling design choice: (1) ensuring that each passenger was selected at random with known probability of inclusion; (2) guaranteeing representative coverage over the different seasons of the year and different days of the week; (3) working within economic constraints, to keep both the researchers' transportation costs and the number of hours required by airport personnel to administer the survey at a reasonable level; and (4) selecting a convenient sample design that would allow appropriate accuracy, close airport personnel supervision, and result in few administrative errors.

Among the alternative sampling methods available, a combination or hybrid design appeared most able to meet both the statistical requirements and practical implications. A year-long survey period was chosen, and an initial survey date (cluster) was randomly selected. All flights during this calendar day were sampled (total enumeration within clusters) rather than distributing questionnaires to individual passengers on scattered individual flights (sampling within clusters). Succeeding survey dates were systematically spaced every 13ᵗʰ day after the initial date. This sample design produced a total of 28 survey days or clusters, 7 in each season of the year, with each day of the week represented once in each quarter. Thus, all seasons were proportionally represented, as were all weekdays and weekend days in each season. By systematically spacing the sample clusters every 13ᵗʰ day, the researchers assumed that no bias was introduced for either the season of the year or day of the week. If a holiday happened (randomly) to be selected as a sample day, it was included in the study. In fact, to exclude a holiday purposely would have introduced an unwanted systematic bias into the analysis.

The overall result of this sample design was a type of *random systematic cluster* sample. Randomness existed because the initial survey date was randomly chosen; surveying passengers every 13ᵗʰ day added a systematic component; and giving all passengers questionnaires on the selected survey dates created a cluster effect.

6.3 Spatial Sampling

The types of sample design considered to this point have not been explicitly spatial in nature. Under certain circumstances, however, spatial sampling is necessary. Spatial sampling is applied when a map of a continuously distributed variable (such as vegetative cover, soil type, or pH of surface water) is being used and a sample of locations is being selected from this map. If a geographer conducting fieldwork must select

sample site locations within a defined target area, spatial sampling is also needed.

Spatial sampling from maps or other spatial sampling frames may involve point samples, line samples (traverses), or area samples (quadrats) (figure 6.6). Of these three types of spatial sampling, geographers use point sampling most frequently, and the various concepts involving spatial sampling procedures can be illustrated most easily through point sampling designs. This

Figure 6.6   Types of Spatial Sampling

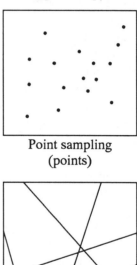

Point sampling
(points)

Line sampling
(traverses)

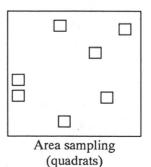

Area sampling
(quadrats)

section focuses on point sampling. Details on quadrat and traverse sampling are covered in more specialized texts.

For all types of spatial sampling, a spatial sampling frame (such as a map) must be constructed. This frame must include a coordinate system that allows clear identification of locations within the sampled area. For point sampling from a map, designation of the (X,Y) coordinate pair will identify a unique location (figure 6.7).

In a **simple random point sample** (figure 6.8, case 1), 2 numbers (an X and Y coordinate pair) are selected to designate each point. To locate *n* sample points in a sampled area for a study, $2(n)$ random numbers must be drawn from a random numbers table or other source. The resulting pattern of points may be uneven, with some portions of the sampled area seemingly underrepresented, and others appearing overrepresented. Locational or spatial unevenness is a natural consequence of the randomization process.

Systematic point sampling is a convenient way to avoid the problems possible with an uneven distribution of points across the study area. Analogous to the regular sampling interval (*k*) from a list of individuals in nonspatial systematic sampling, systematic point sampling uses a "distance interval" (figure 6.8, case 2). A "starting point" is randomly selected, then all other points are located using the distance interval to space them evenly across the entire sampled area. This procedure is easily implemented. Only one point (2 numbers from the random numbers table) must be randomly located; the subsequent placement of all other points in the systematic pattern is automatically determined. This approach also offers representative, proportional coverage of the entire sampled area. Bias will enter the sample design when a spatial regularity or periodicity exists in the distribution of some phenomenon that matches the distance interval. Systematic point sampling is widely used in

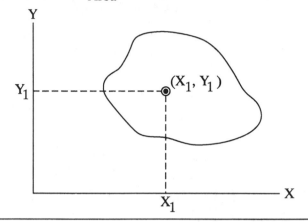

Figure 6.7   Designating a Location within an Area

geographic research, particularly when geographers deal with environmental and resource problems where data are continuously distributed across an area.

Just as in nonspatial cases, **stratified point sampling** may be either proportional (constant-rate) or disproportional (variable-rate). The approach selected depends on the circumstances of the problem. Suppose a geographer is studying environmental degradation in a region that includes particularly vulnerable tidal and nontidal wetlands. In figure 6.8 (cases 3 and 4), stratum 1 shows the non-wetland portion of the spatial sampling frame (comprising 60 percent of the total sampled area), while stratum 2 identifies wetlands (the other 40 percent). Suppose proportional representation of both strata is desired. This would dictate that 60 percent of the sample points be placed in non-wetland locations and 40 percent on wetland sites. If a total of 20 sample points is sufficient, 12 points will be in non-wetland locations, and 8 points in wetlands (figure 6.8, case 3). Suppose instead, particular attention needs to be focused on possible environmental problems in the wetlands area. In this situation, the wetlands stratum needs more detailed monitoring or "oversampling," while maintaining adequate coverage of the

Figure 6.8 Types of Spatial Point Sampling

Case 1:
Simple random
point sample

Case 2:
Systematic point sample
(aligned)

} Distance
interval

Randomly
selected
"starting
point"

Case 3:
Proportional stratified
point sample

Stratum 1
(non-wetland
area)

Stratum 2
(wetland
area)

Case 4:
Disproportional stratified
point sample

Stratum 1
(non-wetland
area)

Stratum 2
(wetland
area)

Case 5:
Random point sample
within clusters
("two stage" cluster sample)

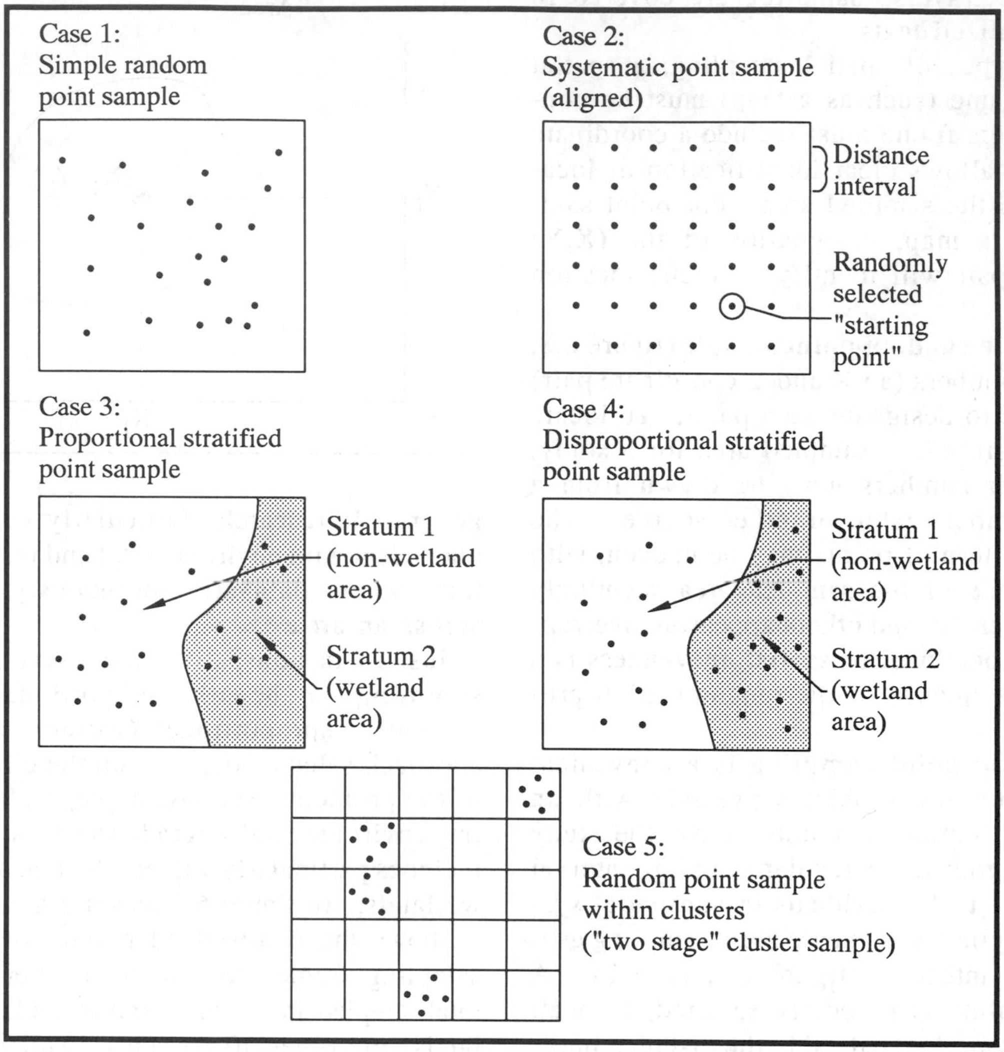

non-wetland portion of the sampled area. The disproportional stratified design (figure 6.8, case 4) shows twice the sample point intensity in the wetland stratum.

In spatial point sampling, a **cluster design** has the important advantage of keeping travel costs and other logistic problems to a minimum because the points of a cluster are necessarily grouped together (figure 6.8, case 5). This feature makes point sampling within clusters par-

ticularly attractive in geographic projects having an extensive study area. In many field studies, a cluster approach may be the only practical alternative.

Although two methods of detailed analysis within clusters are available in nonspatial cluster sampling (figure 6.5), only the "two-stage" cluster approach is appropriate in spatial cluster sampling. In the spatial sampling of a variable continuously distributed across an area, an infinite number of

Figure 6.9 Selected Types of Composite or Hybrid Point Sampling

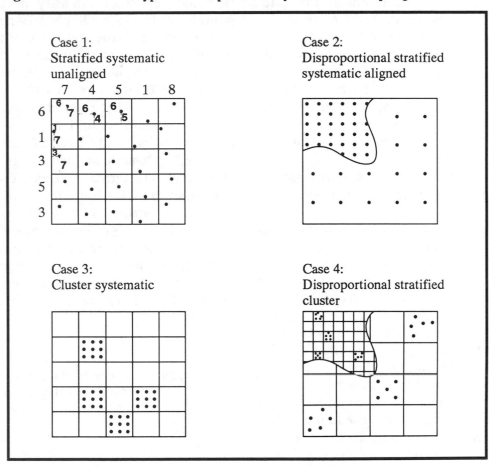

points could be selected in each subarea. This makes the total enumeration approach virtually impossible in a spatial context.

However, a cluster approach to point sampling is sometimes inadvisable. If the phenomenon being studied varies from one portion of the sampled area to another, it may be necessary to ensure that some point locations are selected from all sections or subareas. Cluster sampling excludes substantial parts of the study area, and such uneven areal representation is a disadvantage in many geographic problems. Therefore, the practical advantages of convenience and travel cost reduction often have to be

balanced against this unrepresentative spatial bias.

In addition to the types of simple point sampling, a number of combination or hybrid point sampling designs are also possible (figure 6.9). A composite sample design, with its procedural complexity, is only worthwhile if it is likely to improve the accuracy of the sample in estimating the population.

Perhaps the most widely used hybrid sampling model is the **stratified systematic unaligned point sample.** First suggested by Berry and Baker (1968), this approach has the following steps:

- place a regular grid system over the entire study area, taking care to ensure that grid size is at an appropriate level of spatial resolution for the problem;
- select a random number for each row and column in the gridded area—the row random number specifies the horizontal position of points in that row, whereas the column random number specifies the vertical position of points in that column (figure 6.9, case 1);
- locate a single point in each grid using the appropriate row and column positions.

This procedure provides proportional representation of all segments of the sampled area, yet avoids possible problems with regularities or periodicities in the spatial pattern that can be encountered with an aligned systematic point sample (figure 6.8, case 2). The needed sample points can also be generated fairly easily. As a result of these advantages, a stratified systematic unaligned point sample is a good choice in many realistic geographic situations.

Other combination or hybrid sampling models can be designed, but should be used only if they closely meet specific circumstances of a sampling problem. Geographers try to create models in such areas as climatology, resource management, and migration to replicate or duplicate reality as closely as possible. Similarly, in geographic sampling problems, the sample design must reflect the character of the situation. This allows precise and accurate estimates to be obtained with reasonable effort. Other hybrid point sample design options include: "disproportional stratified systematic aligned" (figure 6.9, case 2); "cluster systematic" (figure 6.9, case 3); and "disproportional stratified cluster" (figure 6.9, case 4).

Are any of these hybrid designs really feasible? Suppose a geographer is concerned with monitoring changes in both the spatial distribution and degree of intensity of nitrogen and phosphorus levels in a bay receiving agricultural runoff. In the narrow estuaries of the bay, runoff problems are likely to be more severe and will need careful monitoring. On the other hand, no portion of the bay should be left totally unmonitored. What type of spatial point sampling procedure should be used in locating monitor buoys in the bay? A disproportional stratified systematic aligned design might be the most practical alternative (figure 6.10). Disproportional stratification places an appropriate density of nutrient monitoring stations in the needed places. Systematic placement allows efficient collection of water samples, since the buoys are aligned regularly for pick-up by boat. This sample design is "tailor-made" to best fit the practical circumstances of this geographic problem.

Figure 6.10   Location of Nutrient Monitoring Stations on a Bay: Disproportional Stratified Systematic Aligned Sample

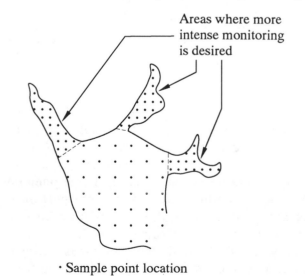

Areas where more intense monitoring is desired

· Sample point location

Key Terms and Concepts

References and Additional Reading

Berry, B. J. L. and A. M. Baker. 1968. "Geographic Sampling," in B. J. L. Berry and D. F. Marble (editors), *Spatial Analysis: A Reader in Statistical Geography*. Englewood Cliffs, NJ: Prentice-Hall, pp. 91-100.

Dixon, C. J. and B. Leach. 1978. *Sampling Methods for Geographical Research*. Norwich, England: Geo Abstracts.

Ferber, R., et. al. 1980. *What Is A Survey?* (informational brochure). Washington, DC: American Statistical Association.

Gregory, S. 1978. *Statistical Methods and the Geographer* (4th edition). London: Longman.

Hauser, P. 1975. *Social Statistics in Use*. New York: Russell Sage Foundation.

Jessen, R. T. 1978. *Statistical Survey Techniques*. New York: John Wiley and Sons.

Kish, L. 1965. *Survey Sampling*. New York: John Wiley and Sons.

McGrew, J. C. and R. A. Rosing. 1979. *Salisbury-Wicomico County Regional Airport Passenger Survey*. Salisbury: Delmarva Advisory Council.

Scheaffer, R. L., W. Mendenhall and L. Ott. 1986. *Elementary Survey Sampling* (3rd edition). Boston: Duxbury.

Sudman, S. 1976. *Applied Sampling*. New York: Academic Press.

Williams, W. H. 1978. *A Sampler on Sampling*. New York: John Wiley and Sons.

CHAPTER 7

ESTIMATION IN
SAMPLING

The primary objective of sampling is to make inferences about the population from which a sample is taken. More specifically, sample statistics are used to estimate population parameters such as the mean, total, and proportion. When sample statistics represent the larger population accurately, they are considered unbiased estimators.

This chapter considers both the theory and practice of estimating from samples. The basic terminology and the theoretical concepts that underlie sample estimation are discussed in section 7.1. The distinction is made between point estimation and interval estimation, and the nature of the sampling distribution of a statistic is explained. The importance of the central limit theorem when inferring sample results to a population is also discussed. Confidence intervals, which indicate the level of precision for an estimate, can be determined for any desired sample statistic. Section 7.2 includes a variety of confidence interval equations. Material is organized by type of sample (random, systematic, stratified) and by parameter (mean, total, proportion). A series of examples illustrates the procedure for constructing confidence intervals around each parameter using the different sampling methods.

To save time and effort, geographers often want to know—*before* taking a full sample—just how large a sample is needed for a research problem. Issues and methods related to determining sample size in advance of taking a full sample are discussed in section 7.3. Examples illustrate how sample size is established for problems using the mean, total, and proportion.

7.1 Basic Concepts in Estimation

Point Estimation and Interval Estimation

In statistical estimation, a basic distinction is made between **point estimation** and **interval**

estimation. The concept of point estimation is relatively straightforward. A statistic is calculated from a sample and then used to estimate the corresponding population parameter. With probability sampling, the best (unbiased) point estimate for a population parameter is the corresponding sample statistic (table 7.1). For example, the best point estimate for the population mean (μ) is the sample mean ($\overline{X}$); the best point estimate for the population standard deviation (σ) is the sample standard deviation (s).

Note that the denominator for calculating sample standard deviation is ($n-1$) rather than (n). This slight adjustment makes the sample standard deviation a truly unbiased estimator of the population standard deviation, particularly for smaller samples. With larger sample sizes

Table 7.1 Point Estimators of Population Parameters

Descriptive Statistic	Population Parameter	Sample Statistic*	Calculating Formula
mean	μ	$\overline{X}$	$\dfrac{\sum\limits_{i=1}^{n} X_i}{n}$
standard deviation	σ	s	$\sqrt{\dfrac{\sum\limits_{i=1}^{n} (X_i - \overline{X})^2}{n-1}}$
total	τ	T	$N(\overline{X}) = N\dfrac{\sum X_i}{n}$
proportion**	ρ	p	$\dfrac{x}{n}$

*Best point estimate
**x = Number of units sampled having particular characteristic
n = Total number of units sampled

(above 30 or so), the difference between dividing by (n) or dividing by ($n-1$) is insignificant. However, using (n) with a smaller sample would result in an estimate that underrepresents the magnitude of the true standard deviation in the population.

How precise are sample point estimators? How close (or distant) from the true population parameter is the calculated sample statistic? Because probability sampling involves some uncertainty, it is unlikely that a sample statistic will equal the true population parameter exactly. It is more valuable to know the likelihood that a sample statistic is within a certain range or interval of the population parameter. The determination of this range forms the basis of interval estimation. In this section, the procedures for placing a confidence interval or bound about a sample statistic arc explaincd. A **confidence interval** represents the level of precision associated with the sample estimate. Its width is determined by: (1) the sample size; (2) the amount of variability in the sample data; and (3) the probability level or level of confidence selected for the problem. These ideas are explored in detail, with the mean as the illustrative measure.

The Sampling Distribution of a Statistic

Suppose a random sample of size (n) is drawn from a population, and the mean of that sample ($\overline{X}$) is calculated. A second independent sample of size (n) could also be drawn and its mean calculated. If this process were repeated for many similar-sized independent samples in a population, the frequency distribution of this set of sample means could be graphed (figure 7.1). This curve is referred to as a **sampling distribution of sample means**.

A sampling distribution can be developed for *any* statistic, not just the mean. After many independent samples of size (n) are drawn from a population, the statistic of interest (mean, total, proportion) can be calculated for each sample, and the distribution of the sample statistics graphed.

The resulting frequency distribution has a shape, a mean, and a certain amount of variability (as reflected by the standard deviation and variance). These general characteristics of sampling distributions are important, but for now the focus is on particular characteristics of sample means.

The Central Limit Theorem

When the sample statistic is the mean, its frequency distribution has a particular set of features that are of vital theoretical importance in sampling and general statistical inference. These features are summarized in the **central limit theorem.**

Central Limit Theorem

Suppose all possible random samples of size (n) are drawn from an infinitely large, normally distributed population having a mean = μ and a standard deviation = σ. The frequency distribution of these sample means ($\overline{X}$'s) will have:

1. a normal distribution;
2. a mean of μ (the population mean); and
3. a standard deviation of $\sigma/\sqrt{n}$

This particular set of characteristics applies only to the distribution of sample means and not to other sample statistics.

Figure 7.1   Sampling Distribution of Sample Means and Frequency Distribution of Population Values

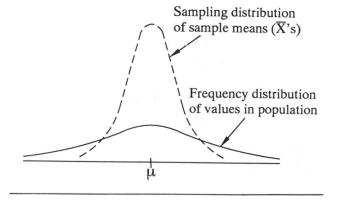

Sampling distribution of sample means ($\overline{X}$'s)

Frequency distribution of values in population

μ

The central limit theorem offers several important insights. Given the effect of randomness in drawing samples, some sample means will be above the population mean, while others will be below it. Therefore, it seems logical that the mean of a set of sample means is μ. It also follows that a frequency distribution of sample means ($\overline{X}$'s) is normally distributed and centered on the population mean (μ). The likelihood that a sample mean differs only slightly from μ is greater than the likelihood that a sample mean differs greatly from μ. When a sample is taken, large values in the population are generally counterbalanced by small values. The sample mean is likely to be quite close to the population mean, especially for a large sample.

The central limit theorem also provides insight into the variability of the sample means. According to this theorem, the standard deviation of the sampling distribution of means is equal to the population standard deviation divided by the square root of the sample size. This measure of standard deviation is called the **standard error of the mean** ($\sigma_{\overline{x}}$):

$$\sigma_{\overline{x}} = \frac{\sigma}{\sqrt{n}} \qquad (7.1)$$

The standard deviation indicates the typical or standard amount that a value is likely to differ from the mean of a set of values. In a similar way, the standard error of the mean indicates the typical or standard amount that a sample mean will differ from the true population mean. Quite simply, standard error is a basic measure of the amount of **sampling error** in a problem.

The magnitude of sampling error is influenced by the sample size and the population standard deviation. First, the larger the sample size (n), the smaller the amount of sampling error ($\sigma_{\overline{x}}$) (figure 7.2). This has a certain appeal—the mean of a larger sample tends to be closer to the true population mean than the mean of a smaller sample. Second, the larger the standard deviation of the population (σ), the larger the amount

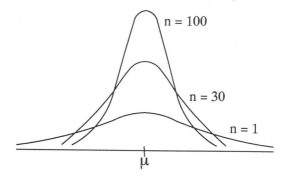

Figure 7.2 Frequency Distribution of Sample Means with Different Sample Sizes

of sampling error. This too seems logical, for a sample taken from a population containing a large amount of variability should have more error than a similarly sized sample taken from a population containing a small amount of variability.

How can the central limit theorem have such general applicability, if the population from which samples are drawn is supposed to be normally distributed and infinitely large? Actually, these seemingly restrictive qualifications are not as severe as they may first appear.

In reality, many populations are not normally distributed, and some frequency distributions of interest to geographers are highly skewed and extremely non-normal. However, whatever the shape of the frequency distribution of the underlying population, the frequency distribution of sample means is approximately normal if the sample size is large. This is a very important result. Generally, a sample size of 30 or more is sufficient to guarantee an approximately normal distribution of sample means, regardless of how the population is distributed.

Many problems of concern to geographers involve taking a sample from a finite-sized population. However, the central limit theorem is completely true only for infinitely large populations. With a finite population, a **finite population correction factor (fpc)** may be incorporated into the estimation process:

$$\text{fpc} = \sqrt{\frac{N-n}{N-1}} \qquad (7.2)$$

where: fpc = finite population correction factor
$\quad\quad\quad$ N = population size
$\quad\quad\quad$ n = sample size

This factor reduces the amount of sampling error slightly, and increases the precision of the sample estimate. When using the finite population correction factor, the standard error of the mean is calculated as:

$$\sigma_{\bar{x}} = \frac{\sigma}{\sqrt{n}} \sqrt{\frac{N-n}{N-1}} \qquad (7.3)$$

The finite population correction is particularly important in the estimation process when the **sampling fraction**—defined as the ratio of sample size to population size (n/N)—is large. If a relatively large sample is taken from a relatively small population, the sampling fraction will be large, and the finite population correction should be included in the sampling error formula to provide a more precise estimate. Conversely, if a relatively small sample is taken from a very large population, the sampling fraction will be small, and the fpc will be very close to 1. In this situation, the correction factor can be excluded from the standard error formula. If a sample of size 25 is taken from a population of 10,000, the finite population correction factor is calculated as:

$$\text{fpc} = \sqrt{\frac{N-n}{N-1}} = \sqrt{\frac{10,000-25}{10,000-1}} = .9988$$

Including the correction factor in the standard error equation in this case will have no meaningful effect on the precision of the sample estimate. In general, the fpc value is included in sample estimation equations if the sampling fraction is greater than .05, which occurs whenever the sample size is more than 5 percent of the population size ($n/N > .05$).

In summary, the central limit theorem provides important information about the frequency distribution of sample means. Such a distribution is considered normal, has a mean of μ (the population mean), and a standard deviation of $\sigma/\sqrt{n}$ (called the standard error of the mean, $\sigma_{\bar{x}}$). However, the practical focus of geographic research is usually on a *single* sample mean ($\bar{X}$) drawn from a population having a mean of μ and standard deviation of σ. Although this single sample mean falls somewhere within the normal frequency distribution of sample means, its actual location is not known.

How precise or exact is this single sample mean? The central limit theorem allows this question to be answered. A confidence interval is placed about the sample mean estimate, and the probability of the true population mean falling within this interval can be calculated. That is, a confidence interval is positioned around the sample mean, and an investigator has a measurable level of confidence that the true population mean lies within that interval.

7.2 Confidence Intervals and Estimation

Suppose a geographer wants to estimate a confidence interval about a sample mean with a 90 percent certainty that the interval range contains the population mean. Since the central limit theorem indicates that the distribution of sample means is normal, the probability distribution of Z-values from the normal table (appendix, table A) can be used to define the confidence interval. The general format of a confidence interval is:

$$\bar{X} \pm Z\sigma_{\bar{x}} \qquad (7.4)$$

where: X = sample mean
$\quad\quad\quad$ Z = Z-value from normal table
$\quad\quad\quad$ $\sigma_{\bar{x}}$ = standard error of the mean

If the desired probability is .90, the corresponding Z-value from the normal table is 1.65. In this case, the researcher is 90 percent confident that the true population mean lies within the confidence interval defined by:

$$\overline{X} \pm 1.65 \sigma_{\overline{x}}$$

The confidence interval contains expressions that are added to and subtracted from the mean to define the **upper and lower bounds** of the interval:

$$\text{Upper bound} = \overline{X} + 1.65 \sigma_{\overline{x}}$$
$$\text{Lower bound} = \overline{X} - 1.65 \sigma_{\overline{x}}$$

This confidence interval is shown graphically in figure 7.3. The shaded area, representing the confidence interval, encompasses 90 percent of all the sample means that theoretically could be taken. Of the 10 sample means shown ($\overline{X}_1$ to $\overline{X}_{10}$), 9 of them (90 percent) are within the confidence interval. This probabilistic result is expected because the interval was constructed with a 90 percent chance of containing μ. Of course, when taking a single sample, the location of μ is not known. However, a .90 probability exists that the interval or bound placed around that single sample mean does, in fact, contain the true population mean. Conversely, a .10 probability exists that the confidence interval placed around the sample mean does *not* include μ. For example, by chance, the confidence interval around $\overline{X}_5$ fails to encompass the true population mean, μ.

The probabilities that an unusually large sample mean (well above μ) or an unusually small sample mean (well below μ) could be drawn are represented graphically by the two unshaded tails of the sampling distribution (figure 7.3). The sample mean $\overline{X}_5$ is too far below the true population mean μ to lie within the confidence interval.

Several terms are used when making interval estimates in sampling. The **confidence level** refers to the probability that the interval sur-

Figure 7.3 Distribution of Sample Means and the Confidence Interval Concept

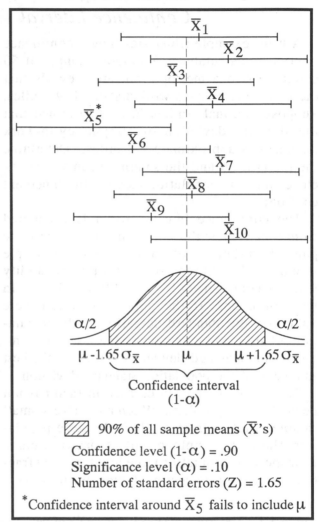

90% of all sample means ($\overline{X}$'s)

Confidence level $(1 - \alpha) = .90$
Significance level $(\alpha) = .10$
Number of standard errors $(Z) = 1.65$

*Confidence interval around $\overline{X}_5$ fails to include μ

rounding a sample mean *encompasses* the true population mean. This confidence level probability is defined as $1 - \alpha$. The **significance level** refers to the probability that the interval surrounding a sample mean *fails to encompass* the true population mean. The significance level is denoted by α and represents the total sampling error. Because error is equally likely in either direction from μ, the probability of the sample mean falling into either tail of the distribution is $\alpha/2$.

Example 7.1 Procedure for Constructing a Confidence Interval

A brief example illustrates how a confidence interval is calculated. A random sample of 50 commuters in a metropolitan area reveals that the average journey-to-work distance is 9.6 miles. Suppose national studies have determined that the standard deviation of all journey-to-work distances is approximately 3 miles. Determine the interval around the sample mean which encloses the true population mean with 90 percent confidence.

The confidence interval for μ is calculated from four values: the sample mean ($\overline{X} = 9.6$), the population standard deviation ($\sigma = 3$), the sample size ($n = 50$), and the Z-value for the probability level or confidence ($1 - \alpha = .90$; $Z = 1.65$). In most problems, if the population mean were unknown, the standard deviation of the population would also be unknown. In those situations, the sample standard deviation is used as the best estimator of the population standard deviation.

The finite population correction factor is not needed in this problem. When a relatively small sample is taken from a relatively large population (here, a sample of 50 is drawn from a metropolitan population), the low sampling fraction makes the correction factor unnecessary.

The single best estimate of the true population mean (μ) is the sample mean ($\overline{X} = 9.6$), the statistic around which the confidence interval is placed. Using equations 7.3 and 7.4, the confidence interval is defined as:

$$\overline{X} \pm Z\sigma_{\overline{x}} = \overline{X} \pm Z\frac{\sigma}{\sqrt{n}}$$

$$9.6 \pm 1.65\frac{3}{\sqrt{50}} = 9.6 \pm 0.70$$

This interval ranges from a lower bound of 8.90 to an upper bound of 10.30, indicating a .90 probability (90 percent chance) that the mean journey-to-work distance for all commuters is between 8.90 and 10.30 miles.

A confidence interval can be created for any desired level of confidence. For certain practical geographic problems, being 90 percent certain of a sample result may not be appropriate. A commonly accepted and widely used confidence level is .95. This "conventional" probability is generally considered sufficiently rigorous to remove most doubt about the conclusion reached. In addition, it usually produces a relatively narrow confidence interval, which allows a precise estimate of the population parameter. In table 7.2, the relationships are shown between confidence level, significance level, number of standard errors, and confidence inter-

Table 7.2 The Relationship Between Confidence Level and Confidence Interval for the Journey-to-Work Example

Confidence Level $(1 - \alpha)$	Significance Level (α)	Number of Standard Errors (Z)	Confidence Interval $(X \pm Z\sigma_{\overline{x}})$	Confidence Interval Range	Width
.80	.20	1.28	9.6 ± .49	9.11 to 10.09	0.98
.90*	.10	1.65	9.6 ± .70	8.90 to 10.30	1.40
.95	.05	1.96**	9.6 ± .75	8.85 to 10.35	1.50
.98	.02	2.33	9.6 ± .89	8.71 to 10.49	1.78
.99	.01	2.58	9.6 ± .99	8.61 to 10.59	1.98

*Calculations for this confidence level are detailed in example 7.1.
**A Z-value of 2.00 is often used here for convenience.

val characteristics for alternative versions of the journey-to-work distance problem. Notice that if a researcher wants a higher level of confidence in the estimation, a wider confidence interval and a less precise estimate will result. Generally, the investigator has the responsibility of deciding how to handle the trade-off between confidence level and confidence interval for any particular study.

The exact equation used to place a confidence interval or bound around a sample estimate depends on two factors: the type of sample taken, and the population parameter being estimated by the sample statistic. The equations presented in this section are organized around three simple types of probability sampling—simple random and systematic sampling (table 7.3) and stratified sampling (table 7.4). For each of these sample designs, confidence interval equations are provided for three population parameters that often need to be estimated from sample statistics: the mean, total, and proportion. More advanced texts dealing exclusively with sampling also consider estimates for other sampling designs, such as cluster, composite, or hybrid samples. Interested readers should refer to the additional readings at the end of the chapter.

Two important matters require brief discussion before confidence interval estimates are applied to examples. First, the population variance (σ^2) is assumed to be unknown in the confidence interval equations of tables 7.3 and 7.4. In most problems of interest to geographers, this is not a restrictive assumption. In fact, if the population variance were known, the population mean would almost certainly be known as well, and sample estimation would serve no purpose. With an unknown population variance, the sample variance (s^2) provides the best estimator for the standard error. Algebraically, one convenient way to include the sample variance in the confidence interval equations is to alter slightly the finite population correction portion of the equations:

Table 7.3 Confidence Interval Equations for Random and Systematic Samples[*]

Part 1: Population Parameter—Mean (μ)

Best point estimate:
$$\hat{\mu} = \overline{X} = \frac{\Sigma X_i}{n}$$

Standard error of the point estimate (sampling error):
$$\hat{\sigma}_{\overline{x}} = \sqrt{\frac{s^2}{n}\left(\frac{N-n}{N}\right)} \text{ where}$$

$$s^2 = \frac{\Sigma\left(X_i - \overline{X}\right)^2}{n-1}$$

Confidence interval (bound) around the point estimate:
$$\overline{X} \pm Z\hat{\sigma}_{\overline{x}} = \overline{X} \pm Z\sqrt{\frac{s^2}{n}\left(\frac{N-n}{N}\right)}$$

Part 2: Population Parameter—Total (τ)

Best point estimate:
$$\hat{\tau} = T = N\overline{X} = \frac{N\Sigma X_i}{n}$$

Standard error of the point estimate (sampling error):
$$\hat{\sigma}_T = \sqrt{N^2\left(\frac{s^2}{n}\right)\left(\frac{N-n}{N}\right)}$$

Confidence interval (bound) around the point estimate:
$$T \pm Z\hat{\sigma}_T = T \pm Z\sqrt{N^2\left(\frac{s^2}{n}\right)\left(\frac{N-n}{N}\right)}$$

Part 3: Population Parameter—Proportion (ρ)

Best point estimate:
$$\hat{\rho} = p = \frac{x}{n}$$

Standard error of the point estimate (sampling error):
$$\hat{\sigma}_p = \sqrt{\frac{p(1-p)}{n-1}\left(\frac{N-n}{N}\right)}$$

Confidence interval (bound) around the point estimate:
$$p \pm Z\hat{\sigma}_p = p \pm Z\sqrt{\frac{p(1-p)}{n-1}\left(\frac{N-n}{N}\right)}$$

[*]Exclude the finite population correction from the confidence interval equations if $n/N < .05$
Replace Z with the corresponding t if $n < 30$

Table 7.4 Confidence Interval Equations for Stratified Samples*

Part 1: Population Parameter—Mean (μ)

Best point estimate:
$$\hat{\mu} = \overline{X} = \frac{1}{N}\sum_{i=1}^{m} N_i\overline{X}_i = \frac{1}{N}\left[N_1\overline{X}_1 + \ldots + N_m\overline{X}_m\right]$$

Standard error of the point estimate (sampling error):
$$\hat{\sigma}_{\overline{x}} = \sqrt{\frac{1}{N^2}\sum_{i=1}^{m} N_i{}^2\left(\frac{s_i{}^2}{n_i}\right)\left(\frac{N_i - n_i}{N_i}\right)}$$

$$= \sqrt{\frac{1}{N^2}\left[N_1{}^2\left(\frac{s_1{}^2}{n_1}\right)\left(\frac{N_1 - n_1}{N_1}\right) + \ldots + N_m{}^2\left(\frac{s_m{}^2}{n_m}\right)\left(\frac{N_m - n_m}{N_m}\right)\right]}$$

Confidence interval (bound) around the point estimate:
$$\overline{X} \pm Z\hat{\sigma}_{\overline{x}} = \overline{X} \pm Z\sqrt{\frac{1}{N^2}\sum_{i=1}^{m} N_i{}^2\left(\frac{s_i{}^2}{n_i}\right)\left(\frac{N_i - n_i}{N_i}\right)}$$

Part 2: Population Parameter—Total (τ)

Best point estimate:
$$\hat{\tau} = T = \sum_{i=1}^{m} N_i\overline{X}_i = N_1\overline{X}_1 + \ldots + N_m\overline{X}_m$$

Standard error of the point estimate (sampling error):
$$\hat{\sigma}_T = \sqrt{\sum_{i=1}^{m} N_i{}^2\left(\frac{s_i{}^2}{n_i}\right)\left(\frac{N_i - n_i}{N_i}\right)}$$

Confidence interval (bound) around the point estimate:
$$T \pm Z\hat{\sigma}_T = T \pm Z\sqrt{\sum_{i=1}^{m} N_i{}^2\left(\frac{s_i{}^2}{n_i}\right)\left(\frac{N_i - n_i}{N_i}\right)}$$

Part 3: Population Parameter—Proportion (ρ)

Best point estimate:
$$\hat{\rho} = p = \frac{1}{N}\sum_{i=1}^{m} N_i p_i = \frac{1}{N}\sum_{i=1}^{m} N_i\frac{x_i}{n_i} = \frac{1}{N}\left[N_1\frac{x_1}{n_1} + \ldots + N_m\frac{x_m}{n_m}\right]$$

Standard error of the point estimate (sampling error):
$$\hat{\sigma}_p = \sqrt{\frac{1}{N^2}\sum_{i=1}^{m} N_i{}^2\left(\frac{p_i(1-p_i)}{n_i - 1}\right)\left(\frac{N_i - n_i}{N_i}\right)}$$

Confidence interval (bound) around the point estimate:
$$p \pm Z\hat{\sigma}_p = p \pm Z\sqrt{\frac{1}{N^2}\sum_{i=1}^{m} N_i{}^2\left(\frac{p_i(1-p_i)}{n_i - 1}\right)\left(\frac{N_i - n_i}{N_i}\right)}$$

*Exclude the finite population correction from the confidence interval equations if $n/N < .05$
Replace Z with the corresponding t if $n < 30$
m = number of strata; subscript i refers to stratum i (N_i = size of population stratum i, n_i = size of sample from stratum i, $\overline{X}_i$ = sample mean of stratum i, s_i^2 = variance of stratum i, p_i = sample proportion with characteristic of stratum i); N = total population of all strata ($N = N_1 + N_2 + \ldots + N_m$)

$$\sigma_{\bar{x}}^2 = \frac{\sigma^2}{n}\left(\frac{N-n}{N-1}\right) = \frac{s^2}{n}\left(\frac{N-n}{N}\right) \qquad (7.5)$$

where: $\sigma_{\bar{x}}^2$ = square of the standard error

s^2 = sample variance

$$= \frac{\Sigma(X_i - \overline{X})^2}{n-1}$$

Second, confidence intervals need to be altered if the sample size is small. The standard normal deviate (Z) is included in all of the equations shown in tables 7.3 and 7.4. However, with small sample sizes ($n < 30$), the standard Z-value must be replaced by the corresponding value from the student's t distribution (appendix, table D). When $n < 30$, the formula $\overline{X} \pm t\sigma_{\bar{x}}$ replaces $\overline{X} \pm Z\sigma_{\bar{x}}$ in the confidence interval equation. Like the normal distribution, the t distribution is symmetric and bell-shaped. The exact shape of the t distribution depends on the sample size; as (n) approaches 30, the value of t approaches the standard normal (Z) value.

Two pieces of information are needed to use the t table:

1. the desired significance level (α). Common levels are α = .10, .05, and .01.
2. the number of degrees of freedom (df), where df is defined as one less than the sample size (df = $n - 1$).

Suppose, for example, the researcher wants to place a 95 percent confidence interval (α = .05) around a sample estimate where n = 20. The t table value (α = .05 and df = 19) is 2.09. This table value is similar in magnitude to the standard normal Z-value of 1.96 when α = .05.

For sample sizes less than 30, the t value will always be slightly larger than the Z-value at the corresponding level of significance, resulting in a slightly larger confidence interval or bound on the error of estimation. This larger confidence interval for t is to be expected because the smaller the sample, the greater the uncertainty that the sample precisely represents the population from which it is drawn, and the larger the sampling error.

Geographic Examples of Confidence Intervals

Nearly 10 years have passed since the last census of population was taken in Middletown, and local planners wish to update selected demographic statistics. Tax records show a total of 3,500 households in the community. However, precise estimates of the following variables are wanted:

1. The *mean* number of people per household. Planners suspect this figure has declined sharply over the recent intercensal period;
2. The *total* number of people in the community, based on the average number of people per household estimated from item ($\hat{1}$);
3. The *proportion* of households with one or more children aged less than 18.

Middletown planners decide to use either a simple random or systematic sample to calculate point estimates and enclose these parameters within confidence intervals. The procedures they followed are illustrated for the sample mean (example 7.2), total (example 7.3), and proportion (example 7.4). The corresponding equations are found in table 7.3.

Example 7.2   Random or Systematic Sample Estimate (Mean)

To estimate and place bounds on the average number of people per household, Middletown planners take a simple random or systematic sample of 25 households. From the sample data, the sample mean and variance are: $\overline{X}$ = 2.73 and s^2 = 2.6. They wish to determine the confidence interval that contains the true average household size of Middletown, with a 90 percent certainty ($1 - \alpha$ = .90). The relevant values for this problem are:

$\overline{X}$=2.73 s^2=2.6 N=3,500 n=25 α=.10

Since $n < 30$, t rather than Z should be used in the confidence interval equation:

$$\bar{X} \pm t \sqrt{\frac{s^2}{n}\left(\frac{N-n}{N}\right)}$$

Since $\alpha = .10$ and df $= n - 1 = 24$, the value from the t table is 1.711. Because the sampling fraction (n/N) is less than .05, the finite population correction can be ignored. In this problem, the confidence interval is calculated as:

$$2.73 \pm 1.711\sqrt{\frac{2.6}{25}} = 2.73 \pm 0.552$$

Middletown planners are 90 percent certain that the true mean household size is in the interval from 2.178 to 3.282.

A confidence interval of this width may not provide precision sufficient to allow practical policy decisions. One of two strategies is available to narrow the interval: lower the confidence level from .90, or increase the sample size above 25. The latter strategy requires more effort, but the larger sample size will permit the confidence interval to be narrowed.

Suppose the decision is made to increase the number of households surveyed from 25 to 250. Keeping the level of confidence at .90, a narrower, more precise, confidence interval results. Suppose the mean and variance of this larger survey of households are calculated from the sample data as $\bar{X} = 2.68$ and $s^2 = 4.3$. With $n = 250$, two changes occur that help narrow the width of the confidence interval. First, since $n > 30$, Z rather than t is used in the confidence interval formula. With $\alpha = .10$, the value from the normal table becomes $Z = 1.65$. Second, the finite population correction factor (fpc) needs to be included, because the sampling fraction n/N is greater than .05. The revised confidence interval around the sample mean $(\bar{X})$ is calculated as:

$$\bar{X} \pm Z\hat{\sigma}_{\bar{x}} = \bar{X} \pm Z\sqrt{\frac{s^2}{n}\left(\frac{N-n}{N}\right)}$$

$$= 2.68 \pm 1.65\sqrt{\frac{4.3}{250}\left(\frac{3,500-250}{3,500}\right)}$$

$$= 2.68 \pm 0.209$$

With this larger sample, planners are now 90 percent certain that the true mean household size in Middletown, μ, is in the narrower interval from 2.471 to 2.889.

Example 7.3   Random or Systematic Sample Estimate (Total)

The Middletown planners can use this calculated average number of people per household to estimate the total number of people in their community. The best estimate of the population total (τ) is the sample total (T), which is the sample mean $(\bar{X})$ multiplied by the population size (N):

$$\hat{\tau} = T = N\bar{X}$$

where: $\hat{\tau}$ = best estimate of true population total
  T = sample total

For the Middletown problem, the best estimate of the population total is:

$$\hat{\tau} = T = 3,500(2.68) = 9,380$$

Using the equations in table 7.3, part 2, the confidence interval around the sample total is determined by:

$$T \pm Z\sqrt{N^2\left(\frac{s^2}{n}\right)\left(\frac{N-n}{N}\right)}$$

When $\alpha = .10$ and $Z = 1.65$ and the confidence interval for the total population in Middletown is calculated as:

$$9,380 \pm 1.65 \sqrt{3,500^2 \left(\frac{4.3}{250}\right)\left(\frac{3,500-250}{3,500}\right)} =$$

$$9,380 \pm 729.84$$

Planners are 90 percent certain that the true total number of people in the community, τ, is in the interval from 8,650.16 to 10,109.84.

Example 7.4 Random or Systematic Sample Estimate (Proportion)

The best estimate of the proportion of all households in Middletown with one or more children less than 18 years of age is the sample proportion. A survey of 250 households in Middletown reveals that 105 of them have children. The best point estimate of the population proportion is:

$$\hat{\rho} = p = \frac{x}{n}$$

where: $\hat{\rho}$ = best estimate of true population proportion

p = sample proportion

x = number in sample having specified characteristic

The best estimate of the proportion of Middletown households having children is:

$$\hat{\rho} = p = \frac{x}{n} = \frac{105}{250} = .42$$

Using the equations in table 7.3, part 3, the confidence interval around the sample proportion is calculated as:

$$p \pm Z \sqrt{\frac{p(1-p)}{n-1}\left(\frac{N-n}{N}\right)}$$

To estimate the true population proportion with 90 percent confidence $(1 - \alpha = .90$ and $Z = 1.65)$, the confidence interval is:

$$.42 \pm 1.65 \sqrt{\frac{(.42)(.58)}{250-1}\left(\frac{3,500-250}{3,500}\right)} = .42 \pm .050$$

Middletown planners can conclude with 90 percent certainty that ρ, the true proportion of households with children, is in the interval from .370 to .470.

Many geographic problems use sampling designs other than simple random or systematic. When geographers work with stratified samples, alternative equations are needed to place confidence intervals around point estimates (table 7.4). Problems follow that show calculation procedures for a stratified sample mean (example 7.5), total (example 7.6), and proportion (example 7.7).

Example 7.5 Stratified Sample Estimate (Mean)

School officials wish to evaluate the "geographic competencies" of all high school juniors in a large school district. One group of students has completed an introductory geography course taught in several of the district's high schools, while another group of students has not. School officials suspect that the geographic competencies of these two groups of juniors are quite different. If a simple random sample was taken of all juniors, either of the groups could be over- or under-represented in the sample. Therefore, a stratified sample design, with proportional representation of each group of students, is appropriate.

To evaluate students' geographic skills, school officials decide to use the Secondary-Level Geography Test, designed by the National Council for Geographic Education (NCGE). This test evaluates a student's basic geographic knowledge in three areas—geographic skills, physical geography, and human geography. Of 2,160 total juniors in the school district, 480 have completed the geography course, and 1,680 have not. School officials have only 90 test booklets from NCGE and must restrict their sample to 90 students.

The procedure for estimating the mean test score from a stratified sample of students is shown in table 7.5.

Example 7.6 Stratified Sample Estimate (Total)

In a county with 430 farms, the agricultural extension agent wants to estimate the total number of uncultivated farm acres in woodlots or fallow. The agent believes that the number of uncultivated acres will vary by size of farm, so a simple random sample could result in certain farm sizes being over- or under-represented. Taking a stratified sample that proportionally represents each of the farm size categories is a better sample design. In the county, 172 farms are less than 50 acres, 107 farms are 50-99 acres, 65 farms are 100-299 acres, 42 are 300-499 acres, and 44 are over 500 acres. The procedure for estimating the total number of uncultivated farm acres from a stratified sample of farms is shown in table 7.6. Note that the relatively small sample sizes and the extreme differences in variability within the different strata combine to produce a very wide (imprecise) confidence interval.

Example 7.7 Stratified Sample Estimate (Proportion)

An urban park and recreation department wants to investigate the public support for a bond issue to finance various park improvements. They want to estimate the proportion of the population in several neighborhoods adjacent to a city park who favor a proposed tax levy. Three different neighborhoods (with residents likely to have different attitudes about the bond issue) are situated adjacent to the park. Neighborhood 1 contains 155 households, neighborhood 2 has 62 households, and neighborhood 3 has 95. Park and recreation officials decide a stratified sample is appropriate to ensure proportional representation of each neighborhood in the sample. Suppose

Table 7.5 Estimate of Stratified Sample Mean— Example 7.5

Task: Estimate the mean test score of all high school juniors on the Secondary-Level Geography Test, based on a stratified sample, and place a 95% ($\alpha = .05$) confidence interval around the estimate.

Stratum 1	Stratum 2
Students Having Taken Introductory Geography Course	Students Who Have Not Taken Introductory Geography Course

$N_1 = 480 \quad n_1 = 20$ $\qquad N_2 = 1680 \quad n_2 = 70$

$X_1 = 64.36 \quad s_1^2 = 65.61$ $\qquad \overline{X}_2 = 51.18 \quad s_2^2 = 90.25$

The best estimate of the true population mean (μ) is the stratified sample mean ($\overline{X}$):

$$\hat{\mu} = \overline{X} = \frac{1}{N} \sum_{i=1}^{m} N_i \overline{X}_i$$

$$= \frac{1}{2160} \left[480(64.36) + 1680(51.18) \right] = 54.11$$

The confidence interval around $\overline{X}$ is:

$$\overline{X} \pm Z\hat{\sigma}_{\overline{x}} = \overline{X} \pm Z \sqrt{\frac{1}{N^2} \sum_{i=1}^{m} N_i^2 \left(\frac{s_i^2}{n_i} \right) \left(\frac{N_i - n_i}{N_i} \right)}$$

Exclude the finite population correction $\left(\dfrac{N_i - n_i}{N_i} \right)$ as (n/N) < .05 for each stratum:

$$\overline{X} \pm Z \sqrt{\frac{1}{N^2} \sum_{i=1}^{m} N_i^2 \left(\frac{s_i^2}{n_i} \right)}$$

$$= 54.11 \pm 1.96 \sqrt{\frac{1}{2160^2} \left[(480)^2 \left(\frac{65.61}{20} \right) + (1680)^2 \left(\frac{90.25}{70} \right) \right]}$$

$$= 54.11 \pm 1.90$$

With 95% certainty, the mean test score of all high school juniors, μ, is within the interval from 52.21 to 56.01

Table 7.6   Estimate of Stratified Sample Total—Example 7.6

Task: Estimate the total number of uncultivated farm acres in woodlots or fallow in the county, based on a stratified sample, and place a 90% ($\alpha = .10$) confidence interval (bounds) around the estimate.

Stratum 1 (< 50 acres)				Stratum 2 (50-99 acres)			Stratum 3 (100-299 acres)			Stratum 4 (300-499 acres)		Stratum 5 (> 500 acres)	
6	2	17	4	12	27	16	120	40	85	212	170	220	380
10	20	10	26	50	20	40	10	55	32	40	110	60	80
23	16	15	3	4	10								
5	12	8	40										

$$N_1 = 172 \qquad N_2 = 107 \qquad N_3 = 65 \qquad N_4 = 42 \qquad N_5 = 44$$

$$n_1 = 16 \qquad n_2 = 8 \qquad n_3 = 6 \qquad n_4 = 4 \qquad n_5 = 4$$

$$\overline{X}_1 = 13.56 \qquad \overline{X}_2 = 22.37 \qquad \overline{X}_3 = 57.0 \qquad \overline{X}_4 = 133.0 \qquad \overline{X}_5 = 185.0$$

$$s_1^2 = 102.01 \qquad s_2^2 = 248.69 \qquad s_3^2 = 1576.09 \qquad s_4^2 = 5596.54 \qquad s_5^2 = 21966.20$$

The best estimate of the true population total (τ) is the stratified sample total (T):

$$\hat{\tau} = T = \sum_{i=1}^{m} N_i \overline{X}_i = 172(13.56) + 107(22.37) + 65(57.0) + 42(133.0) + 44(185.0) = 22156.91$$

The confidence interval about T is: $T \pm Z\hat{\sigma}_T = T \pm Z\sqrt{\sum_{i=1}^{m} N_i^2 \left(\frac{s_i^2}{n_i}\right)\left(\frac{N_i - n_i}{N_i}\right)}$

Include the finite population correction $\left(\dfrac{N_i - n_i}{N_i}\right)$ as $(n/N) > .05$ for each stratum:

$$T \pm Z\sqrt{\sum_{i=1}^{m} N_i^2 \left(\frac{s_i^2}{n_i}\right)\left(\frac{N_i - n_i}{N_i}\right)}$$

$$= 22156.91 \pm 1.65\sqrt{(172)^2\left(\frac{102.01}{16}\right)\left(\frac{172-16}{172}\right) + (107)^2\left(\frac{248.69}{8}\right)\left(\frac{107-8}{107}\right) + \ldots}$$

$$\ldots(65)^2\left(\frac{1576.09}{6}\right)\left(\frac{65-6}{65}\right) + (42)^2\left(\frac{5596.54}{4}\right)\left(\frac{42-4}{42}\right) + (44)^2\left(\frac{21966.20}{4}\right)\left(\frac{44-4}{44}\right)$$

$$= 22156.91 \pm 6041.33$$

With 90% certainty, the total number of uncultivated farm acres in woodlots or fallow in the county, τ, is within the interval from 16115.58 to 28198.24

Table 7.7 Estimate of Stratified Sample Proportion—Example 7.7

Task: Estimate the proportion of all households favoring a bond issue to finance various park improvements, based on a stratified sample, and place a 90% (α = .10) confidence interval (bounds) around the estimate.

Stratum 1 (Neighborhood 1)	**Stratum 2** (Neighborhood 2)	**Stratum 3** (Neighborhood 3)
$N_1 = 155$	$N_2 = 62$	$N_3 = 95$
$n_1 = 40$	$n_2 = 16$	$n_3 = 24$
$x_1 = 22$	$x_2 = 12$	$x_3 = 17$
$p_1 = \dfrac{x_1}{n_1} = \dfrac{22}{40} = .55$	$p_2 = \dfrac{x_2}{n_2} = \dfrac{12}{16} = .75$	$p_3 = \dfrac{x_3}{n_3} = \dfrac{17}{24} = .71$

where: x_i = number of households from sample stratum i favoring bond issue

The best estimate of the population proportion (ρ) is the stratified sample proportion (p):

$$\hat{\rho} = p = \frac{1}{N} \sum_{i=1}^{m} N_i p_i = \frac{1}{312}\left[155(.55) + 62(.75) + 95(.71)\right] = .64$$

The confidence interval about p is:

$$p \pm Z\hat{\sigma}_p = p \pm Z \sqrt{\frac{1}{N^2} \sum_{i=1}^{m} N_i^2 \left(\frac{p_i(1-p_i)}{n_i - 1}\right)\left(\frac{N_i - n_i}{N_i}\right)}$$

Include the finite population correction $\left(\dfrac{N_i - n_i}{N_i}\right)$ as (n/N) > .05 for each stratum:

$$p \pm Z \sqrt{\frac{1}{N^2} \sum_{i=1}^{m} N_i^2 \left(\frac{p_i(1-p_i)}{n_i - 1}\right)\left(\frac{N_i - n_i}{N_i}\right)} =$$

$$.64 \pm 1.65 \sqrt{\frac{1}{(312)^2}\left[(155)^2\left(\frac{(.55)(.45)}{40-1}\right)\left(\frac{155-40}{155}\right) + (62)^2\left(\frac{(.75)(.25)}{16-1}\right)\left(\frac{62-16}{62}\right) + (95)^2\left(\frac{(.71)(.29)}{24-1}\right)\left(\frac{95-24}{95}\right)\right]}$$

$$= .64 \pm 1.65(.0021) = .64 \pm .0035$$

With 90% certainty, the proportion of all households favoring a bond issue to finance park improvements, ρ, is within the interval from .6365 to .6435

officials also decide that a total sample size of 80 households is sufficiently large and practical. The procedure for estimating the proportion of residents in favor of the bond issue from a stratified sample of neighborhoods is shown in table 7.7. The very narrow interval around the estimated sample proportion is caused by the relatively large sampling fractions from all strata and the similar estimated proportions from each stratum. Of course, a very precise sample estimate is highly desirable.

7.3 Sample Size Selection

In geographic problems using sampling, researchers often want to determine the minimum sample size needed to make sufficiently precise estimates—*before* the complete sample is actually taken. Taking a larger sample than necessary wastes both time and effort. Some of the major factors to consider in selecting sample size are:

1. The type of sample (random, stratified, etc.);
2. The population parameter being estimated (mean, total, proportion);
3. The degree of precision (width of confidence interval that can be tolerated); and
4. The level of confidence to be obtained for the estimate.

Key trade-offs occur as sample size is increased. At a particular confidence level (.95, for example), increasing the sample size provides greater precision and narrows the confidence interval width around the sample estimate. Similarly, at a particular level of precision (e.g., estimating the sample proportion within .03 of the true population proportion), increasing the sample size will raise the level of confidence that the sample estimate is within the selected interval.

Unfortunately, a larger sample generally requires more time and effort. In many practical sampling problems, the investigator faces conflicting objectives: (1) take a sample large enough to achieve the desired precision level and confidence interval width; but (2) avoid taking too large a sample, which wastes time and effort and provides estimates more precise than needed for the problem. Considerable extra effort might yield only marginal benefit.

This section illustrates how an appropriate sample size is determined if random sampling is

used. The procedure is shown for each of the three population parameters (mean—μ; total—τ; and proportion—ρ).

Suppose the task is to determine the minimum sample size needed to place the sample mean estimate within a desired confidence interval or bound around the true population mean. Recall that the confidence interval for μ is $\overline{X} \pm Z\hat{\sigma}_{\overline{x}}$, where $\hat{\sigma}_{\overline{x}}$ is the standard error of the mean (sampling error). Let E (for Error) be used to designate the "tolerable error" of the sample mean estimate from the population mean:

$$E = Z\hat{\sigma}_{\overline{x}} = Z\sqrt{\frac{\sigma^2}{n}} \qquad (7.6)$$

This error is equivalent to one-half the width of the confidence interval. After some algebraic manipulation, the sample size (n) needed to estimate μ with a certain level of precision or tolerable error (E), at a chosen level of confidence (Z), can be expressed as:

$$E = Z\sqrt{\frac{\sigma^2}{n}}$$

$$E^2 = Z^2\left(\frac{\sigma^2}{n}\right)$$

$$nE^2 = Z^2\sigma^2$$

$$n = \frac{Z^2\sigma^2}{E^2} = \left(\frac{Z\sigma}{E}\right)^2 \qquad (7.7)$$

If the population standard deviation (σ) is unknown, the sample standard deviation (s) is substituted:

$$n = \left(\frac{Zs}{E}\right)^2 \qquad (7.8)$$

The required minimum sample size is directly related to the desired level of confidence (Z) and variability in the population (σ), but inversely related to the degree of error (E) the investigator is willing to tolerate.

In geographic research, the population standard deviation is almost never known. A sample standard deviation value is usually substituted to determine an appropriate sample size. A value for (s) is best derived from a pretest or preliminary sample, but might also be obtained from some previous study or could be an educated guess. Since a population parameter is being estimated, at least 30 observations should be included in a pretest or preliminary sample. The geographer may return to the general population to draw additional sample units if needed, which may then be combined with the original sample units to create a single larger sample to meet the size requirement. If this "return" procedure is used, it is considered a **two-stage sampling design**.

The minimum sample size needed to estimate a population total within a certain tolerable error level (E) can also be determined in advance. The procedure for a total is directly analogous to that used for the mean. The confidence interval for τ is $T \pm Z\hat{\sigma}_T$, where $\hat{\sigma}_T$ is the sampling error or standard error of the total:

$$E = Z\hat{\sigma}_T = Z\sqrt{N^2\left(\frac{\sigma^2}{n}\right)} \qquad (7.9)$$

Algebraic manipulation isolates (n), the minimum sample size needed to estimate τ, with a selected level of tolerable error (E) at a chosen confidence level (Z):

$$n = \left(\frac{NZ\sigma}{E}\right)^2 \qquad (7.10)$$

If the population standard deviation (σ) is unknown, the sample standard deviation (s) is substituted:

$$n = \left(\frac{NZs}{E}\right)^2 \qquad (7.11)$$

Once again, the population standard deviation is seldom known and must be estimated with the sample standard deviation, which generally requires a pretest or preliminary sample.

When estimating a population proportion with a certain allowable level of error (E), the minimum sample size can also be calculated in advance of full sampling. Again, a pretest or preliminary survey may be used to estimate the population proportion (ρ) from the sample proportion (p). The confidence interval for ρ is $p \pm Z\hat{\sigma}_p$, where $\hat{\sigma}_p$ is the sampling error or standard error of the proportion:

$$E = Z\hat{\sigma}_p = Z\sqrt{\frac{p(1-p)}{n-1}} \qquad (7.12)$$

The researcher can isolate the minimum sample size (n) algebraically as:

$$n = \frac{Z^2\rho(1-\rho)}{E^2} \qquad (7.13)$$

If the population proportion (ρ) is unknown, the sample proportion (p) is substituted:

$$n = \frac{Z^2 p(1-p)}{E^2} \qquad (7.14)$$

When estimating a population proportion, a pretest or preliminary survey is not necessary. The numerator of equation 7.14 contains the product $p(1-p)$. Consider the values this product will take given different values of p:

p	0.1	0.2	0.3	0.4	0.5	0.6	0.7	0.8	0.9
$p(1-p)$	.09	.16	.21	.24	.25	.24	.21	.16	.09

The maximum product $p(1 - p) = .25$ occurs when $p = 0.5$ and $(1 - p) = 0.5$. If this largest possible product of $p(1 - p)$ is inserted into equation 7.14, the result will be the largest minimum sample size needed under conditions of maximum uncertainty and represents a "worst case scenario":

$$n = \frac{Z^2 p(1-p)}{E^2} = \qquad (7.15)$$

$$\frac{Z^2(.5)(.5)}{E^2} = \frac{Z^2(.25)}{E^2} = \frac{Z^2}{4E^2}$$

To illustrate the procedures for estimating minimum sample size, planning examples from Middletown are continued (see examples 7.2–7.4).

Example 7.8 Sample Size Selection (Mean)

Middletown planners wish to estimate the mean number of people per household and be 90 percent confident that their estimate will be within 0.3 persons of the true population mean, μ. What is the minimum number of households that must be surveyed to ensure this degree of precision at this selected confidence level?

Suppose a preliminary sample of 30 households is drawn and (s) is calculated as 1.25. Since the population standard deviation is unknown, equation 7.8 is used:

$$n = \left(\frac{Zs}{E}\right)^2 = \left(\frac{(1.65)(1.25)}{0.3}\right)^2 = 47.26$$

This result suggests that a random sample of at least 48 households should be taken to ensure the desired degree of precision at the 90 percent confidence level. It is always better to include an extra observation or two above the absolute minimum necessary sample size, so rounding up to 48 households is appropriate. Since this result was obtained from a preliminary sample of 30 households, only 18 additional households in a two-stage sampling design need be contacted to complete the study satisfactorily.

Example 7.9 Sample Size Selection (Total)

Middletown planners also want to estimate the total number of people in the community. Tax records show a total of 3,500 households, but the number of people per household is not known. Suppose the planners wish to estimate total community population within 1,000 people and be 90 percent confident with that level of precision and sampling error (E). What is the minimum number of households that must be surveyed?

From a pretest or preliminary survey of 30 households, the standard deviation of the number of people per household (s) is calculated as 2.05. Because the population standard deviation is not known, the sample size needed to meet the requirements of the problem is calculated from equation 7.11:

$$n = \left(\frac{NZs}{E}\right)^2 = \left(\frac{(3,500)(1.65)(2.05)}{1,000}\right)^2 = 140.156$$

A random sample of at least 141 households should be taken to estimate the total population within 1,000 people, and be 90 percent confident in that level of precision. Since 30 households were already surveyed in the preliminary survey, another 111 households should be contacted in the second stage of the survey design.

Example 7.10   Sample Size Selection (Proportion)

Middletown planners wish to estimate the proportion of households having one or more children less than 18 years of age and want to be 90 percent certain their sample estimate is within .04 (4%) of the true population estimate. To determine in advance the minimum sample size needed when estimating a population proportion (ρ), two options are possible:

1. Suppose a preliminary survey reveals that 36 percent of Middletown households contain one or more children. Since the population proportion (ρ) is unknown, $p = .36$ is used with equation 7.14 to determine the necessary sample size:

$$n = \frac{Z^2 p(1-p)}{E^2} = \frac{(1.65)^2(.36)(.64)}{(.04)^2} = 392.0$$

Thus, a random sample of 392 households should be taken to estimate the proportion of Middletown households with children within .04 (4%) and be 90 percent confident in a result that precise. Since 30 households have already been surveyed, another 362 observations must be taken in the second stage of the survey design.

2. If no preliminary survey is taken, the maximum product $p(1 - p) = .5(.5) = .25$ may be used in a "worst case" situation (equation 7.15):

$$n = \frac{Z^2}{4E^2} = \frac{(1.65)^2}{4(.04)^2} = 425.4$$

In this case, a random sample of 426 households should be taken to achieve the desired level of precision. The slightly larger minimum sample size required with this option is to be expected. The worst possible situation has been assumed, and a larger sample size is likely to be needed to reduce the sampling error (E) to the desired level of precision (.04).

Key Terms and Concepts

References and Additional Reading

Dixon, C. J. and B. Leach. 1978. *Sampling Methods for Geographical Research.* Norwich, England: Geo Abstracts.

Ferber, R., et. al. 1980. *What Is A Survey?* (informational brochure). Washington, DC: American Statistical Association.

Gregory, S. 1978. *Statistical Methods and the Geographer* (4th edition). London: Longman.

Hauser, P. 1975. *Social Statistics in Use.* New York: Russell Sage Foundation.

Jessen, R. T. 1978. *Statistical Survey Techniques.* New York: John Wiley and Sons.

Kish, L. 1965. *Survey Sampling.* New York: John Wiley and Sons.

Scheaffer, R. L., W. Mendenhall and L. Ott. 1986. *Elementary Survey Sampling* (3rd edition). Boston: Duxbury.

Sudman, S. 1976. *Applied Sampling.* New York: Academic Press.

Williams, W. H. 1978. *A Sampler on Sampling.* New York: John Wiley and Sons.

PART IV

INFERENTIAL PROBLEM SOLVING IN GEOGRAPHY

CHAPTER 8

ELEMENTS OF INFERENTIAL STATISTICS

The previous chapter introduced various concepts concerning estimation in sampling. Sample estimation involves inference. A primary objective of sampling is to infer some characteristic of the population based on statistics derived from a sample in that population. Sample statistics are used to make point estimates of population parameters, such as the mean, total, and proportion. In addition, to determine the level of precision of these point estimates, a confidence interval or bound is placed around the sample statistic, making it possible to state the likelihood that a sample statistic is within a certain range or interval of the population parameter. In this chapter, these ideas are extended to a form of statistical inference known as **hypothesis testing**. Geographers can apply inferential hypothesis testing to help reach conclusions for a wide variety of geographic problems.

The practical application and value of inferential statistics are best understood in the context of the scientific research process. As outlined and discussed in chapter 1 (figure 1.1), the formulation of hypotheses and their testing through inferential statistics often play a central role in the development of the science of geography. For example, hypothesis evaluation may lead to the refinement of spatial models and the development of laws and theories. In addition, conclusions from inferential testing often contribute toward the advancement of scientific research in geography.

Methods of hypothesis testing are introduced in this chapter with examples of one sample difference tests. A properly created sample is a key to applying inferential statistics successfully. Geographic researchers often need to verify whether a particular sample is truly representative of the population from which it is drawn. As discussed in chapter 6, simple random sampling procedures are often employed to guarantee that each member of the population has the same chance of being selected in the sample, reducing the possibility of inaccuracy or bias.

Even with the use of random sampling, however, sampling error could cause an unrepresentative result.

Researchers can test for possible bias in a sample by comparing a sample statistic to the comparable population parameter. For example, a sample mean could be compared to a population mean, or a sample proportion could be compared to a population proportion. These are examples of **one sample difference tests**.

If the population parameter and the sample statistic are not significantly different, the researcher has confidence that the sample is truly representative of the target population. In such cases, the chosen sample is adequate for further analysis. On the other hand, if the population parameter and sample statistic are significantly different, the sample may be inaccurate or otherwise deficient. Excessive sampling error may produce an unrepresentative sample. Human error may occur in one of the steps taken to produce the sample.

Alternatively, a particular sample may represent the situation described in figure 7.3, where the mean of sample 5 ($\overline{X}_5$) falls into a tail of the sampling distribution of means. That illustration demonstrates how a sample statistic calculated from a properly drawn random sample has a 10 percent chance of being quite different from the actual population parameter value.

An example helps clarify the need for testing the difference between a sample and population mean. Suppose a study is undertaken to determine the attitudes of community residents about constructing a neighborhood swimming pool. The advisory committee wants to select a 5 percent sample of families at random for a telephone survey. However, because families with children are more likely to desire this public facility, the committee wants to make sure that their sample accurately represents the family structure in the community.

One way to measure the success of the committee's objective is to determine the number of children in each family selected in the sample. Using these data, the mean family size for the sample could be computed. This sample mean could then be compared to the average family size for the target population found in either recent census data for the community or perhaps in information from a school census. If the mean family size for the sample matches the mean for the target population, then the committee can have confidence that their sample is truly representative of family structure in the area, and the critical results of the survey can then be analyzed with the knowledge that the sample is not biased.

In this chapter, two complementary methods of hypothesis testing are presented. The classical/traditional approach (section 8.1) provides a solid, logical foundation for understanding hypothesis testing. The *p*-value or prob-value method of hypothesis testing (section 8.2) builds on this solid foundation, yet provides additional useful information concerning the research problem. An example problem from the previous chapter is used to illustrate these methods. A difference of means Z test uses the concepts and procedures developed in chapter 7 on sampling estimation and provides a transition from the placement of confidence intervals around sample estimates to the use of inferential statistics in hypothesis testing.

In section 8.3, some related difference tests are presented. Finally, section 8.4 discusses the circumstances or geographic problems for which inferential testing is appropriate. In addition, the various issues that influence the selection of the proper statistical test are discussed.

8.1 Classical/Traditional Hypothesis Testing

Classical hypothesis testing involves a formal multistep procedure that leads from the statement of hypotheses to a conclusive statement (decision) regarding the hypotheses (table 8.1). Hypothesis testing has the general goal of making inferences about the magnitude of one or more population parameters, based on sample estimates of these parameters. Hypotheses regarding a population parameter are evaluated using sample information, and a conclusion is reached (at some preselected significance level) about the hypotheses. Because of the nature of sampling, a measurable probability can always be assigned to the conclusions reached through statistical hypothesis testing. This logic should sound very familiar, as it is based directly on the themes discussed in chapter 7 on sampling estimation.

Recall in example 7.2, planners in the city of Middletown wanted to update selected demographic statistics. More specifically, community officials estimated the mean number of people per household from two random samples—a smaller sample having only 25 households and a larger sample of 250 households. Both samples were taken from a population of 3,500 Middletown households, and a bound or confidence interval was placed around the sample estimates. In the smaller, 25-household sample,

---

Table 8.1 Steps in Classical/Traditional Hypothesis Testing

Step 1: Statement of Null and Alternate Hypotheses

Step 2: Select Appropriate Statistical Test

Step 3: Select Level of Significance

Step 4: Delineate Regions of Rejection and Nonrejection of Null Hypothesis

Step 5: Calculate the Test Statistic

Step 6: Make Decision Regarding Null and Alternate Hypotheses

---

the sample mean was 2.73 people per household, with a 90 percent confidence interval of .552. In the larger, 250-household sample, the sample mean was 2.68 people per household, with an associated 90 percent confidence interval of .209:

smaller sample ($n = 25$): $\overline{X} \pm t\hat{\sigma}_{\overline{x}} = 2.73 \pm .552$

larger sample ($n = 250$): $\overline{X} \pm Z\hat{\sigma}_{\overline{x}} = 2.68 \pm .209$

Suppose that the planners need additional information. From the 1990 Census of Population, Middletown officials have learned that the average number of people per household in the United States is 2.61 ($\mu = 2.61$), a continuation of the general national trend in declining household size (3.33 in 1960; 3.14 in 1970; and 2.76 in 1980). Middletown planners want to see if household size in their community is typical, or representative, of this national figure. Is mean household size in their community similar to the national mean or significantly different? To answer this question, a hypothesis testing procedure is established to determine how closely the sample of Middletown households compares with the national average household size of 2.61. To test this hypothesis, Middletown officials decide to use the larger, 250-household sample ($\overline{X} = 2.68$ persons per household) to make the comparison, since it is more precise than the smaller sample of 25 households.

The Middletown mean household size example is used to discuss the steps of the classical hypothesis testing process. Appropriate terminology and concepts regarding statistical testing are introduced as needed.

Step 1:   Statement of Null and Alternate Hypotheses

Two complementary hypotheses of interest are the **null hypothesis** (denoted H_0) and the **alternate** or **alternative hypothesis** (denoted H_A). Consider the formulation of hypotheses concerning the mean of a population (μ). The typical claim is that μ is equal to some value, μ_H (for **hypothesized mean**). This claim of equality is called the null hypothesis, and takes the general form:

$$H_0: \mu = \mu_H$$

The null hypothesis can also be stated:

$$H_0: \mu - \mu_H = 0$$

In the latter form, attention is focused on H_0 as a statement of "no difference" between μ and μ_H, which is the case if ($\mu - \mu_H$) equals zero (or null). The null hypothesis statement always includes the equals sign.

The converse of the null hypothesis is the alternate hypothesis, H_A. The alternate hypothesis expresses the conditions under which H_0 is to be rejected and can be viewed as a positive statement of difference. The two hypotheses are mutually exclusive, for if H_0 is rejected, H_A is accepted. The alternate hypothesis takes one of two forms, depending on how the research problem is structured. In some problems, the form of H_A is **nondirectional**, while in others it is **directional**, but H_A always consists of an inequality indicating the conditions under which H_0 is rejected. So, for example, each of the following is a valid form of H_A:

$$H_A: \mu \neq \mu_H$$
$$H_A: \mu < \mu_H$$
$$H_A: \mu > \mu_H$$

The selection of a specific form of H_A depends on how the hypothesized difference is stated. The directional and nondirectional formats offer two possibilities. In the Middletown household size example, the alternate hypothesis can be stated in a nondirectional form:

$$H_0: \mu = 2.61$$
$$H_A: \mu \neq 2.61$$

These statements hypothesize that mean household size of the sample of Middletown households is no different from the mean national household size of 2.61. If the sample mean is close to this population mean, the likely conclusion is that the null hypothesis *should not* be rejected. Conversely, if the sample mean is not close to the population mean, the likely conclusion is that H_0 *should* be rejected. By expressing H_A in this form, the direction of difference between household size of Middletown and the United States is not important. That is, the Middletown household size could be either greater than *or* less than the national figure. If H_0 is rejected, the *only* conclusion that can be drawn is that the difference in household sizes is significant; no conclusion on the direction of that difference is possible.

For some situations, the alternate hypothesis can provide more specific information about the hypothesized direction of difference:

$$H_0 : \mu = 2.61$$

$$H_A : \mu < 2.61$$

or

$$H_A : \mu > 2.61$$

In addition to determining whether a significant difference exists, the direction of that difference (expressed in H_A) is also specified. If $H_A: \mu < 2.61$, then rejection of H_0 indicates that Middletown households are smaller in size than the national average. This would be a reasonable alternate hypothesis to test if Middletown has many single persons, young couples without children, or elderly residents. If $H_A: \mu > 2.61$, then rejection of H_0 indicates that Middletown household sizes are larger than the national average. This might be a logical alternate hypothesis if there are many families with children residing in the community.

With classical hypothesis testing, Middletown planners must decide which form of H_A to use. The critical issue is whether *a priori* knowledge

about any direction of difference between Middletown and nationwide household sizes exists. If Middletown planners have no preconceived rationale for believing their household size is larger or smaller than the national average, the nondirectional format would be appropriate.

In classical hypothesis testing, the conclusion is to either reject or not reject the null hypothesis (i.e., choose between the null and alternate hypotheses). Because this decision is based on a single sample, there is a measurable chance or probability of making an incorrect decision or reaching a wrong conclusion. Error comes from two possible sources (table 8.2).

Type I Error

A decision could be made to reject the null hypothesis as false when, in fact, it is true. For the Middletown example, it could be concluded that the difference between the national average household size and the sample of Middletown households is significant, when actually no difference exists. The likelihood of this sort of error occurring is equivalent to the significance level (α), discussed in chapter 7 (see figure 7.3).

Type II Error

Conversely, a decision could be made to not reject the null hypothesis when it is actually false. For the Middletown example, it could be concluded that their household size does not

Table 8.2 Possible Decisions in Classical/Traditional Hypothesis Testing

Decision From Hypothesis Testing	Null Hypothesis in Reality	
	True	False
Reject H_0 as False	Type I Error (probability = α)	Correct Decision (prob. = $1 - \beta$)
Do Not Reject H_0	Correct Decision (prob. = $1 - \alpha$)	Type II Error (prob. = β)

differ from the national average, when a significant difference really exists. The likelihood of this error occurring is beta (β).

The logic of hypothesis testing operates on much the same principle as judicial decision making in a court of law (table 8.3). In hypothesis testing, the null hypothesis is presumed correct until rejected or proven otherwise. Similarly, in court, a defendant is presumed innocent until proven guilty. In the judicial system, convicting a person who is truly not guilty is considered a more serious error than freeing a guilty person. Similarly, in hypothesis testing, rejecting H_0 as false when it is actually true (a Type I error) is considered more serious than not rejecting H_0 when it is actually false (a Type II error). In a court of law, when reasonable doubt exists about the guilt of a defendant, the jury reaches a verdict of not guilty. The defendant is *not* declared "innocent"; the verdict is "not guilty." Similarly, if reasonable doubt exists about rejecting the null hypothesis, the researcher should not reject it. For this reason, the significance level (α) of most problems is kept relatively low (at a level such as .05 or .01), to minimize the chances of a serious Type I error.

Step 2: Select Appropriate Statistical Test

The second step in classical hypothesis testing is selecting the most appropriate statistical test to examine the research problem. A logical and convenient way to categorize the many statistical tests available is by the type of question being asked.

1. Does a sample statistic differ significantly from a population parameter? Does a significant difference exist between two or more samples? Inferential statistics offers a wide variety of difference tests, including "goodness-of-fit" tests, specially designed "dependent" sample tests, and tests with special application to explicit spatial data—such as point and area patterns. Many of the tests discussed in chapters 9, 10, and 11 are examples of difference tests.
2. Does a significant association or relationship exist between two or more variables, samples, or groups? The direction and strength of relationship can be measured through correlation, the subject of chapter 12.
3. Does a significant nonrandom form of relationship occur between variables, samples, or groups? The form or nature of the relationship between variables is measured through regression analysis, the focus of chapter 13.

The appropriate statistical test for the Middletown household size problem is the **difference of means Z test**:

$$Z = \frac{\overline{X} - \mu}{\sigma_{\overline{x}}} = \frac{\overline{X} - \mu}{\sigma / \sqrt{n}} \tag{8.1}$$

where: Z = test statistic

$\overline{X}$ = sample mean

μ = population mean

σ = population standard deviation

n = sample size

Because a single sample mean is being compared to a population mean, this test is sometimes referred to as a *one sample* difference of

Table 8.3   Possible Decisions in a Court of Law

Decision from Jury	True Situation Not Guilty	(Unknown) Guilty
Guilty	Incorrect Decision	Correct Decision
Not Guilty	Correct Decision	Incorrect Decision

means test to differentiate it from tests that compare two or more samples for differences.

In chapter 5 on probability, the concept of a Z-score or standard score was introduced. The one sample difference of means Z test is structurally similar to the Z-score equation (table 8.4). The standardized Z-score of a value in a set of data may be interpreted as the number of standard deviations a value lies above or below the mean. Using similar logic, the Z-score difference of means test statistic measures the number of standard errors a sample mean lies above or below the hypothesized population mean.

For the Middletown household size example, the population standard deviation (σ) is not known, a common occurrence in scientific research and hypothesis testing. Recall from chapter 7 that the sample standard deviation, s, is a proper estimator of σ, provided the sample size (n) is greater than 30. Thus, substituting s for σ in equation 8.1 gives:

$$Z = \frac{\overline{X} - \mu}{s/\sqrt{n}} \qquad (8.2)$$

where: $n > 30$

One Sample Difference of Means Z Test
(Large Sample)

Primary Objective: Compare a random sample mean to a population mean for difference

Requirements and Assumptions:
(1) Random sample
(2) Population from which sample is drawn is normally distributed
(3) Variable measured at interval or ratio scale

Hypotheses:
H_0: $\mu = \mu_H$ (where μ_H is the hypothesized mean)
H_A: $\mu \neq \mu_H$ (two-tailed)
H_A: $\mu > \mu_H$ (one-tailed) or
H_A: $\mu < \mu_H$ (one-tailed)

Test Statistic:
If sample size is greater than 30: $Z = \dfrac{\overline{X} - \mu}{\sigma_{\overline{x}}}$

Step 3: Select Level of Significance

The next task in classical hypothesis testing is to place a probability on the likelihood of a sampling error. As mentioned earlier, committing a Type I error and rejecting a null hypothesis as false when it is actually true is generally considered serious. Therefore, the usual procedure is to select a fairly low significance level (α) such as .05 or .01. Then, the conclusion reached is specified in terms of the level of significance of the result. In classical hypothesis testing, a null hypothesis may be rejected at the .05 level, which is the same as saying the statistical test is significant at the .05 level. This would mean that the chance that a Type I error has occurred is only 5 percent (1 in 20), and it is only 5 percent likely that the null hypothesis has been improperly rejected because of random sampling error.

For many geographic research problems, a highly demanding or extremely stringent significance level may not be necessary. In general, the choice of significance level depends on the nature of the problem and the effects of the decision. With some geographic problems, the consequences of sampling error may be severe, and in these instances, a very low significance level is required. With the Middletown household size example, using a commonly accepted "conventional" significance level such as $\alpha = .05$ provides sufficient precision.

Table 8.4 Similar Structure: Z-Score of a Data Value and Difference of Means Z Test

Z-Score of a Value (i) in a Set of Data	Z-Value of a Sample Mean in a Frequency Distribution of Sample Means
$Z_i = \dfrac{X_i - \overline{X}}{s}$	$Z = \dfrac{\overline{X} - \mu}{\sigma_{\overline{x}}}$ where $\sigma_{\overline{x}} = \dfrac{\sigma}{\sqrt{n}}$

Step 4: Delineate Regions of Rejection and Nonrejection of Null Hypothesis

Once a significance level has been selected, this value is used to create the **regions of rejection and nonrejection** of the null hypothesis (figure 8.1). The total area in which H_0 is rejected, as represented by the significance level, encompasses 5 percent ($\alpha = .05$) of the area under the curve. This area of rejection can be distributed in one of two ways. In case 1, the alternate hypothesis is nondirectional, so the

shaded rejection area of H_0 is distributed equally between the two tails of the curve. With this **two-tailed** format and $\alpha = .05$, 2.5 percent ($\alpha/2 = .025$) of the total area is in each of the rejection regions or tails of the distribution. In case 2, the alternate hypothesis is directional, so the shaded rejection region of H_0 is placed entirely on one tail of the distribution. In a **one-tailed** format shown in case 2, the rejection region happens to be on the right tail, but the placement of the rejection region depends on the form of H_A. In both the two-tailed and one-tailed cases, the unshaded area under the curve delineates test statistic values where H_0 is not rejected. This area of nonrejection is 95 percent ($1 - \alpha = .95$) of the total area under the curve.

The next task is to determine the critical Z-score values that delimit the boundaries separating the rejection and nonrejection regions. When $\alpha = .05$ and H_A is nondirectional, each of the two tails contains $\alpha/2$ or .025 of the area under the curve (figure 8.2). This leaves .475 of the area on *each* side of the distribution ($1 - \alpha = .95$ in total) in the nonrejection region. In the normal curve table (appendix, table A), the Z-value that corresponds with the .475 probability for the upper side of the distribution is 1.96. Therefore, in this problem, a

Figure 8.1 General Regions of Rejection and Nonrejection of Null Hypothesis: Significance Level (α) = .05

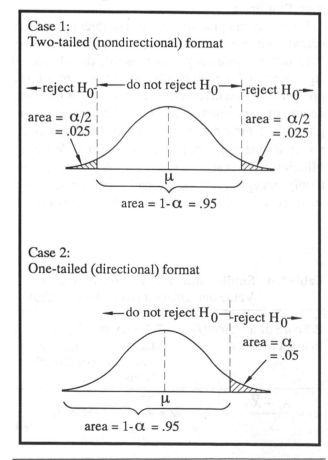

Figure 8.2 Normal Table Probability Values Associated with a Significance Level (α) = .05: Two-tailed Case

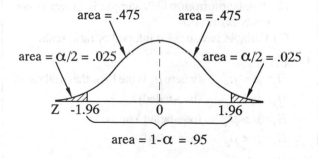

Z-value of 1.96 defines the boundary between the rejection and nonrejection regions, and a calculated Z-value (test statistic) must be less than or equal to 1.96 to keep the null hypothesis from being rejected. Given the symmetrical nature of a normal curve, a similar-magnitude Z-score of –1.96 is the critical value delineating the boundary separating H_0 and H_A at the lower tail.

The following decision rule identifies and summarizes the regions of rejection and nonrejection of the null hypothesis for α = .05 and H_A nondirectional:

Decision rule:

If $Z < -1.96$ or if $Z > 1.96$, reject H_0.

Conversely, if $-1.96 \le Z \le 1.96$, do not reject H_0.

Step 5: Calculate the Test Statistic

At this step of the hypothesis testing procedure, sample data are evaluated using the test statistic. In the Middletown sample, mean household size was 2.68 persons, with a sample standard deviation of 2.07 persons (from example 7.2, s^2 = 4.3 and s = 2.07). Substituting these sample statistics into the difference of means Z test gives the following:

$$Z = \frac{\overline{X} - \mu}{s/\sqrt{n}} = \frac{2.68 - 2.61}{2.07/\sqrt{250}} = 0.53$$

The Middletown sample mean is 0.53 standard errors *above* the U.S. average. If household size in Middletown had been less than the national norm of 2.61, the calculated test statistic would have been negative. This calculated value is now compared with the critical Z-values determined in step 4 to reach a final decision.

Step 6: Make Decision Regarding Null and Alternate Hypotheses

All the information needed to make a decision regarding rejection or nonrejection of the null hypothesis is now available. The calculated test statistic is $Z = 0.53$. The critical Z-values are –1.96 and 1.96. Since the test statistic lies between the critical values, $(-1.96 \le Z \le 1.96)$, the null hypothesis should *not* be rejected. The conclusion is that mean household size in Middletown does not differ from the mean household size nationally. The steps in the classical hypothesis testing procedure for the Middletown example are summarized in table 8.5.

8.2 *P*-Value or Prob-Value Hypothesis Testing

The formal multistep procedure of classical hypothesis testing provides a logical basis and excellent theoretical underpinning for inferential decision making. However, the usefulness of the results from classical analysis is limited in some important ways. First, a specific significance level must be selected to delineate the regions of rejection and nonrejection of the null hypothesis. This *a priori* selection of α is often arbitrary and may lack a clear theoretical basis. A significance level of .05 or .01 is often chosen because they are the "conventional" probabilities commonly provided in statistical tables. Second, the final decision regarding the null and alternate hypotheses is binary in nature: either

Table 8.5   Summary of Classical Hypothesis Testing: Middletown Household Size Example

---

Step 1: H_0: μ = 2.61 and H_A: $\mu \neq 2.61$

Step 2: One sample difference of means Z test selected as test statistic

Step 3: α = .05

Step 4: If $Z < -1.96$ or if $Z > 1.96$, reject H_0
　　　　If $-1.96 \le Z \le 1.96$, do not reject H_0

Step 5: Calculate Z (from random sample) = 0.53

Step 6: Since $-1.96 \le Z \le 1.96$, do not reject H_0

H_0 is rejected or not rejected at that arbitrary significance level. This type of conclusion provides only limited information about the calculated test statistic. Geographers rarely use classical hypothesis testing today for statistical problem solving; instead, the *p*-value approach is commonly employed in geographic research.

The more flexible *p*-value method of hypothesis testing provides additional valuable information. With this approach, the *exact* significance level associated with the calculated test statistic value is determined. That is, the **p-value** is the exact probability of getting a test statistic value of a given magnitude, *if* the null hypothesis of no difference is true. This can generally be interpreted as the probability of making a Type I error. Preselection of a "standard" significance level (α) is avoided, and decisions to reject or not reject H_0 at that arbitrary level need not be made.

In the Middletown example, the decision rule was to reject the null hypothesis if $Z < -1.96$ or $Z > 1.96$ and to not reject H_0 if $-1.96 \leq Z \leq 1.96$. The calculated test statistic value of $Z = 0.53$ led to the conclusion that the mean size of Middletown households was no different than mean national household size, and the null hypothesis was *not* rejected.

This conclusion is of limited use. All that has been decided is that the Middletown household size is no different from the national average household size. However, knowing the exact significance level associated with the specific sample mean or calculated test statistic value of 2.68 would be more informative. If the null hypothesis had been rejected on the basis of the 2.68 sample mean, what would be the exact significance level and likelihood that a Type I error had been made? The probability of making a Type I error cannot be determined with the classical approach.

The Middletown sample can be viewed in a more informative way using the *p*-value approach (figure 8.3). The critical values that separate the regions of rejection and nonrejection of the null

---

Figure 8.3   *P*-Values in Two-tailed Hypothesis Test: Middletown Mean Household Size *(X)*

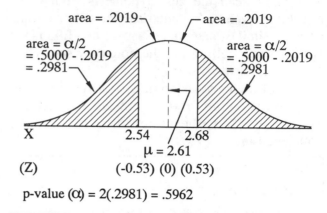

p-value (α) = 2(.2981) = .5962

---

hypothesis are now based on the location of the particular sample mean ($\overline{X} = 2.68$) relative to the population mean ($\mu = 2.61$). The unshaded region of nonrejection is centered on $\mu = 2.61$, and the difference between $\overline{X}$ and μ ($\overline{X} - \mu = .07$) is used to establish the width of the nonrejection interval on either side of μ. Thus, the nonrejection region has an upper bound of 2.68 ($\mu + .07$) and a lower bound of 2.54 ($\mu - .07$). The rejection regions occupy the extremes of the distribution, outside the upper and lower bounds (shaded areas).

In the *p*-value approach to hypothesis testing, the area within the rejection regions represents the *p*-value. This *p*-value area is calculated in four steps: (1) the test statistic for Z is calculated, as in the classical approach; (2) the probability or relative area under the normal curve is determined for that Z-value; (3) the shaded, rejection area is determined by subtracting the probability (of step 2) from .5000; and (4) this area is doubled if a nondirectional (two-tailed) alternate hypothesis is used. The resulting probability is the *p*-value.

This procedure is illustrated with the Middletown example of mean household size. The difference of mean Z test statistic is calculated:

$$Z = \frac{\overline{X} - \mu}{s/\sqrt{n}} = \frac{2.68 - 2.61}{2.07/\sqrt{250}} = 0.53$$

The resulting Z-value is then used to determine a probability or area under the normal curve. When $Z = .53$, the table of normal values lists a probability of .2019. This probability represents the nonrejection or unshaded area between $Z = 0$ and $Z = +.53$ (figure 8.3). The area in the shaded *upper* tail of the normal distribution is found by subtraction: $.5000 - .2019 = .2981$. Since this problem is nondirectional (two-tailed), a second rejection region of area $= .2981$ is located at the *lower* tail below $\overline{X} = 2.54$. Thus, the total area in the rejection region is twice .2981 or .5962. This result defines the p-value associated with the null hypothesis ($p = .5962$).

How can this p-value be interpreted? The significance level is the total area in the rejection region and indicates the likelihood of making a Type I error. If the decision is made to reject H_0, the significance level equals the p-value or .5962, representing a 59.62 percent chance of a Type I error. In this situation, the likelihood of making an error is very high, so the clear and logical decision is not to reject the null hypothesis.

The p-value provides a measure of the belief or conviction that the decision not to reject the null hypothesis is correct. For example, a p-value relatively close to 1 indicates a high degree of trust in the validity of the null hypothesis, whereas a p-value relatively close to 0 suggests that little faith should be placed in the null hypothesis. In the Middletown example, the p-value of .5962 indicates that the null hypothesis should not be rejected, and the Middletown mean household size is not different from the nationwide average household size.

Suppose the null and alternate hypotheses for the Middletown household size example are:

$$H_0: \mu = 2.61$$
$$H_A: \mu > 2.61$$

This is a one-tailed or directional test that asks whether the mean household size of Middletown is greater than the national average. This alternate hypothesis would be appropriate if planners thought that Middletown has an atypically large number of households with children. The p-value associated with $\overline{X} = 2.68$ is .2981 ($p = .2981$) (figure 8.4). In the earlier two-tailed example, this value was doubled because H_0 was rejected if $\overline{X}$ was different from μ, regardless of the direction of that difference. In this example, no such doubling is necessary. The critical region of rejection of H_0 is located entirely in one tail of the normal distribution (in this case, the upper or right tail).

The p-value from a one-tailed test is interpreted in much the same way as a two-tailed p-value. That is, if the null hypothesis is rejected in this problem, the chance of a Type I error is 29.81 percent. Since this p-value is not at all close to any "conventional" significance level (such as .05), the decision should be against rejecting the null hypothesis. The conclusion in this one-tailed test is that mean household size in Middletown is not larger than the nationwide average household size.

Note how the one-tailed p-value reduced the likelihood of making a Type I error by one-half.

Figure 8.4   *P*-Value in One-tailed Hypothesis Test: Middletown Mean Houshold Size (X)

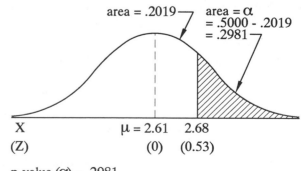

p-value (α) = .2981

Therefore, whenever a prior rationale exists for doing so, a one-tailed (as opposed to two-tailed) hypothesis testing format should be applied.

The *p*-value method provides more information than the classical method. The exact significance level and probability of making a Type I error is calculated for any test statistic value. The geographer can then evaluate the particular situation revealed by the sample data and report the *p*-value results without having to reach a decision to reject or not to reject a null hypothesis based on an arbitrary significance level. Since virtually all statistical analysis in geography is now done with the aid of statistical computer packages, *p*-values are calculated automatically. Furthermore, the *p*-value method contains all of the information derived from classical testing. The decision to reject the null hypothesis when *p* < .05 is equivalent to a classical test using a significance level of .05.

The ease of deriving multiple *p*-values also permits geographic studies to emphasize comparative, investigative analysis. Through multiple applications of a statistical test, for example, trends in spatial data over time can be examined or different regions can be compared with one another at a particular time. In other instances, the application of different test statistics to the same set of spatial data may provide new geographic insights.

Use of the *p*-value approach, however, does not provide an excuse to interpret results subjectively or avoid making decisions. Just because the computer can produce multiple *p*-values quickly, the researcher must still be responsible for interpreting the statistical and geographic meaning of the *p*-values in a consistent and rational way. *P*-values report what the sample data reveal about the credibility of a null hypothesis, but still demand the same stringent rules of inference as required in classical hypothesis testing.

Since the *p*-value method offers many advantages over classical hypothesis testing, *all statistical tests presented in the remainder of this text will use the p-value method*. In the statistical analysis of the geographic problems that follow, all associated *p*-values will be reported without showing their manual calculation. This approach provides all necessary information to reach conclusions in a succinct and consistent manner. Subsequent chapters emphasize investigative analysis and the comparative use of multiple *p*-values.

8.3 Related One Sample Difference Tests

This section introduces other examples of one sample difference tests. The first problem is a direct variation of the difference of means *Z* test, applied to small samples. The second example is a difference of proportions test, comparing a sample proportion to a hypothesized population proportion.

Difference of Means *t* Test

In the example from the previous section, the sample size was fairly large ($n = 250$), and the *Z*-score one sample difference test was appropriate. In some situations, however, a large sample is neither practical nor feasible. For a sample size less than 30, the Student's *t* distribution is used instead of the normal distribution (*Z*).

The *t* distribution was used in chapter 7 to estimate confidence intervals or bounds around sample estimates for small samples. Compared with the standard normal (*Z*) distribution, applying the *t* distribution to confidence interval estimation widened the bound on the error of estimation slightly. The smaller sample resulted in greater uncertainty about the precision of the estimate. Similar logic is applied to an inferential difference test. If the sample size is small ($n < 30$), the difference of means test is adjusted. Either of the following options is procedurally correct:

Option 1: use $t = \dfrac{\overline{X} - \mu}{s/\sqrt{n-1}}$ if $s = \sqrt{\dfrac{\Sigma(X - \overline{X})^2}{n}}$ (8.3)

Option 2: use $t = \dfrac{\overline{X} - \mu}{s/\sqrt{n}}$ if $s = \sqrt{\dfrac{\Sigma(X - \overline{X})^2}{n-1}}$ (8.4)

With either of these small-sample options, the results of the difference of means t test are comparable to Z.

The one sample t test is illustrated with another example from Middletown. Suppose Middletown contains a new planned unit development (PUD) that is predominantly residential. The development includes condominiums, townhouses, and apartments, but also contains some detached single family homes. Planners in Middletown believe that average household size in this PUD is not representative of national household size ($\mu = 2.61$), but is instead significantly smaller. This hypothesis is reasonable if planners believe this development has been particularly attractive to older singles and couples (including retirees) or to young professionals without children. In this case, a directional or one-tailed alternate hypothesis is logical:

$$H_0 : \mu = 2.61$$

$$H_A : \mu < 2.61$$

Rejection of H_0 will indicate that PUD household size is smaller than the national average.

Because time and personnel are limited, only 25 households in the PUD will be sampled. Since $n < 30$, the appropriate difference test is the one sample t test, not the Z test. Suppose the sample of 25 households in this development had a mean household size of 2.03 persons and a standard deviation of 1.50 persons. The resulting test statistic (t) is:

$$t = \frac{\overline{X} - \mu}{s/\sqrt{n-1}} = \frac{2.03 - 2.61}{1.50/\sqrt{25-1}} = -1.90$$

The degrees of freedom in the problem equals $n - 1$ or 24. The p-value associated with $t = -1.90$ at 24 degrees of freedom is .0348, (appendix, table C) as shown in figure 8.5.

In this example, if the decision is made to reject the null hypothesis and conclude that mean house-

---

Figure 8.5 *P*-Value in Small Sample, One-tailed Hypothesis Test: Mean Houshold Size *(X)* in Middletown PUD

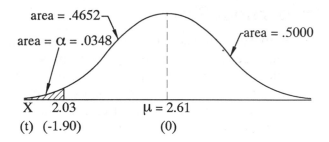

At t = -1.90 and df = 24, table value for t = .4652
p-value (α) = .5000 - .4652 = .0348

One Sample Difference of Means *t* Test (Small Sample)

Primary Objective: Compare a random sample mean to a population mean for difference

Requirements and Assumptions:
   (1) Random sample
   (2) Population from which sample is drawn is normally distributed
   (3) Variable measured at interval or ratio scale

Hypotheses:
   H_0: $\mu = \mu_H$ (where μ_H is the hypothesized mean)
   H_A: $\mu \neq \mu_H$ (two-tailed)
   H_A: $\mu > \mu_H$ (one-tailed) or
   H_A: $\mu < \mu_H$ (one-tailed)

Test Statistic:
   If sample size is less than 30: $t = \dfrac{\overline{X} - \mu}{\sigma_{\overline{x}}}$

hold size in the Middletown PUD is less than mean household size nationally, the probability that a Type I error has been made is only .0348 (or 3.48 percent). With a p-value this close to 0, the decision to reject H_0 is probably correct, since the chance of making an error when rejecting H_0 is so small. Middletown planners should feel quite confident that they are correct in concluding that PUD household size is less than the national average.

One Sample Difference of Proportions Test

Hypotheses can also be formed to study the difference between a sample proportion and a population proportion. The test statistic for a difference of proportions test is sometimes called the Z test for proportions. Like the difference of means Z test, the normal (Z) distribution is used.

Middletown planners estimated the proportion of all households in the community with one or more children less than 18 years of age (example 7.4). Of the 250 families sampled, 105 or 42 percent had children, for a sample proportion (p) and 90 percent confidence interval ($Z\hat{\sigma}_p$) of:

$$p \pm Z\hat{\sigma}_p = p \pm Z\sqrt{\frac{p(1-p)}{n-1}\left(\frac{N-n}{N}\right)} = .42 \pm .05$$

It was 90 percent likely that ρ, the true population proportion of families with children, is in the interval $.37 < \rho < .47$.

This example is now developed further to create a hypothesis test for differences. According to the 1990 Census, 48.6 percent of all families in the United States have one or more children (that is, $\rho = .486$). Is Middletown typical or representative of the national norm, or is Middletown's proportion different?

$$H_O: \rho = .486$$
$$H_A: \rho \neq .486$$

With a nondirectional two-tailed test, H_0 could be rejected if the Middletown proportion is either much greater than or much less than the national proportion.

The appropriate test statistic is the one sample difference of proportions test:

$$Z = \frac{p-\rho}{\sigma_p} \qquad (8.5)$$

where: σ_p = standard error of the proportion

The standard error of the proportion (σ_p) is the standard deviation of the sampling distribution of proportions:

$$\sigma_p = \sqrt{\frac{p(1-p)}{n}} \qquad (8.6)$$

In the Middletown example, the one sample difference of proportions test statistic is calculated as:

$$\sigma_p = \sqrt{\frac{p(1-p)}{n}} = \sqrt{\frac{(.486)(.514)}{250}} = .0316$$

$$Z = \frac{p-\rho}{\sigma_p} = \frac{.42-.486}{.0316} = \frac{-.066}{.0316} = -2.09$$

If the null hypothesis is rejected when $Z = -2.09$, the exact associated significance level and p-value is .0366 (figure 8.6). This indicates a 3.66 percent chance that a Type I error has been made if H_0 is rejected. This fairly low p-value suggests that the null hypothesis should probably be rejected, and the conclusion made that the proportion of Middletown families with children is indeed different from the national proportion.

The title of this section suggests that the various one sample difference tests are related. This can be seen in the similar structure of the one sample test statistics (table 8.6). A general format is common to each of these difference

Figure 8.6 *P*-Values in One Sample Difference of Proportions Test: Proportion *(p)* of Middletown Families with Children

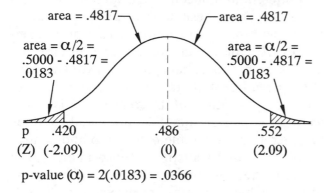

area = .4817 area = .4817

area = $\alpha/2 =$ .5000 - .4817 = .0183

area = $\alpha/2 =$ .5000 - .4817 = .0183

p	.420	.486	.552
(Z)	(-2.09)	(0)	(2.09)

p-value (α) = 2(.0183) = .0366

Table 8.6 The Similar Structure of One Sample Difference Tests

Difference Test	Test Statistic
Difference of means* large sample (n > 30)	$Z = \dfrac{\overline{X} - \mu}{\sigma_{\overline{x}}} = \dfrac{\overline{X} - \mu}{s/\sqrt{n}}$
Difference of means* small sample (n ≤ 30)	$t = \dfrac{\overline{X} - \mu}{\sigma_{\overline{x}}} = \dfrac{\overline{X} - \mu}{s/\sqrt{n-1}}$
Difference of proportions	$Z = \dfrac{p - \rho}{\sigma_p} = \dfrac{p - \rho}{\sqrt{p(1-p)/n}}$

*Population standard deviation (σ) is unknown

Best estimate of $\sigma = s = \sqrt{\dfrac{\Sigma(X_i - \overline{X})^2}{n}}$

General Format of One Sample Difference Test

$$Z \text{ or } t = \frac{\text{sample statistic} - \text{hypothesized population parameter value}}{\text{standard deviation of sample statistic (standard error)}}$$

tests. Note that the numerator of each test statistic is the difference between the sample statistic and the hypothesized population parameter value. The denominator is the standard deviation of the sampling distribution, which is also referred to as the standard error of the sample statistic.

8.4 Issues in Inferential Testing and Test Selection

So far in this chapter, attention has been focused on a related set of one sample difference tests. Numerous other inferential tests can be applied to help solve geographic research problems. The selection of the proper inferential test depends on a series of characteristics or dimensions. Examination of these dimensions provides the framework for categorizing inferential tests.

Geographers have actively debated the circumstances for appropriate use of inferential testing in problem solving. The discussion centers on the difference between artificial and natural sampling. In an **artificial sample**, the investigator draws an unbiased, representative sample from a statistical population and then is able to infer certain characteristics about the population based on the sample data. As discussed in chapters 6 and 7, all representative samples contain an element of randomness, which is incorporated into the sample design procedure. No one questions the appropriateness of inferential techniques in geographic problems or situations in which a proper artificial sample has been taken.

Opinion differs on whether inferential statistics should be applied toward geographic data sets not obtained from artificial sampling. Geographers often wish to analyze spatial patterns or data sets that comprise complete enumerations or total populations. In these situations, are inferential procedures ever appropriate? Some geographers suggest that inferential

---

One Sample Difference of Proportions Test

Primary Objective: Compare a random sample proportion to a population proportion for
difference

Requirements and Assumptions:
(1) Random sample
(2) Variable measured at nominal scale and is dichotomous (binary)

Hypotheses:
H_0: $\rho = \rho_H$ (where ρ_H is the hypothesized
proportion)

H_A: $\rho \neq \rho_H$ (two-tailed)

H_A: $\rho > \rho_H$ (one-tailed) or

H_A: $\rho < \rho_H$ (one-tailed)

Test Statistic:

$$Z = \frac{p - \rho}{\sigma_p}$$

---

statistics may be permitted when using a data set considered a "natural sample."

In **natural sampling**, the "natural" or real-world processes that produce the spatial pattern under analysis contain random components. Suppose a geographer wishes to analyze a data set showing the pattern of all hurricane "land-falls" along a coastline during the last century. The landfall site of a hurricane is partly a function of prevailing global water currents and wind patterns (nonrandom or systematic influences), but is also affected by a complex variety of meteorological processes associated with that particular hurricane (random influences). It is possible to argue, then, that the observed pattern of hurricane landfall sites is actually a "natural sample" from the population of all possible land-fall locations that could have occurred.

The pattern of state-level 1980-90 population change (shown in figure 1.2) is another example of a natural sample. This choropleth map pattern is partly the result of such nonrandom or systematic factors as climate and job opportunities related to the location of natural resources

and partly the result of numerous individual and family decisions to migrate from one state to another (random influences).

If inferential procedures are applied to natural samples such as these, the results must be interpreted with extreme caution. Descriptive summaries of natural samples are appropriate, but the researcher must take care to avoid making improper inferential statements. Issues regarding the application of inferential statistics to natural samples are quite complex and continue to generate considerable discussion and controversy in applied statistics. Those interested in pursuing these arguments further (in the context of geographic research) are directed to the references at the end of this chapter.

Because of the idiosyncratic nature of geographic problems, selecting the single best statistical procedure to solve that problem is often difficult. In addition, the same geographic problem can be structured or organized in different ways, or data can be collected in various ways. Very often, determining the structure of a problem and the organization of data in that problem directs the geographer to a particular technique or set of appropriate techniques.

All inferential techniques have certain general characteristics in common. As stated earlier, the general goal of hypothesis testing is to make inferences about the magnitude of one or more population parameters, based on sample estimates of those parameters. When applying any inferential test statistic, it is assumed that a *random sample* has been drawn from a population. Often, however, other unbiased types of sampling (such as systematic sampling) are also valid for inferential hypothesis testing. If multiple samples are required for a particular problem, it is always assumed that *each sample is drawn separately and independently*. Suppose a geographer wishes to compare the average size of pebbles on two beaches. This would be done by taking a random sample of pebbles from one beach and a separate and independent random sample of pebbles from the other beach. (An exception to the assumption of independence occurs if a matched-pair or dependent sample difference test is used, examples of which will be seen in chapter 9).

The geographer needs to know more than these general characteristics to select the most appropriate inferential test for a particular geographic problem. To make the selection, the many inferential techniques must be placed in a logical organizational framework, and the important characteristics of the individual geographic problem considered. The remainder of this section focuses on the key issues that geographers must consider when determining the appropriate inferential test: type of question, category of test (parametric versus nonparametric), and level of measurement.

Type of Question

The numerous inferential tests can be classified by the type of question under investigation. At the most basic level, statistical inference is concerned with differences between a sample (or set of samples) and a hypothesized value of a population parameter (or set of parameters). However, problems are organized or structured differently according to the type of difference being examined. The example problems earlier in this chapter were concerned with the difference between a single sample and a hypothesized population parameter, referred to as one sample difference tests. Other inferential tests consider the magnitude of difference between two or more independently drawn random samples, known as two sample or k sample difference tests. Thus, the number of samples or groups distinguishes one type of geographic problem from another. In certain special cases, the data set under investigation consists of one sample of observations collected for two or more different variables or at two or more different time periods. In these circumstances, a matched-pair (dependent sample) difference test is appropriate. In other geographic problems, the data are in the form of

frequency distributions or frequency counts by category, and special purpose categorical tests are used. Data may be explicitly spatial—in the form of a point or area pattern—and specially designed tests analyze the point pattern or area pattern for randomness, clustering, or dispersal. If the strength or nature of relationship between two or more variables is a concern, then inferential tests of correlation or regression may be appropriate.

All of these examples indicate that the test selection process is somewhat detailed. However, upcoming chapters will include examples of each of the above situations. With experience, proper techniques can be selected for solving geographic problems.

Parametric and Nonparametric Tests

Statisticians have traditionally divided inferential techniques into two categories: parametric and nonparametric. Inferential tests that require knowledge about population parameters and make certain assumptions about the underlying population distribution are termed **parametric tests**. For example, in the one sample difference tests for Z and t, population parameters such as μ, σ, and ρ were included in the formulas for the test statistic. A commonly applied assumption is that the population be normally distributed with mean μ and standard deviation σ. Other assumptions that apply for some multisample parametric techniques include equal variance in each population (the so-called homoscedasticity assumption) and a linear relationship between variables.

Another group of statistical tests require no such knowledge about population parameter values and have fewer restrictive assumptions concerning the nature of the underlying population distribution. These tests are termed **nonparametric** or **distribution-free** tests. The issues regarding selection of a parametric or nonparametric test to solve a geographic problem are discussed further as specific problems are introduced in upcoming chapters.

Level of Measurement

Another closely related dimension in the test selection process is the level of measurement (nominal, ordinal, interval/ratio) at which the data have been collected and structured for the problem. Parametric tests require sample data measured on an interval/ratio scale, because their test statistics utilize parameters such as the mean or standard deviation—descriptive statistics that can only be calculated from interval/ratio data. Nonparametric tests, on the other hand, do not require such "interval/ratio" statistics. Some nonparametric tests are specifically designed to be used with ordinal data, whereas others are designed to be applied effectively to nominal or categorical data.

For data measured at a nominal or ordinal scale, *only* a nonparametric test can be applied. However, nonparametric tests can be used when interval-ratio data are converted to an ordinal or nominal scale. A set of interval/ratio data may need to be converted to an ordinal or nominal scale if one or more parametric test assumptions (normality, homoscedasticity, linearity) are moderately or severely violated. This "downgrading" strategy can be productive in conducting comparative investigative geographic analysis. In a geographic problem with interval/ratio level data, additional insights may be gained by running *both* a parametric and comparable nonparametric test on the same data and comparing the resulting *p*-values. This strategy of "pairing" a parametric and nonparametric test to solve the same geographic problem is demonstrated several times in later chapters.

The number of geographic problems that can be examined using inferential statistics is limitless. In the remainder of the text, inferential techniques are applied to a number of real-world geographic situations. Each inferential technique or set of techniques is presented using a common format: (1) The rationale, purposes, and objectives of the technique are discussed. A set of appropriate geographic problems that could be solved using the technique are listed, and an overview of the key assumptions and required conditions is provided. (2) The basic formulas and computational procedures associated with the technique are presented. (3) The geographic problem under examination is discussed, and the inferential technique applied. (4) The use of the technique for a particular geographic problem is evaluated, including discussion of both inferential issues and geographic factors that might have affected the test results.

Key Terms and Concepts

References and Additional Reading

Barber, G. M. 1988. *Elementary Statistics for Geographers.* New York: The Guilford Press.

Cliff, A. D. 1973. "A Note on Statistical Hypothesis Testing." *Area* 5(3):240.

Court, A. 1972. "All Statistical Populations Are Estimated From Samples." *Professional Geographer* 24:160-161.

Gould, P. R. 1970. "Is Statistix Inferens the Geographical Name for a Wild Goose?" *Economic Geography* (Supplement) 46:439-450.

Hays, A. 1985. "Statistical Tests in the Absence of Samples: A Comment." *Professional Geographer* 37:334-338.

Marzillier, L. F. 1990. *Elementary Statistics.* Dubuque, IA: Wm. C. Brown Publishers.

Meyer, D. R. 1972a. "Geographical Population Data: Statistical Description Not Statistical Inference." *Professional Geographer* 24:26-28.

Meyer, D. R. 1972b. "Samples and Populations: Rejoinder to 'All Statistical Populations Are Estimated From Samples'." *Professional Geographer* 24:161-162.

Ott, R. L., C. Rexroat, R. Larson, and W. Mendenhall. 1992. *Statistics: A Tool for the Social Sciences* (5th edition). Boston: PWS-Kent Publishing Co.

Silk, J. 1979. *Statistical Concepts in Geography.* London: George Allen and Unwin.

Summerfield, M. 1983. "Population, Samples, and Statistical Inference in Geography." *Professional Geographer* 35:143-149.

CHAPTER 9

MULTIPLE SAMPLE
DIFFERENCE TESTS

The previous chapter introduced the process of hypothesis testing from both the classical approach, accepting or rejecting the null hypothesis at a given significance level, and the more informative and practical *p*-value method. Both approaches were illustrated using one sample difference tests, which compare a sample statistic to a population parameter using either means or proportions. Geographers confront many other problems in their research where the objective is to determine whether significant differences exist between variables or samples. This chapter extends the concept of hypothesis testing to include additional applications of difference tests in geography.

The focus now is on several types of difference tests that are commonly applied by geographers. The chapter is organized by the relationship between the samples (independent or dependent) and the number of samples being tested (two or more than two). Section 9.1 discusses methods for testing differences between two independent samples. The choice of two sample difference tests is based primarily on the measurement scale of the data, and methods are presented to test two sample means (interval/ratio), two rank sums (ordinal), and two proportions (nominal) for significant differences.

Geographers often want to determine the differences between two sets of values defined for one group of individuals or spatial locations. In this case, the samples are considered dependent rather than independent. In section 9.2, both parametric and nonparametric versions of hypothesis tests are used to analyze so-called matched pairs or dependent sample problems.

Many geographic problems involve three or more samples. Testing for significant differences in more than two independent samples is commonly known as "analysis of variance." The methodology to test hypotheses of multiple sample problems at both the interval/ratio and ordinal scale is presented in section 9.3.

9.1 Two Sample Difference Tests

In many geographic problems, the research question concerns whether the data found in two independent samples are significantly different. If the sample differences are determined to be significant, statistical inference allows the geographer to conclude that the samples were drawn from truly different populations. However, if sample differences are not found to be significant, the logical conclusion is that the samples have been drawn from a single population rather than two distinct populations.

Three primary tests can be applied to analyze two independent samples for differences depending on the level of data measurement (interval/ratio, ordinal, nominal). When data are measured using an interval/ratio scale, the *means* for each sample can be computed and tested for a significant difference. This parametric procedure, referred to as the two sample difference of means test, is analogous to the one sample difference of means test using either the Z or *t* distribution (see section 8.3). With data in *rank* or ordinal format, a nonparametric equivalent to the difference of means test, known as the Wilcoxon or Mann-Whitney test, can be applied. For problems using data at the nominal scale, only the frequency of values for each category can be determined for both samples. When the absolute frequencies are converted to relative frequencies (sample proportions or percentages), a difference of proportions test is used to determine if the corresponding *proportions* differ significantly. In each of the three cases, the two sample tests of significant difference require two independent samples from two distinct populations.

Suppose a physical geographer is interested in studying surface runoff levels in two different environments: (1) a natural setting with no modification by human activity and (2) an urban setting, which has been highly modified through development. Because urban areas contain higher percentages of non-natural surfaces (e.g., concrete and asphalt) and lower percentages of natural open space, the amount of surface runoff should be greater in urban areas. The natural landscape in rural areas allows more moisture to soak directly into the ground.

To test for significant differences in runoff, spatial random samples could be taken from both the natural and urban settings. At each sample location, the runoff volume from precipitation could be measured over a particular period of time. If the data were measured as interval/ratio, a parametric difference of means test could be applied to see whether the mean runoff from the urban area exceeds that in the natural setting. Alternatively, a Wilcoxon difference of ranked means test could be used if runoff data were ordered from high to low within the two sampled areas. If runoff data were coded into two categories (e.g., high and low volume), a difference of proportions test could be used to see if the proportions differ significantly between the samples.

Two Sample Difference of Means Test

The most widely used two sample difference test compares the means of two independent random samples. With data measured on an interval/ratio scale, the difference of means test is applied to determine whether two sample means are significantly different. As in other inferential situations, the researcher is actually interested in extending the results to infer differences between the two populations being studied.

The null hypothesis for problems testing two sample means for significant differences has the form H_0: $\mu_1 = \mu_2$, which is equivalent to H_0: $\mu_1 - \mu_2 = 0$. The alternate hypothesis is stated as H_A: $\mu_1 \neq \mu_2$, when the direction of difference is not hypothesized (two-tailed format) or as H_A: $\mu_1 > \mu_2$ (or H_A: $\mu_1 < \mu_2$), when the

direction of difference is hypothesized (one-tailed format).

Like the one sample difference of means procedure discussed in section 8.3, the parametric two sample difference of means test has several forms depending on the size of the samples and nature of the population variances. If the populations are normally distributed with known variances and the sample sizes are large (n_1 and $n_2 > 30$), the sampling distribution for the difference of means follows the normal (Z) distribution, and the test statistic is:

$$Z = \frac{\overline{X}_1 - \overline{X}_2}{\sigma_{\overline{x}_1 - \overline{x}_2}} \qquad (9.1)$$

where: $\overline{X}_1$ = mean of sample 1

$\overline{X}_2$ = mean of sample 2

$\sigma_{\overline{x}_1 - \overline{x}_2}$ = standard error of the difference of means

and: $$\sigma_{\overline{x}_1 - \overline{x}_2} = \sqrt{\frac{\sigma_1^2}{n_1} + \frac{\sigma_2^2}{n_2}} \qquad (9.2)$$

Following the general difference test format, the numerator of the two sample test statistic in equation 9.1 denotes the difference between the two sample statistics, in this case the sample means. Recall from chapter 7 that $\overline{X}$ is the best estimate of μ, the population mean. Similarly, in this situation involving two samples, $(\overline{X}_1 - \overline{X}_2)$ is the best estimate of the difference of population means $(\mu_1 - \mu_2)$. The denominator of equation 9.1 represents the standard error of the difference in means, a measure of variability in the two populations (equation 9.2).

In most geographic problems, however, variances for the two populations are not known, which suggests that the test statistic for Z in equation 9.1 is often not appropriate to test for differences in means. In situations where population variances are unknown, the standard error must be estimated from sample variances (s^2). Here, the difference of means test requires the t distribution, and the test statistic is:

$$t = \frac{\overline{X}_1 - \overline{X}_2}{\sigma_{\overline{x}_1 - \overline{x}_2}} \qquad (9.3)$$

In addition to its usefulness for problems with unknown population variances, the t distribution also provides more valid results than Z when sample sizes are less than 30.

Two methods exist for using sample data to estimate the standard error in the denominator of equation 9.3. Selection of the appropriate method depends on the assumed relationship between the two population variances:

1. Pooled Variance Estimate if ($\sigma_1^2 = \sigma_2^2$)
2. Separate Variance Estimate if ($\sigma_1^2 \neq \sigma_2^2$)

When the population variances are assumed to be *equal*, a **pooled variance estimate (PVE)** is calculated as the weighted average of the two sample variances:

$$PVE = \sqrt{\frac{s_1^2 (n_1 - 1) + s_2^2 (n_2 - 1)}{n_1 + n_2 - 2}} \qquad (9.4)$$

The denominator of equation 9.4 represents the degrees of freedom in the problem. When the two population variances are unknown but assumed equal, the standard error estimate for equation 9.2 can be written:

$$\sigma_{\overline{x}_1 - \overline{x}_2} = PVE \sqrt{\frac{1}{n_1} + \frac{1}{n_2}} \qquad (9.5)$$

In problems where the population variances are assumed to be *unequal*, the pooled estimate is not appropriate. In such cases, the sample variances are substituted directly into the standard error portion of equation 9.2 as best estimates of the respective population variances. A

Two Sample Difference of Means Test

Primary Objective: Compare two independent random sample means for differences

Requirements and Assumptions:
1. Two independent random samples
2. Each population normally distributed
3. Variables measured at interval or ratio scale

Hypotheses:
H_o: $\mu_1 = \mu_2$
H_A: $\mu_1 \neq \mu_2$ (two-tailed)
H_A: $\mu_2 > \mu_2$ (one-tailed) or
H_A: $\mu_1 < \mu_2$ (one-tailed)

Test Statistics:
If sample sizes are greater than 30 and population variances are known:

$$Z = \frac{\overline{X}_1 - \overline{X}_2}{\sigma_{\overline{x}_1 - \overline{x}_2}}$$

If sample sizes are less than 30, or population variances are unknown:

$$t = \frac{\overline{X}_1 - \overline{X}_2}{\sigma_{\overline{x}_1 - \overline{x}_2}}$$

---

separate variance estimate (SVE) is then calculated:

$$SVE = \sigma_{\overline{x}_1 - \overline{x}_2} = \sqrt{\frac{s_1^2}{n_1} + \frac{s_2^2}{n_2}} \qquad (9.6)$$

Since geographers rarely know population parameters (like μ and σ), they must rely heavily on sample data to estimate population characteristics. This task supports one of the central goals of inferential statistics discussed earlier— to use samples to produce unbiased estimates of population parameters. The previous discussion stated that the method chosen to estimate variance—pooled vs. separate—depends on whether the population variances are *assumed* to be equal.

In practice, researchers usually decide equality or inequality of population variances by testing the corresponding sample data. Sample variances are considered the best estimates of population variances.

To determine whether the variances of the two samples are significantly different, an analysis of variance (ANOVA) or F-test (see section 9.3) can be applied. If the sample variances are found to be different, the population variances are assumed to be unequal, and the separate variance method should be used to test for difference in the means. However, if the sample variances are not significantly different, the pooled variance estimate should be applied.

Wilcoxon Rank Sum W Test

For some two sample problems, data are available only in ordinal or ranked form and, as a result, sample means and variances cannot be calculated. In these cases, the Wilcoxon rank sum W test, a nonparametric alternative to the difference of means test, is appropriate. In other situations, the researcher may have interval/ratio data, but cannot prove the assumption of normality of population distributions, raising validity questions about the use of a parametric difference of means test. In these instances, the Wilcoxon W test is again useful for analyzing differences between two samples and making inferences to the corresponding populations.

A slightly different form of the Wilcoxon procedure, known as the Mann-Whitney U test, is also available for researchers and is popular for geographic research. The focus here is on the Wilcoxon rank sum W test because of its ease of computation and application to the p-value approach.

The Wilcoxon rank sum W test uses either ranked data directly or interval/ratio data downgraded to the ordinal equivalent. Like the difference of means t test, which looks for significant differences between sample means, the Wilcoxon procedure tests for significant differ-

ences in ranked positions between two samples. Although no distributional form is assumed for the two populations, the procedure requires that the two distributions be similar in shape. This characteristic makes the Wilcoxon rank sum W test especially useful for problems with small samples drawn from populations that are not necessarily normal.

The objective of the two sample Wilcoxon procedure is similar to that for the parametric two sample difference of means test: compare differences in two samples drawn independently from two separate populations. The null and alternate hypotheses for the two inferential tests, however, usually are stated differently. In parametric applications like the difference of means, the hypotheses assume that a parameter (such as μ) takes on a certain value in one or more populations (such as $\mu_1 = \mu_2$). Sample data are used to calculate a test statistic and to evaluate the likelihood of the null and alternate hypotheses.

Unlike parametric procedures for testing hypotheses, nonparametric methods generally do not involve sample or population parameters. Researchers can choose among several different formats for defining hypotheses in problems using ordinal data. In some cases, hypotheses are phrased in general statements about the population(s) without referring to parameters. Using this format with the Wilcoxon procedure, the null hypothesis assumes that identical distributions exist for the two populations being studied. As a second alternative, some researchers identify the median of a distribution as the relevant parameter for describing a distribution of ranks and for evaluating similarity or difference of ranks. For this alternative procedure, the null hypothesis for the two sample Wilcoxon test assumes the population median of the first distribution equals that for the second. Other researchers base their statement of hypotheses on a common method used to compare the ranks of two sample distributions: the mean rank. Using this format, the null hypothesis for the Wilcoxon procedure assumes the mean rank of the first population equals that for the second. Each of these strategies is used in formulating hypotheses for geographic problem solving. In the boxed summary for the Wilcoxon test, the hypotheses are general written statements about the population(s), without reference to any parameter.

The Wilcoxon rank sum W procedure combines the data from two samples and places them in a single ranked order. When two or more values are tied for a particular rank, the average rank value is assigned to each tied position. The samples are then considered separately, and the sum of ranks (W) is determined for each sample. The sum of ranks for sample 1 is called W_1, and the sum of ranks for the second sample, W_2. If the two samples are of equal size, W_1 and W_2 should be identical when no differences exist in the samples. This would confirm the null hypothesis. However, as the sample differences become more extreme, the two rank sums should also show greater disparity, making the null hypothesis less likely. The Wilcoxon rank sum W test uses a variation of the Z test to see if the sum of sample ranks is significantly different from what it should be if the two samples are actually drawn from the same population. With this procedure, sample sizes do not need to be the same.

The test statistic (Z_W) for the two sample Wilcoxon procedure is:

$$Z_W = \frac{W_i - \overline{W_i}}{s_W} \qquad (9.7)$$

where:

W_i = sum of ranks of sample i

$\overline{W_i}$ = mean of W_i

$$= n_i \left(\frac{n_1 + n_2 + 1}{2} \right) \qquad (9.8)$$

s_W = standard deviation of W

$$= \sqrt{n_1 n_2 \left(\frac{n_1 + n_2 + 1}{12} \right)} \qquad (9.9)$$

Wilcoxon Rank Sum W Test

Primary Objective: Compare two independent random sample rank sums for differences

Requirements and Assumptions:
1. Two independent random samples
2. Both population distributions have the same shape
3. Variables measured at the ordinal scale or interval/ratio scale variables downgraded to ordinal

Hypotheses:

H_o: The distribution of measurements for the first population is equal to that of the second population

H_A: The distribution of measurements for the first population is not equal to that of the second population (two-tailed)

H_A: The distribution of measurements for the first population is larger (or smaller) than that for the second population (one-tailed)

Test Statistic:

$$Z_W = \frac{W_i - \overline{W_i}}{s_W}$$

$\overline{W_i}$ and s_W represent the theoretical mean and standard deviation of W_i, respectively, and are determined totally by the sample sizes. As shown in equation 9.9, only one standard deviation exists for W. However, because rank sums (W_1 and W_2) can be determined for each sample, two means can also be calculated using equation 9.8—one for sample 1 ($\overline{W_1}$) and another for sample 2 ($\overline{W_2}$).

In the Wilcoxon rank sum procedure, two complementary test statistics (Z_{W_1} and Z_{W_2}) can be calculated. Z_{W_1} uses the rank sum and the mean from the first sample, while Z_{W_2} uses the corresponding values for the second sample. The two test statistics differ only in sign: one will be negative and the other positive.

Application to Geographic Problem

An examination of global demographic and economic development patterns reveals that significant differences exist between countries and that certain regional generalizations exist. Differences in development appear most pronounced when less developed counties (LDCs) are contrasted with more developed countries (MDCs). International differences can be noted for such important variables as birth rate, gross national product, percentage of elderly population, and degree of urbanization. Moreover, the degree of disparity in some of these indices should change over time as the process of development continues.

Several geographic questions can be proposed. For which key demographic and economic indicators is the disparity between LDC and MDC countries most pronounced? For which indicators does it appear that the gap between LDCs and MDCs is widening (or narrowing)? What trends in the disparity of these indices occur as the development process continues?

To answer these questions, information is analyzed for selected variables from both the 1980 and 1990 *World Population Data Sheet* published by the Population Reference Bureau. The following 11 variables are tested for significant differences between LDCs and MDCs:

- birth rate,
- death rate,
- annual rate of natural increase,
- infant mortality rate,
- total fertility rate,
- percentage of population under age 15,
- percentage of population over age 64,
- life expectancy at birth,
- percentage urban population,
- per capita gross national product, and
- percentage of married women using contraception.

According to the United Nations classification of countries, the more developed group consists of all nations in Europe and the individual countries of Canada, United States, Australia, New Zealand, Japan, and the former USSR. All other nations are placed into the LDC category. From the 142 LDCs represented in the *1990 World Population Data Sheet,* 28 nations are selected for analysis through a simple random sample. Of the 32 nations on the *Data Sheet* classified as MDCs, a second simple random sample of 7 countries is generated. Because the Population Reference Bureau did not include data on contraception in the 1980 *Data Sheet,* this variable is not analyzed for 1980. Also, because 3 of the countries appearing in the list of sampled LDCs for 1990 (French Polynesia, New Caledonia, and Antigua and Barbuda) did not appear in the *Data Sheet* for 1980, the sample size in 1980 is reduced from 28 to 25. Missing data for some countries reduces the sample size slightly for some of the variables. Table 9.1 shows the nations selected in the two independent samples, the original birth rate data in ratio form, and the equivalent ordinal ranks.

Two statistical methods are applied to look for significant differences between the two groups of countries for the variables selected. Since the original data are measured on an interval/ratio scale, sample means for each variable can be calculated for the countries in the LDC and MDC groups and tested for significant differences using a difference of means test. The populations from which each sample is selected are assumed to be normally distributed. (The method of testing sample data for normality is discussed in section 10.1.)

Because the population variances are unknown, the test statistic for t, rather than Z, is used. The sample variances are first checked for equality, and either the pooled variance or separate variance estimate is used to calculate the standard error. The calculations for testing birth rate differences between the LDC and MDC samples in 1990 are presented in table 9.2. The resulting mean birth rates (34.36 for LDC and 12.43 for MDC) are shown to be significantly different with $t = 9.63$ and $p = .000$.

The interval/ratio scale data are downgraded to ordinal ranks, and the Wilcoxon rank sum W test is used to look for significant differences between

Table 9.1 Number of Births per 1,000 Population and Rank of Values for Less Developed (LDC) and More Developed (MDC) Country Samples, 1990

	Sample	Birth Rate (X)	Rank
Antigua and Barbuda	LDC	15	7.5
Austria	MDC	12	3.5
Bangladesh	LDC	39	23
Burundi	LDC	48	33
Cameroon	LDC	42	26
Chile	LDC	22	13
China	LDC	21	12
Comoros	LDC	47	32
Cuba	LDC	18	10
Dominica	LDC	26	16
Fiji	LDC	27	17
France	MDC	14	6
French Polynesia	LDC	31	20
Grenada	LDC	37	22
Guadeloupe	LDC	20	11
Hungary	MDC	12	3.5
Iraq	LDC	46	29.5
Italy	MDC	10	1
Luxembourg	MDC	12	3.5
Madagascar	LDC	46	29.5
Malawi	LDC	52	35
Morocco	LDC	35	21
Nepal	LDC	42	26
New Caledonia	LDC	25	15
Reunion	LDC	24	14
Saint Lucia	LDC	28	18
Saudi Arabia	LDC	42	26
Senegal	LDC	46	29.5
Solomon Islands	LDC	41	24
South Korea	LDC	16	9
Swaziland	LDC	46	29.5
Switzerland	MDC	12	3.5
Tanzania	LDC	51	34
Turkey	LDC	29	19
Yugoslavia	MDC	15	7.5

			Sample Statistics			
Sample	n	$\overline{X}$	s	F^*	p-value	W
1 (LDC)	28	34.36	11.61			601.5
				51.5	0.000	
2 (MDC)	7	12.43	1.62			28.5

*Test for significant differences in sample variances

Source: Population Reference Bureau. 1990. *World Population Data Sheet, 1990.*

the samples. Calculations for the ordinal birth rate variable (1990) are shown in table 9.3. With highly differing ranked sums for LDC ($W_1 = 601.5$) and for MDC ($W_2 = 28.5$), the differences between the two samples are significant at the $p = 0.000$ level ($Z = 4.02$). Note that the test statistics (Z_{W_1} and Z_{W_2}) based on calculations for the two samples differ only in sign (table 9.3). This result suggests that the sum of ranks for the LDC group is too large (or the sum of ranks for the MDC group too small) for the two samples to have been drawn from one population. Therefore, differences in birth rate

Table 9.2 Work Table for Difference of Means t Test: Birth Rates, 1990

$$H_0: \quad \mu_1 = \mu_2$$
$$H_A: \quad \mu_1 \neq \mu_2$$

$$t = \frac{\overline{X}_1 - \overline{X}_2}{\sigma_{\overline{x}_1 - \overline{x}_2}}$$

$$\sigma_{\overline{x}_1 - \overline{x}_2} = \sqrt{\frac{\sigma_1^2}{n_1} + \frac{\sigma_2^2}{n_2}}$$

(A) When $\sigma_1 = \sigma_2$: $\sigma_{\overline{x}_1 - \overline{x}_2} = \text{PVE}\sqrt{\frac{1}{n_1} + \frac{1}{n_2}}$

$$\text{PVE} = \sqrt{\frac{s_1^2(n_1 - 1) + s_2^2(n_2 - 1)}{n_1 + n_2 - 2}}$$

(B) When $\sigma_1 \neq \sigma_2$: $\sigma_{\overline{x}_1 - \overline{x}_2} = \sqrt{\frac{s_1^2}{n_1} + \frac{s_2^2}{n_2}}$

Since s_1^2 is significantly different from s_2^2 ($p = .0000$), the separate variance estimate is used for $\sigma_{\overline{x}_1 - \overline{x}_2}$ (see table 9.1)

$$\sigma_{\overline{x}_1 - \overline{x}_2} = \sqrt{\frac{s_1^2}{n_1} + \frac{s_2^2}{n_2}} = \sqrt{\frac{(11.61)^2}{28} + \frac{(1.62)^2}{7}} =$$

$$\sqrt{\frac{134.79}{28} + \frac{2.62}{7}} = \sqrt{4.814 + 0.375} = 2.278$$

Therefore: $t = \dfrac{34.36 - 12.43}{2.278} = \dfrac{21.93}{2.278} = 9.63$

p-value $= 0.000$

Table 9.3 Work Table for Wilcoxon Rank Sum W Test: Birth Rates, 1990

$$Z_{W_i} = \frac{W_i - \overline{W}_i}{s_W}$$

where: W_i = sum of ranks for sample i

$$\overline{W}_i = n_i\left(\frac{n_1 + n_2 + 1}{2}\right)$$

$$s_W = \sqrt{n_1 n_2 \left(\frac{n_1 + n_2 + 1}{12}\right)}$$

From table 9.1: $W_1 = 601.5$

$W_2 = 28.5$

$$\overline{W}_1 = 28\left(\frac{28 + 7 + 1}{2}\right) = 504$$

$$\overline{W}_2 = 7\left(\frac{28 + 7 + 1}{2}\right) = 126$$

$$s_W = \sqrt{(28)(7)\left(\frac{28 + 7 + 1}{12}\right)} = 24.25$$

$$Z_{W_1} = \frac{601.5 - 504}{24.25} = 4.02$$

$$Z_{W_2} = \frac{28.5 - 126}{24.25} = -4.02$$

p-value $= 0.000$

between the two groups of countries are highly significant.

The results from significance tests for differences between LDCs and MDCs for the selected variables in 1990 appear in table 9.4. The sample with the larger mean is identified, and the results of testing the sample variances for equality are shown as either a pooled estimate (no significant differences) or separate estimate (significant differences). In all variables except percentage of population aged 64 and over, sample variances differ significantly ($p < .05$) between the LDC and MDC groups. After applying the appropriate difference of means t test, all variables except death rate show significant differences between the two samples at p-values less than or equal to 0.01. Differences in death rates between the two groups are not statistically significant, as indicated by the low value of t (-0.24) and

the corresponding high p-value ($p = .813$). Less than a 19 percent chance exists that the differences in death rates between the LDCs and MDCs are significant. To conclude that LDC death rates are different from MDC death rates, an 81.3 percent chance exists that a Type I error has been made.

Results from the nonparametric Wilcoxon rank sum W analysis correspond closely to those from the difference of means t test for 1990 (table 9.4). Again, death rate is the only variable that shows an unacceptably high level of significance and weak p-value ($p = 0.319$). With the data in ordinal form, differences in percentage urban population between the LDCs and MDCs are significant only at the weaker $p = 0.061$ level. These results verify the general similarity between the two methods for testing significant differences in two-sample problems.

Table 9.4   Difference of Means t and Wilcoxon Ranked Sum Results for Two Sample Difference Tests Between Less Developed (LDC) and More Developed (MDC) Countries, 1990

	Larger Mean	s^{2*}	t	p-value (t)	W	Z	p-value (Z)
Birth Rate	LDC	S	9.63	0.000	28.5	-4.03	0.000
Death Rate	MDC	S	-0.24	0.813	150.0	-1.00	0.319
Natural Increase	LDC	S	12.16	0.000	28.0	-4.05	0.000
Infant Mortality Rate	LDC	S	6.04	0.000	40.0	-3.55	0.000
Total Fertility Rate	LDC	S	7.60	0.000	35.5	-3.73	0.000
% Population < 15	LDC	S	12.04	0.000	28.0	-4.05	0.000
% Population > 64	MDC	P	-12.01	0.000	224.0	-4.13	0.000
Life Expectancy	MDC	S	-6.02	0.000	207.5	-3.38	0.001
% Urban Population	MDC	S	-2.81	0.010	161.5	-1.87	0.061
Per capita GNP	MDC	S	-3.71	0.010	176.0	-3.62	0.000
% Contraception Use	MDC	S	-6.32	0.000	124.0	-3.27	0.001

Sample sizes   % Urban Population: LDC = 26; MDC = 7   % Contraception Use: LDC = 18; MDC = 6
                Per capita GNP: LDC = 22; MDC = 7   All other variables: LDC = 28; MDC = 7

See table 9.1 for a list of countries in each sample

*method to estimate variance: S = Separate (Prob of $F < .05$)
                                P = Pooled   (Prob of $F > .05$)

Results using the 1980 data closely match those from 1990 for most variables (table 9.5). Birth rate, natural increase, infant mortality rate, total fertility rate, percentages of population under 5 and over 64, life expectancy, percent urban, and per capita GNP all produce differences between the LDC and MDC groups that are significant at $p < 0.01$ for both inferential procedures. Again, only the death rate showed weak levels of significance between the two samples of countries in both tests.

Summary Evaluation

Using a two sample difference of means t test and Wilcoxon rank sum W test with the same set of data allows the two techniques to be compared directly. Results for all variables during the two time periods are generally consistent. With the exception of the death rate variable, p-values from the t tests closely match the corresponding p-values for the Wilcoxon Z test.

The results of the analysis, however, may be influenced by the operational definitions of "less developed country" and "more developed country." Because the United Nations sometimes defines these groups on a broad regional basis, several countries may be placed in the improper development category. For example, all European countries, including former Eastern Bloc countries like Hungary and Albania, are considered "more developed," whereas some countries thought to be more advanced (such as Israel, South Africa, South Korea, and Argentina) are listed as "less developed." Such operational definitions may create errors in validity.

Table 9.5   Difference of Means t and Wilcoxon Ranked Sum Results for Two Sample Difference Tests Between Less Developed (LDC) and More Developed (MDC) Countries, 1980

	Larger Mean	s^{2*}	t	p-value (t)	W	Z	p-value (Z)
Birth Rate	LDC	S	9.12	0.000	28.5	− 3.97	0.000
Death Rate	LDC	S	1.49	0.146	105.5	− 0.46	0.647
Natural Increase	LDC	P	7.81	0.000	28.0	− 3.99	0.000
Infant Mortality Rate	LDC	S	6.92	0.000	32.5	− 3.78	0.000
Total Fertility Rate	LDC	S	5.09	0.000	49.0	− 3.03	0.002
% Population < 15	LDC	S	14.99	0.000	28.0	− 3.96	0.000
% Population > 64	MDC	P	−15.00	0.000	189.0	− 4.02	0.000
Life Expectancy	MDC	S	− 7.50	0.000	195.0	− 3.63	0.000
% Urban Population	MDC	P	− 3.33	0.002	180.0	− 2.94	0.003
Per capita GNP	MDC	S	− 3.94	0.006	197.0	− 3.72	0.000

Sample sizes   % Population < 15:   LDC = 23; MDC = 7
                  % Population > 64:   LDC = 23; MDC = 7
                  All other variables: LDC = 25; MDC = 7

See table 9.1 for a list of countries in each sample; no data are available for French Polynesia, New Caledonia, and Antigua and Barbuda as of 1980, and they are omitted from the analysis.

Data on contraception use was not available for 1980.

*method to estimate variance: S = Separate (Prob of F < .05)
                                P = Pooled   (Prob of F > .05)

Two Sample Difference of Proportions Test

Geographers often work with categorical data measured on a nominal scale, where interval values or ordinal positions are unknown. For example, questionnaires often produce data that are defined for two or more categories (e.g., sex [male or female] or type of location [urban or rural]). When data are measured nominally, only the frequency or proportion of responses in each category can be calculated. Because such sample statistics as the mean and standard deviation cannot be determined for nominal data, the difference of means test is not appropriate. The Wilcoxon rank sum W test cannot be applied either because nominal data cannot be placed in ordinal or rank format. However, the difference of proportions test provides an alternative procedure to test for significant differences between two independent samples when the data are dichotomous (binary).

For many problems, the variable being analyzed has exactly two categories or classes, such as male and female. In cases where a nominal variable contains three or more classes, it can often be reduced to two categories so that the difference of proportions test can be used. For example, a variable measuring marital status might include the following classes: married, single, divorced, and widowed. A dichotomous variable can be generated by reassigning data into two categories: married and unmarried. The difference of proportions test could then examine whether significant differences exist between the proportion of married persons in the two samples being considered.

In problems testing for significant differences with a dichotomous variable, one of the two variable categories is selected as the focus for the analysis. In the example from the previous paragraph, the focus category was married persons. The proportion of the sample in the focus category is termed p, a decimal value between 0 and 1. The proportion of the sample that is not in this category is termed q, where $q = 1 - p$. The objective of the difference test is to determine whether the proportion of *population 1* (ρ_1) having the focus attribute differs significantly from the corresponding proportion of *population 2* (ρ_2). As in other inferential tests, data from samples are used to make inferences about the populations.

The null hypothesis for the problem is $H_0: \rho_1 = \rho_2$. As in other difference tests, the alternate hypothesis (H_A) can take two forms. If no direction is hypothesized for difference in proportion between the two samples, a two-tailed procedure is applied where $H_A: \rho_1 \neq \rho_2$. On the other hand, if one sample is expected to exceed the other in proportion, a one-tailed test is appropriate. In these instances, the alternate hypothesis is: $H_A: \rho_1 > \rho_2$ or $H_A: \rho_1 < \rho_2$.

The test statistic for the difference of proportions procedure (Z_P) has similar form to other two sample difference tests:

$$Z_P = \frac{p_1 - p_2}{\sigma_{p_1 - p_2}} \qquad (9.10)$$

where: p_1 = proportion of sample 1 in the category of focus

p_2 = proportion of sample 2 in the category of focus

$\sigma_{p_1 - p_2}$ = standard error of the difference of proportions

Just as the sample proportion p is the best estimate of the population proportion, ρ, ($p_1 - p_2$) is the best estimate of the difference in population proportions ($\rho_1 - \rho_2$). The denominator of equation 9.10 represents the standard error of the difference in proportions and can be estimated as follows:

$$\sigma_{p_1 - p_2} = \sqrt{\hat{p}(1 - \hat{p})\left(\frac{n_1 + n_2}{n_1 n_2}\right)} \qquad (9.11)$$

where: $\hat{p}$ = pooled estimate of the focus category for the population

Two Sample Difference of Proportions Test

Primary Objective: Compare two independent random sample proportions for differences

Requirements and Assumptions:
1. Two independent random samples
2. Variables measured at nominal scale and are dichotomous (binary)

Hypotheses:
H_o: $p_1 = p_2$
H_A: $p_1 \neq p_2$ (two-tailed)
H_A: $p_2 > p_2$ (one-tailed) or
H_A: $p_1 < p_2$ (one-tailed)

Test Statistic:

$$Z_p = \frac{p_1 - p_2}{\sigma_{p_1 - p_2}}$$

The pooled estimate, $\hat{p}$, is the proportion in the focus category if the two samples were combined into one sample. Operationally, the pooled estimate is the weighted proportion from the two samples:

$$\hat{p} = \sqrt{\frac{n_1 p_1 + n_2 p_2}{n_1 + n_2}} \qquad (9.12)$$

Application to Geographic Problem

Political geographers often want to determine whether regional differences in voting behavior occur within a country or state. To illustrate the difference of proportions procedure, voting results from the 1990 congressional elections are tested for significant variation between geographic regions of the United States. Random samples of voters are examined to see whether regional differences in the percentage choosing a Democratic member of the House of Representatives are significant. The variable is nominal with three categories (Democrat, Republican, Independent), and separate random samples are taken from distinct regional populations, making

the difference of proportions test appropriate for this problem. To create a dichotomous variable, voting results from exit polls are tabulated as Democrat (the focus variable) and non-Democrat. Samples consist of randomly selected, registered voters in the census regions of the East, South, Midwest, and West.

The voting results are summarized by census region in table 9.6. The highest Democratic vote occurred in the South, where 53 percent of the sampled voters chose a Democrat, while the lowest percentage vote for Democrats occurred in the Midwest (44 percent). Given this regional variation in voting trends, do the proportions of voters choosing Democratic representatives differ significantly from one region to another?

To answer this question, a separate difference of proportions test is applied to each pair of census regions. The calculation procedure for the difference of proportions test is illustrated in table 9.7 using the two regional samples with the greatest disparity in Democratic vote (Midwest and South). The Democratic vote for Representatives in these regions differs by .09 (9 percent). The pooled estimate ($\hat{p}$) is determined from the sample sizes (n_1 and n_2) and the proportion of Democratic votes in the two samples. Since the number of voters sampled in the Midwest and South is equal, the pooled estimate is simply the average of the two regional proportions (0.485). The resulting difference of proportions value

Table 9.6 Voting Behavior by Census Region for House of Representatives, November, 1990

Sample	Region	n	% Democrat (p)	% Non-Democrat ($1 - p$)
1	East	2267	50	50
2	Midwest	2550	44	56
3	South	2550	53	47
4	West	2078	46	54

Source: Modified from data published in *Washington Post*, November 8, 1990.

$(Z_P = 6.429)$ and its corresponding p-value (0.0000) suggest that voters in these two geographic regions differed significantly in their voting behavior during Congressional elections in November, 1990.

Table 9.8 presents the difference of proportion test results of differences in Democratic vote between all regional pairs. Highly significant differences occurred between most regional pairs. The least significant difference was found between the Midwest and West ($Z_P = 1.429; p = 0.1530$). Analysis of this voting data suggests that regional differences are significant among major geographic divisions of the nation. Note that low p-values, indicating strong statistical significance, emerge even with rather small differences in proportions, because of the large sample sizes.

Summary Evaluation

Since voting preference is a nominal variable, the difference of proportions test seems appropriate to test for regional differences in percentage Democratic vote for Representatives. The assumptions and requirements of the test appear to be met. However, political party distinctions often are blurred as voters become concerned more with the personality of the politician than the party he or she represents. Because of this validity problem, the results may not allow extension to conclusions regarding more meaningful political differences such as liberal versus conservative attitudes among the voters.

Table 9.7 Work Table for Difference of Proportions Test: Democratic Vote in the South and Midwest

$$H_0: \quad \rho_1 = \rho_2$$
$$H_A: \quad \rho_1 \neq \rho_2$$

where: ρ_1 = proportion Democratic vote in the South
ρ_2 = proportion Democratic vote in the Midwest

Step 1: Calculate the pooled estimate ($\hat{p}$) of p

$$\hat{p} = \sqrt{\frac{n_1 p_1 + n_2 p_2}{n_1 + n_2}} = \sqrt{\frac{2550(.53) + 2550(.44)}{2550 + 2550}} =$$

$$\sqrt{\frac{2473.5}{5100}} = 0.485$$

Step 2: Calculate the standard error of the estimate of the difference of proportions

$$\sigma_{p_1-p_2} = \sqrt{\hat{p}(1-\hat{p})\left(\frac{n_1 + n_2}{n_1 n_2}\right)} =$$

$$\sqrt{.485(.515)\left(\frac{2550 + 2550}{2550(2550)}\right)} =$$

$$\sqrt{.250}\sqrt{.00078} = .014$$

Step 3: Calculate the difference of proportions value

$$Z_p = \frac{p_1 - p_2}{\sigma_{p_1-p_2}} = \frac{.53 - .44}{0.014} = 6.429$$

p-value = 0.0000

9.2 Matched Pairs (Dependent Sample) Difference Tests

Geographers often want to determine if two sets of values defined for one group of individuals or spatial locations differ. For example,

Table 9.8 Difference of Proportion Tests for Democratic Vote by Region for U.S. Representatives, 1990

Regional Pair	Z_p	p-value
Midwest and South	6.429	0.0000
South and West	4.667	0.0000
East and Midwest	4.286	0.0000
East and West	4.000	0.0000
East and South	2.000	0.0456
Midwest and West	1.429	0.1530

suppose a researcher wishes to compare the number of migrants moving into a set of counties with the number of migrants leaving the same set of counties. The number of in-migrants and the number of out-migrants could be determined for a sample of geographic areas and the differences tested for statistical significance. In another example, residents of an area could be sampled to determine the miles traveled per week for noncommuting purposes. Following a major hike in the price of gasoline, as happened during the Persian Gulf conflict of the summer and fall of 1990, the same individuals could be surveyed a second time to determine changes in their discretionary travel behavior. The researcher may want to determine whether the change in gasoline price led to decreased automobile use for drivers within a particular area.

In both examples, the data under investigation consist of one set of observations (locations or individuals) collected for two different variables or at two different time periods. At first, a difference of means or Wilcoxon W may seem most appropriate to test differences for significance. However, recall from section 9.1 that these two difference tests require *two* **independent samples.** In the situations proposed above, only *one* sample is drawn, and data are collected for two variables (in- and out-migrants) or for one variable at two time periods (miles traveled before and after a price increase for gasoline). Thus, neither the two sample *t* test nor Wilcoxon rank sum W procedure is appropriate for these problems.

When two sets of data are collected for one group of observations, the samples are termed **dependent**, and a **test of matched pairs** is considered the proper inferential procedure to examine differences between the samples. As the name implies, each observation or sample member has two values, termed a "matched pair." The differences in the set of matched pairs are tested for statistical significance and results inferred to the population from which the dependent samples were drawn.

Two common inferential procedures for testing differences in dependent samples are the matched pairs *t* test and the Wilcoxon matched pairs signed-ranks test. The matched pairs *t* test is a parametric test requiring interval/ratio level data and a normally distributed population. Like the Wilcoxon two sample difference test, the related matched pairs signed-ranks test is a distribution-free, nonparametric method that uses either ranked data directly or interval/ratio data downgraded to its ordinal equivalent.

Matched Pairs *t* Test

Rather than comparing the means of two sets of data (as is done in the two sample difference of means test), the matched pairs *t* test considers the difference between the values for each matched pair. The greater this difference (d), the more dissimilar are the results of the two values within the matched pair. The mean of the difference values ($\bar{d}$) is determined from the set of all matched pairs in the sample. If the differences within the matched pair values are small, this average difference value will be close to zero. However, given a problem in which the matched pair differences are large, $\bar{d}$ will also be large, suggesting significant differences between the two sets of data being studied.

The null hypothesis in the matched pair problem states that the mean difference for all matched pairs in the population (δ) equals zero. The best estimate of the population matched pair mean difference (δ) is the sample matched pair mean difference ($\bar{d}$). The alternative hypothesis can be either nondirectional or directional, depending on *a priori* information about the expected matched pair differences.

The test statistic for the matched pairs *t* test is defined as follows:

$$t_{mp} = \frac{\bar{d}}{\sigma_d} \qquad (9.13)$$

where: t_{mp} = matched pairs *t*

$\bar{d}$ = mean of matched pair differences (d)

σ_d = standard error of the mean difference

Matched Pairs _t_ Test

Primary Objective: Compare matched pairs from a random sample for difference

Requirements and Assumptions:
1. Random sample
2. Data collected for two different variables or at two different time periods
3. Population is normally distributed
4. Variables measured at interval or ratio scale

Hypotheses:
H_0: $\delta = 0$
H_A: $\delta \neq 0$ (two-tailed)
H_A: $\delta > 0$ (one-tailed) or
H_A: $\delta < 0$ (one-tailed)

Test Statistic:

$$t_{mp} = \frac{\overline{d}}{\sigma_d}$$

The numerator of equation 9.13 is the mean of matched pair differences (_d_):

$$\overline{d} = \frac{\Sigma d_i}{n} \qquad (9.14)$$

where: d_i = difference for matched pair i

n = number of matched pairs

The difference (d_i) is found by subtracting corresponding values of the second variable (Y) from those of the first variable (X):

$$d_i = X_i - Y_i \qquad (9.15)$$

The mean of the difference values ($\overline{d}$) can also be calculated from the means of the two variables:

$$\overline{d} = \overline{X} - \overline{Y} \qquad (9.16)$$

Like other difference tests, the denominator of equation 9.13 contains a standard error mea-

sure, in this case, the standard error of mean difference in matched pairs:

$$\sigma_d = \frac{s_d}{\sqrt{n}} \qquad (9.17)$$

where: $$s_d = \sqrt{\frac{\Sigma(d_i - \overline{d})}{n-1}} \qquad (9.18)$$

Standard deviation of the matched pair differences can also be derived from the computational formula (see table 3.3):

$$s_d = \sqrt{\frac{\Sigma d_i^{\,2} - \dfrac{(\Sigma d_i)^2}{n}}{n-1}} \qquad (9.19)$$

Wilcoxon Matched Pairs Signed-Ranks Test

In some geographic problems a dependent sample matched pair test may be appropriate, but the sample data for analysis are measured at the ordinal level. In other situations, the sample data may not be drawn from a normally distributed population. In the first instance, the parametric matched pair _t_ test cannot be applied because the measurement scale is not appropriate. In the second case, the parametric test may produce biased results. The Wilcoxon signed-ranks test is the nonparametric equivalent for dependent sample or matched pair problems and is the appropriate procedure in these situations.

The Wilcoxon signed-ranks test uses matched pair differences ranked from lowest (rank 1) to highest. The matched pair data can come either from direct ordinal measurement or from interval/ratio differences downgraded to ranks. The _absolute_ difference between the two variables (rather than the positive or negative difference) is used to determine the rank for each matched pair. When the difference for any matched pair is zero, the data are ignored and the sample size reduced accord-

---

Wilcoxon Matched Pairs Signed-Ranks Test

Primary Objective: Compare matched pairs from a random sample for difference

Requirements and Assumptions:
1. Random sample
2. Data collected for two different variables or at two different time periods
3. Variables measured at ordinal scale or interval/ratio scale variables downgraded to ordinal

Hypotheses:

H_o: The ranked matched pair differences of the populations are equal

H_A: The ranked matched pair differences of the populations are not equal (two-tailed)

H_A: The ranked matched pair differences of the populations are positive or negative (one-tailed)

Test Statistic:

(if $n > 10$)

$$Z_w = \frac{T - \frac{n(n+1)}{4}}{\sqrt{\frac{n(n+1)(2n+1)}{24}}}$$

---

sums (either the positive or negative differences) will be large and the other small.

The Wilcoxon test for dependent samples uses only one of the two possible rank sum values. The decision of which rank sum to test depends on whether the alternative hypothesis (H_A) is directional (one-tailed) or nondirectional (two-tailed). If no direction of difference between the two variables is hypothesized, a two-tailed test is applied, and the *smaller* of T_p and T_n is chosen for testing. In this instance, the researcher is hypothesizing only a difference between the two variables under study and not which is the largest.

The second possibility involves a directional hypothesis and a one-tailed procedure. In this case, the researcher is hypothesizing that either the positive or negative differences for the matched pairs are expected to dominate. The value of T corresponding to the *smaller number* of hypothesized differences (either positive or negative) is selected for testing. Thus, if more differences are expected to be positive, T_n, the sum of the negative differences, is used.

When the number of matched pairs exceeds 10, the rank sum (T) value can be converted to a Z statistic (Z_W) and tested using the distribution of normal values:

$$Z_w = \frac{T - \frac{n(n+1)}{4}}{\sqrt{\frac{n(n+1)(2n+1)}{24}}} \qquad (9.20)$$

where: n = number of matched pairs ($n > 10$)
$\quad\quad\quad T$ = rank sum

Application to Geographic Problem

The matched pairs t test and Wilcoxon signed-ranks tests are illustrated with an example from urban transportation. Over the last 20 years, central business districts of most American cities have experienced significant decline in economic activity. Following World War II, rapid suburbanization, aided by increasing demand for single-family housing and extensive metropoli-

---

ingly. When differences for matched pairs are tied for a particular rank position, the average rank is assigned to each such pair. The null hypothesis for the signed-ranks problem assumes the population of matched pair differences (in ranks) from which the sample is drawn is zero.

Two sums can be calculated from the set of ranked matched pairs: T_p, the sum of ranks for positive differences (variable one greater than variable two); and T_n, the sum of ranks for negative differences (variable two greater than variable one). If the two variables measured for the single sample show no difference, T_p should be about equal to T_n. However, for a problem in which the differences between the two variables are large, the disparity between T_p and T_n will also be large. In these situations, one of the rank

tan freeway construction, led to large-scale retail development outside the central city. These social, economic, and spatial processes have caused retailing activity in the downtown areas of most American cities to decline dramatically. Modest growth of other economic activities located in the central city (e.g., office buildings and public services) have generally failed to counterbalance this decline in retailing, and most central business districts have lost their dominant role in metropolitan life. An important consequence of these changes has been increased traffic flows in the outer city and diminished traffic volumes in downtown areas.

Using data collected by the Akron Metropolitan Transportation Study (AMATS) from 1974 to 1989, traffic volumes at a sample of 11 intersections within the downtown (CBD) of Akron, Ohio, are tested for significant differences over time (figure 9.1). The relative change in traffic flow is compared for two time periods: (a) 1974-1979 — labeled "the 1970s" and (b) 1982-1988 — "the 1980s." Since one sample of locations provides the sites for data collection at two time

Figure 9.1   Sampled Intersections in Downtown Akron

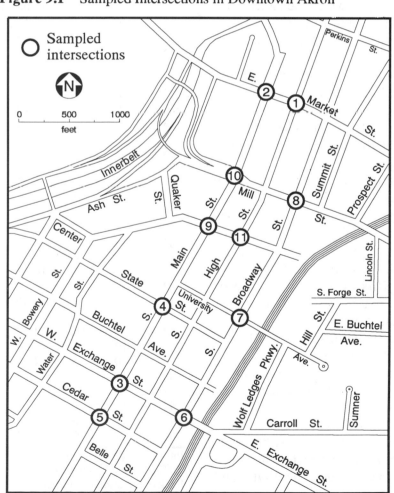

periods, two matched pairs procedures are used to test for decline in traffic volume. First, the matched pairs *t* test is used to analyze the original ratio scale traffic counts for the two decades. The data are then downgraded to their ordinal equivalent by ranking the differences, and the Wilcoxon signed-ranks test is applied to the sum of negative and positive ranks.

Due to financial and practical limitations, AMATS does not collect traffic data at the same time for all intersections they monitor. Thus, the volume of traffic flow at each intersection represents a count from a traffic survey conducted at an early date (1970s) and again at a later date (1980s). Even though the selected traffic counts within either the early or later periods do *not* represent flows at the same date, they are close enough in time to provide a general indication of traffic volume for that particular period of time.

To obtain an adequate sample of traffic data, the flow through each intersection was measured from each of four directions—north, east, south, and west. Thus, the total sample consists of 44 traffic counts taken at the 11 central city locations (table 9.9).

The average difference in traffic flow at the 44 locations surveyed is approximately 1,700 vehicles per day between the two time periods (table 9.10). The mean daily traffic volume for the 1970s period is about 13,000 vehicles, whereas in the 1980s average traffic volume dropped to 11,350. This average difference of about 1,700 vehicles per location per day produces a matched pair *t* value of 4.30, which represents a statistically significant decline in traffic volume ($p = 0.000$).

Using the data in ordinal rather than ratio form, the Wilcoxon signed-ranks test considers either the sum of the negative ranks (surveyed sites showing an increase in traffic volume over time) or the sum of the positive ranks (sites having a drop in traffic volume). Since traffic counts are hypothesized to decline between the 1970s and 1980s, the alternative hypothesis is directional and a one-tailed test is applied (table 9.11).

Table 9.9 Traffic Volume Data for Akron: Matched Pairs Tests

Matched Pair	Inter- section[1]	Direc- tion	1970 Volume[2]	1980 Volume[2]	*d*	Rank[3]
1	1	N	3.94	8.27	−4.32	39
2	1	E	24.69	25.11	−0.42	7
3	1	S	8.15	10.41	−2.26	25
4	1	W	25.69	26.07	−0.38	6
5	2	N	16.50	10.33	6.17	41
6	2	E	25.90	26.42	−0.52	10
7	2	S	19.85	11.76	8.09	43
8	2	W	27.46	26.87	0.58	11
9	3	N	9.58	6.42	3.15	30
10	3	E	16.46	16.56	−0.10	3
11	3	S	8.39	4.99	3.40	34
12	3	W	13.78	14.66	−0.89	16
13	4	N	12.55	9.29	3.27	32
14	4	E	3.07	2.23	0.84	14
15	4	S	11.65	8.85	2.79	29
16	4	W	4.16	3.94	0.22	4
17	5	N	8.64	4.83	3.81	35
18	5	E	13.71	11.36	2.35	26
19	5	S	6.39	3.75	2.64	28
20	5	W	16.20	12.12	4.07	38
21	6	N	15.03	13.99	1.04	18
22	6	E	27.10	19.00	8.11	44
23	6	S	22.05	22.51	−0.46	9
24	6	W	21.07	14.72	6.35	42
25	7	N	18.05	14.00	4.05	37
26	7	E	11.84	5.93	5.92	40
27	7	S	15.05	13.20	1.84	23
28	7	W	7.76	4.61	3.15	31
29	8	N	9.12	10.86	−1.74	22
30	8	E	10.20	9.62	0.59	12
31	8	S	12.14	13.59	−1.45	20
32	8	W	9.32	8.88	0.44	8
33	9	N	16.67	12.81	3.86	36
34	9	E	4.20	4.15	0.05	1.5
35	9	S	12.80	10.59	2.20	24
36	9	W	7.86	6.71	1.15	19
37	10	N	14.19	11.67	2.52	27
38	10	E	9.03	7.43	1.60	21
39	10	S	14.74	11.41	3.33	33
40	10	W	8.69	9.53	−0.84	15
41	11	N	10.62	10.57	0.05	1.5
42	11	E	4.54	4.83	−0.29	5
43	11	S	11.95	11.14	0.81	13
44	11	W	4.62	3.62	1.01	17

[1] Intersection numbers correspond to locations in figure 9.1
[2] Traffic volume is number of vehicles per day (in thousands)
[3] Rank of the absolute differences (*d*)
Source: Akron Metropolitan Area Transportation Study (AMATS), City of Akron, Ohio.

Table 9.10 Work Table for Dependent Sample, Matched Pairs t: Akron Traffic Volumes

$$H_o: \quad \delta = 0$$
$$H_A: \quad \delta > 0 \text{ (one-tailed test)}$$

$$\Sigma d = 75.78 \qquad \Sigma d^2 = 434.65 \qquad n = 44$$

where: d is difference in traffic volume per day (in thousands) (see table 9.9 for data)

$$\bar{d} = \frac{\Sigma d_i}{n} = \frac{75.78}{44} = 1.72$$

or

$$\bar{d} = \bar{X} - \bar{Y} = 13.07 - 11.35 = 1.72$$

$$s_d = \sqrt{\frac{\Sigma d_i^2 - \frac{(\Sigma d_i)^2}{n}}{n-1}} = \sqrt{\frac{434.65 - \frac{75.78^2}{44}}{43}} =$$

$$= \sqrt{\frac{434.65 - 130.51}{43}} = \sqrt{\frac{304.14}{43}} = 2.66$$

$$t_{mp} = \frac{\bar{d}}{s_d/\sqrt{n}} = \frac{1.72}{2.66/\sqrt{44}} = \frac{1.72}{0.40} = 4.30$$

$$p\text{-value} = 0.000 \text{ (one-tailed)}$$

Table 9.11 Work Table for Matched Pair Wilcoxon Signed-Ranks: Akron Traffic Volumes

$$Z_w = \frac{T - \frac{n(n+1)}{4}}{\sqrt{\frac{n(n+1)(2n+1)}{24}}}$$

where: T_n = sum of negative ranks = 177
T_p = sum of positive ranks = 813
n = 44

See table 9.9 for matched pairs ranks

$$Z_w = \frac{177 - \frac{44(44+1)}{4}}{\sqrt{\frac{44(44+1)(2(44)+1)}{24}}}$$

$$= \frac{177 - 495}{\sqrt{\frac{176,220}{24}}} = \frac{318}{85.69} = 3.711$$

$$p\text{-value} = 0.0001 \text{ (one-tailed)}$$

Therefore, T_n should be used because the smaller number of differences are hypothesized to be negative. Given the sum of negative ranks for the problem (177), the resulting test statistic (Z_w) is 3.711. This value represents a significant decline in traffic volume ($p = .0001$). Both the parametric and nonparametric matched pairs tests indicate a significant reduction in traffic volume between the 1970s and 1980s across the 11 intersections surveyed in the central business district of Akron. The low p-values suggest a high degree of confidence that the differences found in the sample traffic data are, in fact, real changes in traffic volume for all traffic at those intersections in downtown Akron.

Summary Evaluation

Several issues associated with data collection and problem structure are of concern when interpreting results for this example. AMATS collects traffic count data sporadically at downtown intersections. Thus, comparing the early (1970s) and late (1980s) periods may produce inconsistencies that are caused by a lack of common dates of data collection at each intersection. Grouping reporting dates together into "1970s" and "1980s" may affect accuracy and produce systematic bias. For example, changes in the economy may affect traffic volumes over short time periods of several years.

The use of the specific set of intersections reflects those locations chosen for traffic counts by AMATS rather than a random sample of all possible traffic count positions in the downtown area. Therefore, the ability of the researcher to extend the results of the specific sample to general downtown traffic changes in Akron may be suspect.

In addition to the way the problem was structured, the accuracy of traffic count data may be affected by many factors, such as road construction and weather conditions. If automated counts are taken, did the equipment work properly? If counts are tabulated manually, were the personnel reliable? These issues not related to sampling error may produce inaccurately collected or missing traffic count data.

9.3 Three or More Sample Difference Tests

Many geographic problems involve comparing three or more independent samples for significant differences. The logic applied to these problems is an extension of the reasoning used with two sample difference tests. If some of the sample differences are significant, statistical inference allows the geographer to conclude that the samples were drawn from at least two different populations. However, if none of the samples are significantly different from one another, the inferential conclusion is that all of the samples are from the same population.

Consider again the geographic problem used in section 9.1 to test for differences between two samples. In that application, a random sample of more developed countries (MDCs) was compared to a second independent random sample of less developed countries (LDCs) over a set of demographic and economic development indicators. Suppose that the research design of this two sample problem is expanded. In past annual editions of the *World Development Report*, The World Bank has divided the countries of the world into 5 categories: (1) low-income, (2) middle-income, (3) industrial market economies,

(4) capital-surplus oil exporters, and (5) non-market (centrally planned) economies. If this more detailed classification scheme is used to test for differences between groups, a two sample procedure, such as the difference of means *t* test or Wilcoxon rank sum *W* test, is no longer appropriate.

Earlier in this chapter, another possible application of two sample difference tests was suggested. In that example, a geographer was interested in studying surface runoff in two different environments: a natural setting with no human landscape modifications and an urban setting, highly modified through development. Suppose that this two sample problem is refined to collect independent random samples from several different environmental settings: (1) a fully natural environment, with no human modification; (2) a rural area with low-density development and only slight human modification; (3) a medium-density suburban area; and (4) a center city that has been substantially altered by unnatural surface features. Once again, the two sample *t* test or Wilcoxon rank sum *W* test cannot be used.

The two inferential procedures used to test three or more samples for differences are **analysis of variance (ANOVA)** and the **Kruskal-Wallis test.** Analysis of variance is a parametric test that requires interval/ratio data drawn from normally distributed populations. The equivalent nonparametric procedure is the Kruskal-Wallis one-way analysis of variance by ranks test (Kruskal-Wallis for short), which uses ordinal data directly or interval/ratio data downgraded to ordinal.

Analysis of Variance (ANOVA)

Variance has already been defined as a descriptive statistic measuring variability about the mean. Analysis of variance (ANOVA) involves the separation of the total variation found in three or more nominal groupings or samples into meaningful components: (1) variability *between* the groups or categories; and (2) variability *within* the groups or categories. Even though the structure of the test uses variation as the key descriptive statistic, it is the means of the samples that are compared for significant differences, making ANOVA a three or more sample difference of means test. The null hypothesis for this parametric procedure takes the form H_0: $\mu_1 = \mu_2 = \ldots = \mu_k$, where k is the number of independent random samples.

Between-group variability focuses on how the sample mean of each group differs from the total or overall mean when all categories are grouped together. If the sample means differ significantly from the total or overall mean, the between-group variability is large. **Within-group variability** measures the variation about the mean of each group and is used to estimate the variation within each group or category in the research problem. When these estimates of variability are totaled for all categories, the result is the total within-group variability.

Sometimes the between-group variation is called the "explained variation," because variability attributed to differences between group or sample means is a measure of how much variation is explained by (or statistically dependent on) the group structure itself. Conversely, internal variation within each of the groups is not "explained" by the grouping or categorization scheme and is thus considered "unexplained" or residual variation.

ANOVA measures and compares the two components of total variation: between-group variability or within-group variability. It determines which is more dominant or pronounced and accounts for a greater proportion of the total variation. Operationally, ANOVA examines the ratio of between-group variability to within-group variability.

The notion of ANOVA as a technique for comparing sample means for differences becomes more apparent when considered graphically (figure 9.2). The null hypothesis asserts that the k independent samples are from the same population, whereas the alternate hypothesis asserts that

the different samples are from separate and distinct populations. In case 1, only a small difference seems to separate the three sample means ($\overline{X}_1$, $\overline{X}_2$, and $\overline{X}_3$), and when ANOVA is applied, the likely decision will be not to reject the null hypothesis. The sample means are inferred to be no more different than would be three independent random samples drawn from the same population. Conversely, in case 2, the large apparent difference between sample means leads to the probable decision to reject H_0 and to conclude that the different samples have been taken from more than one truly different population.

Many possible ANOVA testing structures are available, but only a single-variable model called "one-way analysis of variance" is examined here. The procedure in one-way ANOVA involves the calculation of between-group variability, within-group variability, and the ANOVA test statistic (F), which is the ratio of between-group variability to within-group variability. The ANOVA test statistic (F) is:

$$F = \frac{MS_B}{MS_W} \qquad (9.21)$$

where: MS_B = between-group mean squares
  MS_W = within-group mean squares

To calculate the between-group mean squares (the numerator of the test statistic), the following three-step procedure is followed:

Step 1: Calculate the total or overall mean ($\overline{X}_T$), which is the weighted average of the individual group or sample means ($\overline{X}_i$'s):

$$\overline{X}_T = \frac{\sum_{i=1}^{k} n_i \overline{X}_i}{N} \qquad (9.22)$$

where: $\overline{X}_i$ = mean of sample i
  n_i = number of observations in sample i
  k = number of groups or samples
  N = total number of observations in all samples

Figure 9.2   Null and Alternate Hypotheses in Analysis of Variance (ANOVA)

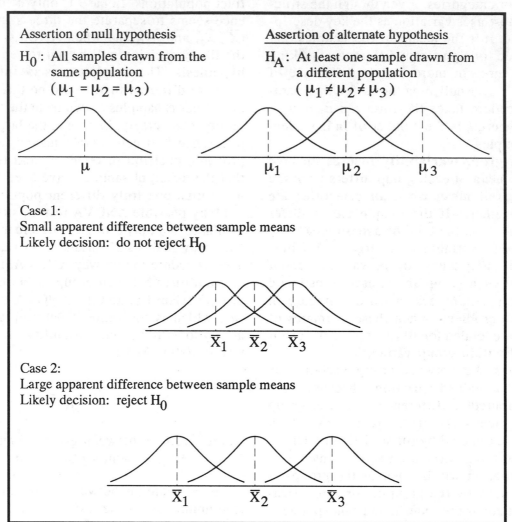

Step 2: Calculate the between-group sum of squares (SS_B):

$$SS_B = \sum_{i=1}^{k} n_1 \left(\overline{X}_i - \overline{X}_T \right)^2 \qquad (9.23)$$

$$= \left(\sum_{i=1}^{k} n_i \overline{X}_i^2 \right) - N \overline{X}_T^2 \qquad (9.24)$$

Equation 9.23 is the definitional formula of the between-group sum of squares, and equation 9.24 is the computational formula.

Step 3: To derive the between-group mean squares (MS_B), the between-group sum of squares value is adjusted by the between-group degrees of freedom (df_B), where df_B depends on the number of groups in the problem:

$$MS_B = \frac{SS_B}{df_B} = \frac{SS_B}{k-1} \qquad (9.25)$$

Derivation of the within-group mean squares (the denominator of the test statistic) is also a multi-step procedure. The first task is to determine the amount of variation in each group and weight the importance of this variation by the sample size of the group. The result is the within-group sum of squares (SS_W):

$$SS_W = \sum_{i=1}^{k} (n_i - 1) s_i^2 \qquad (9.26)$$

The within-group sum of squares is adjusted by the within-group degrees of freedom (df_W) to obtain the within-group mean squares (MS_W):

$$MS_W = \frac{SS_W}{df_W} = \frac{SS_W}{N-k} \qquad (9.27)$$

The sample or group means will fluctuate somewhat, even if they are from a set of independent samples drawn from the same population. This is the expected amount of between-group fluctuation and results from the nature of sampling and sampling distributions. If the null hypothesis is correct, the group means should vary no more than expected if they were a set of independent random samples drawn from the same population, and the between-group variance will be approximately equal to the within-group variance. As a result, the F-ratio value will be approximately equal to one. However, if H_0 is incorrect, and the population means are, in fact, significantly different, then the sample means estimating these population means will vary more than expected from simple random fluctuations of multiple samples from the same population. This will cause the between-group variance to be significantly larger than the within-group variance, and the F-ratio test statistic will be greater than one.

Analysis of Variance (ANOVA) Test

Primary Objective: Compare three or more (k) independent random sample means for differences

Requirements and Assumptions:
  (1) Three or more (k) independent random samples
  (2) Each population is normally distributed
  (3) Each population has equal variance ($\sigma_1^2 = \sigma_2^2 = \ldots = \sigma_k^2$)
  (4) Variables measured at interval or ratio scale

Hypotheses:
  H_0: $\mu_1 = \mu_2 = \ldots = \mu_k$
  H_A: $\mu_1 \neq \mu_2 \neq \ldots \neq \mu_k$ (not all μ_i equal)

Test Statistic:

$$F = \frac{MS_B}{MS_W}$$

Kruskal-Wallis Test

The Kruskal-Wallis test is the nonparametric equivalent of ANOVA. Kruskal-Wallis may be the most appropriate technique in cases where assumptions required for the use of the parametric ANOVA test (normality, equal population variances) are not fully met. Kruskal-Wallis may be considered the nonparametric extension of the Wilcoxon rank sum W test to problems with three or more samples.

In this test, values from all samples are combined into a single overall ranking (as in the Wilcoxon test). The rankings from each sample are summed, and the mean ranks of each sample are then calculated. For example, in sample one the sum of ranks is R_1 and the mean rank is R_1 / n_1, in sample two, the sum of ranks is R_2 and the mean rank is R_2 / n_2, and so on. Because of the random nature of sampling, the sample mean ranks should differ somewhat, even if the samples are drawn from the same population. The Kruskal-Wallis test examines whether the mean rank values are significantly different.

If the multiple (k) samples are from the same population, as asserted by the null hypothesis, their mean ranks should be approximately equal. The best estimate of the mean rank of population i is the sample mean rank (R_i / n_i). Thus, if the null hypothesis is correct:

$$\frac{R_1}{n_1} = \frac{R_2}{n_2} = \ldots = \frac{R_k}{n_k}$$

On the other hand, if the mean ranks differ by more than is likely with chance fluctuations, it may be concluded that at least one of the samples comes from a different population, and the null hypothesis is rejected.

The Kruskal-Wallis H test statistic is:

$$H = \frac{12}{N(N+1)} \sum_{i=1}^{k} \frac{R_i^2}{n_i} - 3(N+1) \qquad (9.28)$$

where: N = total number of observations or values in all samples

$\quad\quad\ = n_1 + n_2 + \ldots + n_k$

$\quad n_i$ = number of observations or values in sample i

$\quad R_i$ = sum of ranks in sample i

In some geographic data sets, the ranks for a number of observations across the k samples may be tied. If this affects more than 25 percent of the values, then a correction factor should be included in the Kruskal-Wallis test statistic. The correction factor for ties is:

$$1 - \frac{\Sigma T}{N^3 - N} \qquad (9.29)$$

where: $T = t^3 - t$

$\quad t$ = number of tied observations in the data

$\quad N$ = total number of observations or values in all k samples

With this addition for correction factor for ties, the H statistic is:

Kruskal-Wallis Test

Primary Objective: Compare three or more (k) independent random sample mean ranks for differences

Requirements and Assumptions:
1. Three or more (k) independent random samples
2. Each population has an underlying continuous distribution
3. Variables measured at ordinal scale or interval/ratio scale variable downgraded to ordinal

Hypotheses:
H_o: The populations from which the three or more (k) samples have been drawn are all identical
H_A: The populations from which the three or more (k) samples have been drawn are not all identical

Test Statistic:

$$H = \frac{12}{N(N+1)} \sum_{i=1}^{k} \frac{R_i^2}{n_i} - 3(N+1)$$

$$H = \frac{\frac{12}{N(N+1)} \sum_{i=1}^{k} \frac{R_i^2}{n_i} - 3(N+1)}{1 - \frac{\Sigma T}{N^3 - N}} \qquad (9.30)$$

The effects of this modification for ties are to increase the value of H, reduce the p-value, and improve the likelihood of discovering significant differences between samples.

Application to Geographic Problem

Both ANOVA and Kruskal-Wallis are now applied to the spatial patterns of housing cost data in the Washington, D.C. metropolitan area. Based on actual sales of homes recorded in local government offices and subsequently summarized in the *Washington Post,* information is available by zip code on the number of sales and median purchase price for 1988 and 1989. Metropolitan

Washington, D.C. contains more than 140 residential zip code areas. These zip codes can be classified by city or county political subdivision or by concentric distance zone from the city center. Several research questions can be explored:

1. Does the purchase price of homes vary significantly by political subdivision? The six

political subdivisions for which data have been published are the District of Columbia, Alexandria City VA, Arlington County VA, Fairfax County VA, Montgomery County MD, and Prince George's County MD (figure 9.3).

2. Does the purchase price of homes vary significantly by distance from downtown

Figure 9.3   Political Subdivisions in the Washington, D.C. Metropolitan Area

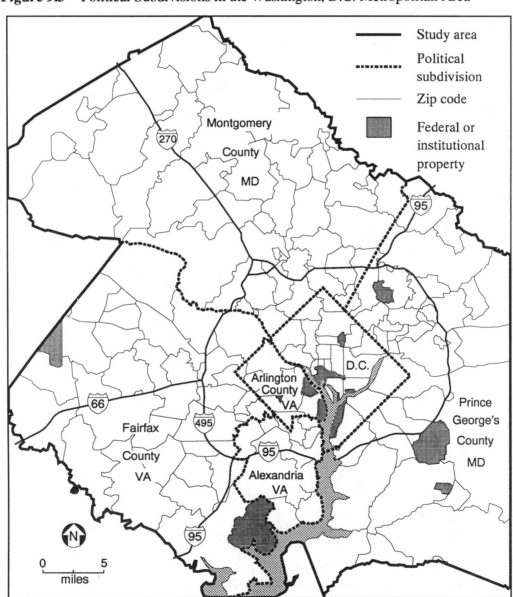

Washington? The four concentric distance zones delineated for this problem are shown in figure 9.4.

3. During the 1988-1989 study period, the total number of home sales in the Washington, D.C. area declined substantially, from a total of 64,040 in 1988 to 54,181 in 1989. What changes in the spatial variability of home prices occurred from 1988 to 1989, for both classification schemes (political subdivision and distance zone from downtown Washington)?

An exploratory research design can be established using both ANOVA and Kruskal-Wallis

Figure 9.4   Concentric Distance Zones in the Washington, D.C. Metropolitan Area

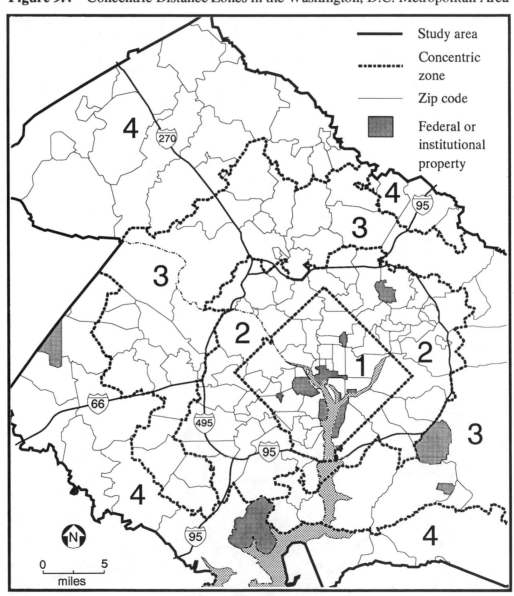

to compare the spatial variation of home purchase prices both by concentric distance zone and by political subdivision for both 1988 and 1989 (table 9.12).

The spatial variability of 1988 home purchase prices is evaluated first by political subdivision. A

Table 9.12 Research Design for Spatial Analysis of Washington, D.C. Metropolitan Area Home Purchase Prices

	1988		1989	
	ANOVA	**Kruskal-Wallis**	**ANOVA**	**Kruskal-Wallis**
Zip codes classified by political subdivision	F =	H =	F =	H =
	p =	p =	p =	p =
Zip codes classified by concentric distance zone	F =	H =	F =	H =
	p =	p =	p =	p =

partial listing of the data and the summary descriptive statistics needed for ANOVA and Kruskal-Wallis are provided in table 9.13 (continued on page 180) for each of the six political subdivisions in metropolitan Washington, D.C. For example, the District of Columbia (group 1) has 20 zip code areas (n_1 = 20) with an average home purchase price ($\overline{X}_1$) of \$166,546.05 and standard deviation (s_1) of \$112,955.81. Furthermore, when these zip code purchase prices are placed in an overall ranking for application of Kruskal-Wallis, the sum of ranks (R_1) is 1188.5 and the mean rank (R_1 / n_1) is 59.4.

According to the summary statistics in table 9.13, the group means ($\overline{X}_i$'s) and mean ranks (R_i / n_i's) are different, but are they significantly different? The work table of calculations for ANOVA is shown in table 9.14. The values in this housing problem are quite large, so most of the ANOVA statistics are expressed in scientific notation. For example, the between-group sum of squares is 203,000,000,000 (in scientific notation, this is 2.03 $E+11$). When the 1988 housing cost data are grouped by political subdivision, the resultant

Table 9.13 Home Purchase Prices in the Washington, D.C. Metropolitan Area: Zip Codes Grouped by Political Subdivision, 1988

District of Columbia (Group 1)				**Prince George's County MD (Group 2)**		
zip code	home purchase price	rank (r)		zip code	home purchase price	rank (r)
20001	\$ 68,421	2		20607	\$ 94,334	22
20002	77,500	7		20613	114,000	39
20003	124,000	46		20623	102,000	28
...	...	...		...	...	...
20036	332,000	137		20903	85,000	13
20037	224,100	116		20912	113,950	38
n_1 = 20				n_2 = 30		

ANOVA statistics: $\overline{X}_1 = 166,546.05$

$s_1 = 112,955.81$

Kruskal-Wallis statistics: $R_1 = \Sigma r = 1188.5$

$R_1 / n_1 = 59.4$

ANOVA statistics: $\overline{X}_2 = 102,750.63$

$s_2 = 20,200.11$

Kruskal-Wallis statistics: $R_2 = 855.5$

$R_2 / n_2 = 28.5$

Table 9.13   (Continued)

| Montgomery County MD (Group 3) | | |
zip code	home purchase price	rank (r)
20814	$ 235,000	118.5
20815	329,293	135
20816	330,000	136
...	...	...
20910	165,000	85
20912	145,000	64.5
$n_3 = 39$		

ANOVA statistics: $\overline{X}_3 = 191,765.08$

$s_3 = 81,415.52$

Kruskal-Wallis statistics: $R_3 = 3202$

$R_3/n_3 = 82.1$

| Fairfax County VA (Group 4) | | |
zip code	home purchase price	rank (r)
22003	$ 180,000	99.5
22015	151,690	73
22020	150,780	71
...	...	...
22181	220,000	114.5
22182	290,000	126
$n_4 = 33$		

ANOVA statistics: $\overline{X}_4 = 206,985.67$

$s_4 = 81,483.05$

Kruskal-Wallis statistics: $R_4 = 3164$

$R_4/n_4 = 95.9$

| Arlington County VA (Group 5) | | |
zip code	home purchase price	rank (r)
22202	$237,000	120
22203	208,120	112
22204	145,000	64.5
...	...	...
22209	205,000	108.5
22213	197,000	106
$n_5 = 8$		

ANOVA statistics: $\overline{X}_5 = 190,815.00$

$s_5 = 37,836.64$

Kruskal-Wallis statistics: $R_5 = 775.5$

$R_5/n_5 = 96.9$

| Alexandria City VA (Group 6) | | |
zip code	home purchase price	rank (r)
22302	$194,600	105
22303	103,500	29
22304	172,900	91
...	...	...
22312	171,250	89
22314	200,000	107
$n_6 = 12$		

ANOVA statistics: $\overline{X}_6 = 165,829.17$

$s_6 = 36,873.52$

Kruskal-Wallis statistics: $R_6 = 967.5$

$R_6/n_6 = 80.6$

Note: data listed for only five zip codes in each political subdivision

Those portions of Fairfax County in Alexandria City are in Group 6

n_i = number of observations (zip code areas) in sample i

$\overline{X}_i$ = mean home purchase price in sample i

s_i = standard deviation of home purchase price in sample i

R_i = sum of ranks in sample i

R/n_i = mean rank in sample i

Source: Modified from data published in *Washington Post*, March 31, April 7, April 14, and April 21, 1990.

Table 9.14 ANOVA Work Table: Home Purchase Prices in the Washington, D.C. Metropolitan Area, With Zip Codes Grouped by Political Subdivision, 1988

$$H_0: \quad \mu_1 = \mu_2 = \ldots = \mu_6$$

$$H_A: \quad \mu_1 \neq \mu_2 \neq \ldots \neq \mu_6$$

Between-Group Variability

Step 1: Total or Overall Mean $(\overline{X}_T)$

$$\overline{X}_T = \frac{\sum_{i=1}^{6} n_i \overline{X}_i}{N} = \frac{20(166,546.05) + \ldots + 12(165,829.17)}{142} = 170,699.12$$

Step 2: Between-Group Sum of Squares (SS_B)

$$SS_B = \left(\sum_{i=1}^{6} n_i \overline{X}_i^{\,2} \right) - N\overline{X}_T = \left(20(166,546.05)^2 + \ldots + 12(165,829.17)^2 \right) - 142(170,699.12)^2 = 203,000,000,000$$

$$= 2.03E + 11$$

Step 3: Between-Group Mean Squares (MS_B)

$$MS_B = \frac{SS_B}{df_B} = \frac{SS_B}{k-1} = \frac{2.03E + 11}{6-1} = 4.06E + 10$$

Within-Group Variability

Step 1: Within-Group Sum of Squares (SS_W)

$$SS_W = \sum_{i=1}^{6} (n_i - 1)\, s_i^{\,2} = 19(112,955.8)^2 + \ldots + 11(36,873.5)^2 = 7.44E + 11$$

Step 2: Within-Group Mean Squares (MS_W)

$$MS_W = \frac{SS_W}{df_W} = \frac{SS_W}{N-k} = \frac{7.44E + 11}{142 - 6} = 5.47E + 9$$

Calculate the Test Statistic (F)

$$F = \frac{MS_B}{MS_W} = \frac{4.06E + 10}{5.47E + 9} = 7.4307 \qquad (p = .0000)$$

Summary Table

Source of Variation	Sum of Squares	Degrees of Freedom	Mean Squares	F Ratio	p
Between Group	2.03 $E + 11$	5	4.06 $E + 10$	7.4301	.0000
Within Group	7.44 $E + 11$	136	5.47 $E + 9$		
Total	9.47 $E + 11$	141			

F statistic is 7.43, and the corresponding p-value is .0000. These results indicate highly significant differences in home purchase prices across political subdivisions in the Washington, D.C. area in 1988.

When the Kruskal-Wallis test is applied to the same 1988 home purchase price data grouped by political subdivision, the resultant H statistic is 52.26, and the corresponding p-value is .0000 (table 9.15). The similar p-values of ANOVA and Kruskal-Wallis in this problem indicate that 1988 home purchase price differs significantly by city or county subdivision in the Washington area.

The analysis of home cost variability is now expanded in two ways: (1) zip codes grouped by distance zone from the city center are compared with zip codes grouped by political subdivision, and (2) data from both 1988 and 1989 are included.

Certain general conclusions seem clear. For both 1988 and 1989, home purchase prices vary significantly ($p = .0000$ in both years for both ANOVA and Kruskal-Wallis) by political subdivision (table 9.16). With both ANOVA and Kruskal-Wallis, test statistic values decline slightly from 1988 to 1989. The ANOVA F statistic declined slightly from 7.432 to 6.227, indicating a decrease in between-group mean squares relative to within-group mean squares. Either differences in home purchase prices are trending somewhat toward equality between cities and counties in the Washington, D.C. area, or variation in home purchase prices within political subdivisions is increasing. An examination of the ANOVA results for the two years indicates that the within-group mean squares increased more dramatically from 1988 to 1989 than did the between-group mean squares value. This suggests home prices varied more within the political subdivisions in 1989 than 1988. In a similar fashion, the Kruskal-Wallis H statistic declined from 52.26 in 1988 to 46.94 in 1989, indicating the mean ranks of the different groups are not quite as dissimilar in 1989 as 1988.

Table 9.15 Kruskal-Wallis Test Work Table: Home Purchase Prices in the Washington, D.C. Metropolitan Area, With Zip Codes Grouped by Political Subdivision, 1988

$$H = \frac{12}{N(N+1)} \sum_{i=1}^{k} \frac{R_i^2}{n_i} - 3(N+1)$$

$$= \frac{12}{142(143)} \left(\frac{(1188.5)^2}{20} + \ldots + \frac{(976.5)^2}{12} \right) - 3(143)$$

$$= 52.26$$

$$(p = .0000)$$

Table 9.16 Results of Spatial Analysis of Washington, D.C. Metropolitan Area Home Purchase Prices: ANOVA and Kruskal-Wallis

	1988		1989	
	ANOVA	**Kruskal-Wallis**	**ANOVA**	**Kruskal-Wallis**
Zip codes classified by political subdivision	F = 7.432 p = .0000	H = 52.26 p = .0000	F = 6.227 p = .0000	H = 46.94 p = .0000
Zip codes classified by concentric distance zone	F = 0.991 p = .3993	H = 4.60 p = .2039	F = 1.449 p = .2311	H = 5.15 p = .1609

However, the slight decline of these test statistic values is occurring within an overall situation where highly significant differences continue to exist in home purchase prices from one political subdivision to another.

No comparably significant differences are found in home purchase price when examined by concentric distance zone from the metro center (table 9.16). The ANOVA F statistics (in 1988, $F = .991$ and $p = .3993$; and in 1989, $F = 1.449$ and $p = .2311$), strongly suggest that distance from the metro center is not a significant determinant of home purchase price. Similarly, the Kruskal-Wallis test results indicate home prices do not vary systematically (nonrandomly) by distance from the metro center. The Kruskal-Wallis H statistics are relatively low, and the resultant p-values are therefore comparatively high (in 1988, $H = 4.60$ and $p = .2039$; and in 1989, $H = 5.15$ and $p = .1609$).

Summary Evaluation

Several inferential issues and general considerations need to be discussed relative to this problem. For an inferential technique to be considered fully valid, the observations included in each sample must be randomly drawn. Since both ANOVA and Kruskal-Wallis require multiple samples, each of these random samples is assumed to be drawn separately and independently. Numerous nonrandom influences are present, however. A group of homes with similar characteristics (and similar prices) is likely to be placed on the market whenever a new residential subdivision or development is completed. Another likely occurrence is the sale of a large number of homes in neighborhoods or communities undergoing rapid demographic or socioeconomic change, and homes in the same neighborhood are likely to have similar prices. These factors make the home sales both nonrandom and nonindependent. Statisticians generally agree that the assumption of independent random samples is very important and should not be relaxed. Thus, the question becomes how seriously have these assumptions of randomness and independence been violated, and does this invalidate the housing variability problem?

As a parametric technique, ANOVA also requires that each population be normally distrib-

uted and have an equal variance. Both of these assumptions are probably violated. If a frequency distribution were drawn showing the number of homes that exist at various purchase prices in the typical metropolitan area, the distribution would very likely be positively skewed (figure 9.5). A large number of housing units are probably available at the low and middle price range. Below a certain price, however, few, if any, units exist. The "long tail" of the distribution, skewing the distribution positively, is toward the upper-income housing market. How severely the data are skewed in this problem is not known. Similarly, equal variance (known as homoscedasticity) most certainly does not exist in each population (political subdivision or concentric distance zone). But, what is the extent of this inequality and has the equal variance assumption of ANOVA been violated? Statisticians have determined that the statistical power of ANOVA is not significantly affected by moderate violations of normality and homoscedasticity, but in this problem the magnitude of these violations is unknown. This suggests that greater confidence might be placed in the Kruskal-Wallis results. This is one of the reasons for presenting Kruskal-Wallis, a nonparametric technique not

Figure 9.5 Positively Skewed Frequency Distribution of Home Purchase Prices

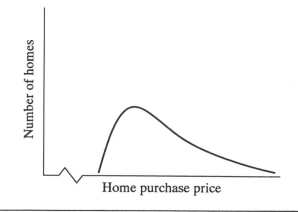

concerned with normality or equal variances, in conjunction with ANOVA in this housing variability problem.

A few other problem characteristics need to be considered. In this problem, a "data value" is the median purchase price of all homes sold in a particular zip code area during the year. The data are spatially aggregated, and a value is not a single home sale price. This does not invalidate the analysis, but is nevertheless a characteristic of the problem that must be recognized.

The observations are not artificial samples (drawn or selected from a population through application of a sample design procedure), but are rather "natural" samples. The data values are natural in the sense that the processes operating to produce the spatial pattern of home sales contain a random component. In this context, the set of homes that happened to be sold during the study period is just one possible sample from the many that could have occurred.

Key Terms and Concepts

References and Additional Reading

The statistical techniques covered in this chapter are discussed in most introductory textbooks. Refer to the list of general references which cover statistical methods in geography, located at the end of the text prior to the tables in the Appendix.

CHAPTER 10

GOODNESS-OF-FIT AND CATEGORICAL DIFFERENCE TESTS

185

In chapter 8, statistics from a single sample were compared with a corresponding population parameter to determine the likelihood of the sample having been drawn from the population. In chapter 9, multiple sample difference tests evaluated whether two or more samples were from the same population or from different populations. In this chapter, the focus is on two special types of difference tests: goodness-of-fit and categorical tests.

Goodness-of-fit tests address geographic research questions in which an actual or **observed frequency** distribution is compared with some **expected frequency** distribution. Goodness-of-fit procedures are sometimes used to test a hypothesis to determine if a set of data fits a particular frequency distribution. In other problems, goodness-of-fit tests attempt to confirm or deny the applicability or validity of a particular geographic model or theory. In these cases, the investigator may expect a certain frequency distribution to exist, based either on theoretical principles or empirical knowledge (such as past observation or experience). In section 10.1, various goodness-of-fit tests are presented, including examples to determine if various data sets have a uniform or equal distribution, a proportional or unequal distribution, or a normal distribution.

In many goodness-of-fit problems, variable responses are organized into categories, and sample observations are assigned to a particular category. The goodness-of-fit procedure analyzes the frequency counts (number of data values) of these categories to determine if the frequency distribution matches some expected distribution. Thus, many goodness-of-fit tests are also considered **categorical difference tests.**

Some geographic research problems extend the single sample or single variable goodness-of-fit procedure to measure the relationship or association between two variables. If data units from both variables are assigned to categories, then the frequency count of each category from one variable may be cross-tabulated or cross-classified with the frequency count of each category for the second variable. This cross-tabulation process is summarized by a contingency table. Contingency table analysis is the topic in section 10.2.

Other geographic problems examine differences between two distributions drawn from independent random samples. If sample values from the variable being studied can be placed in ordinal categories with frequency counts by category, then a two sample difference test can be applied. Section 10.3 discusses the application of a difference test for ordinal categories to solve geographic problems.

10.1 Goodness-of-Fit Tests

In many geographic research problems, goodness-of-fit tests are used to compare an actual or observed frequency distribution with an expected frequency distribution. If a set of sample data closely fits a particular distribution, the population from which the sample has been drawn likely matches that distribution. Goodness-of-fit procedures are sometimes used to test hypotheses that a data set fits a distribution. In other situations, goodness-of-fit tests examine the validity of a geographic model or theory. In this section, applications of goodness-of-fit testing are shown for three different frequency distributions: uniform or equal, proportional or unequal, and normal.

In some problems, the "expected" probability distribution is uniform or equal. For example, suppose an urban geographer wishes to examine Scholastic Aptitude Test (SAT) scores among graduating seniors from five high schools within a metropolitan area. Random samples of 100 students could be taken from each school and frequency distributions produced that show the number of sampled students whose test scores exceed the national median in each of the high schools (figure 10.1). With this measure, the

Figure 10.1   Number of Students with Scholastic Aptitude Test Scores Above National Median: Five High Schools in a Metropolitan Area

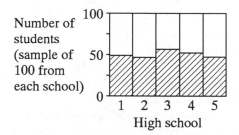

Number of students (sample of 100 from each school)

High school

☐   Observed number of students with SAT score above national median

▨   Observed number of students with SAT score below national median

question of whether SAT scores differ significantly among schools can be examined. A goodness-of-fit procedure can be designed to determine whether the frequency count distribution is uniform or equal for this set of high schools. If the sample frequency distribution is shown to be uniform, it can be inferred that no significant differences are present among the populations of all graduating seniors at the schools studied.

On the other hand, the investigator might expect the distribution of SAT scores to be something other than uniform. Alternative applications of a goodness-of-fit procedure could test for an uneven or proportional distribution. For example, a high school with a low student-teacher ratio or high dollar expenditure per student might be expected to have a larger number of sampled students with SAT scores above the national median. A research hypothesis can be designed in which the "expected" number of students performing well on the SATs is proportional to the student-teacher ratio or dollars spent per pupil at each high school.

Proportional goodness-of-fit tests can also be applied to test the validity of a geographic model.

For example, suppose a recreation planner is studying attendance patterns at a regional park that serves people in several nearby communities. How many park visitors might be expected from each community? To estimate park usage patterns, the potential model of spatial interaction would be useful. This model says that the volume of spatial interaction is directly related to population size and inversely related to distance. In this particular problem, the number of visitors from a community to a park would be directly proportional to the population of the community and inversely related to the distance separating the community and the park. That is:

$$Pot_{ij} = \left(Pop_i / D_{ij}\right) \qquad (10.1)$$

where: Pot_{ij} = potential produced at park j by community i

Pop_i = population of community i

D_{ij} = distance from community i to park j

The recreation planner can compare the observed number of park visitors from each community with the corresponding number of visitors predicted (expected) by the potential model. The extent to which the observed and expected visitor counts from the set of communities match (or differ) can be tested with a goodness-of-fit test.

Goodness-of-fit tests are also used to compare an actual frequency distribution with a theoretical probability distribution such as normal or random (see chapter 5 for a review of these distributions). A frequency distribution often needs to be tested to determine if it meets the assumption of normality, a necessary condition if the data are to be used in a parametric test such as t or ANOVA. Suppose a geomorphologist is studying soil erosion rates of several soil orders (alfisols, aridosols, entisols, etc.) in a large watershed study area. To test for significant differences in erosion rates by soil order, a spatial random sample is taken from each type of soil.

If the samples are drawn from normally distributed populations, a parametric ANOVA could be applied as the difference test. A goodness-of-fit test could be used to determine whether this normality requirement has been met.

In some geographic problems, a frequency distribution should be tested for randomness. In these cases, a Poisson probability distribution would be expected. For example, a goodness-of-fit test for randomness could be applied to the Illinois tornado "touch down" pattern examined in the probability chapter (section 5.4). This example will be explored more fully in the next chapter, which discusses explicitly spatial data.

Chi-square (χ^2) Test

When applied as a goodness-of-fit test, the chi-square statistic compares the observed frequency counts of a single variable (organized into nominal or ordinal categories) with an expected distribution of frequency counts allocated over the same categories. The chi-square test must use absolute frequency counts and cannot be applied if the observations or sampling units are in relative frequency form, such as percentages, proportions, or rates.

Chi-square is a method to determine if a truly significant difference exists between a set of observed frequencies and the corresponding expected frequencies. With this procedure, the focus is on how closely the two frequency counts match, which provides a goodness-of-fit measure. The null hypothesis (H_0) states that the population from which the sample has been drawn fits an expected frequency distribution. Thus, H_0 assumes no difference between observed and expected frequency counts. If the magnitude of difference between frequency counts is small across all categories, the data are likely to be a random sample drawn from the expected frequency distribution, and H_0 is not rejected. Conversely, the alternate hypothesis (H_A) suggests that the magnitude of difference between frequencies is large in at least one category. If H_A is true, the data cannot be considered a ran-

---

Chi-Square (χ^2) Goodness-of-fit Test

Primary Objective: Compare random sample frequency counts of a single variable with expected frequency counts (goodness-of-fit)

Requirements and Assumptions:
1. Single random sample
2. Variable organized by nominal or ordinal categories; frequency counts by category are input to statistical test
3. If two categories, both expected frequency counts must be at least five; if three or more categories, no more than one-fifth of the expected frequency counts should be less than five, and no expected frequency count should be less than two

Hypotheses:
H_0: Population from which sample has been drawn fits an expected frequency distribution (uniform or equal, proportional or unequal, random, etc.); no difference between observed and expected frequencies
H_A: Population does not fit an expected frequency distribution; there is a significant difference between observed and expected frequencies

Test Statistic:

$$\chi^2 = \sum_{i=1}^{k} \frac{(O_i - E_i)^2}{E_i}$$

---

dom sample from the expected frequency distribution, and H_0 is rejected.

The formula for the chi-square test statistic is:

$$\chi^2 = \sum_{i=1}^{k} \frac{(O_i - E_i)^2}{E_i} \qquad (10.2)$$

where: O_i = observed or actual frequency count in the ith category
E_i = expected frequency count in the ith category
k = number of nominal or ordinal categories

If the observed and expected frequency counts for each category are similar, then all of the differences $(O_i - E_i)$ will be slight, χ^2 will be small, the goodness-of-fit will be strong, and H_0 will not be rejected. However, if at least one difference between frequency counts is large, then the χ^2 statistic will be large, the observed frequency counts will not necessarily come from the population or model theorized by the expected frequency counts, and H_0 will be rejected. In later example problems, the methodology for calculating expected frequencies will be discussed.

Chi-square is a very flexible goodness-of-fit test because the number of nominal or ordinal categories may vary. However, minimum size restrictions apply under certain circumstances. For example, if the nominal variable has only two categories, the expected frequency in both should be at least five. For the test to be valid with more than two categories, no more than one-fifth of the expected frequencies should be less than five, and no expected frequency should be less than two. Sometimes categories may need to be combined to increase the expected frequencies to an appropriate size.

Kolmogorov-Smirnov Test

The Kolmogorov-Smirnov statistic is a powerful alternative to chi-square for testing the similarity between two frequency distributions. Chi-square uses frequencies from either nominal or ordinal classes, but Kolmogorov-Smirnov (K-S) requires data measured at the ordinal level or interval/ratio level downgraded to ordinal. Technically, K-S requires continuously distributed variables, but only very slight errors result when the technique is applied to discrete data. In the one sample goodness-of-fit application of Kolmogorov-Smirnov discussed in this section, the observed distribution of the sample data is compared to a particular expected distribution, such as normal or Poisson. In the two sample K-S problem discussed in section 10.3, two observed distributions are tested for difference. In both of these applications of Kolmogorov-

Smirnov, the null hypothesis states that no significant difference exists between the two frequency distributions.

Geographers most often use Kolmogorov-Smirnov to test a data set for normality. If the observed sample distribution matches a theoretical normal distribution, the investigator can infer that the population from which the random sample was drawn is normally distributed. Then, such parametric statistics as analysis of variance and the difference of means t test can be applied with confidence that the assumption of normality has been met.

More specifically, the Kolmogorov-Smirnov test for normality compares the **cumulative relative frequencies** of the observed sample data with the cumulative frequencies expected for a perfectly normal distribution. Recall that cumulative frequencies can be shown graphically as an ogive (section 2.5). If the two sets of cumulative frequencies are closely matched, the theoretical and sample distributions can be considered the same and the population considered normal. Any sizable differences between the two cumulative distributions suggest that the sample data were not taken from a normally distributed population.

To calculate an observed cumulative frequency distribution, individual data are ranked from low to high or aggregated into ordinal classes. The ranked data or classes are then converted to Z-scores and cumulative relative frequencies calculated for each Z-value or class. The appropriate question to ask for each cumulative value is "What proportion of the values are *equal to or less than* this value?"

The corresponding cumulative relative frequency distribution for an expected (perfectly normal) data set having the same mean and standard deviation must also be determined. The comparable question asked here is "If the data were normal, what proportion of the values would lie below (equal to or less than) the given Z-value?" **Cumulative normal values (CNVs)** of the expected distribution can be found for any Z-value using the following method:

if $Z > 0$ (above the mean):   CNV $= .50 + p$ (10.3)

if $Z = 0$ (equal to mean):   CNV $= .50$ (10.4)

if $Z < 0$ (below the mean): CNV $= 1.00 - (p + .50)$

$$(10.5)$$

where: CNV = cumulative normal value

p = probability from normal table (appendix, table A)

Cumulative normal values (CNVs) are graphed for selected Z-values in a perfectly normal expected frequency distribution (figure 10.2). The expected cumulative value corresponding to the mean of a normal distribution is easy to specify (equation 10.4). Since half of the values in a normal distribution are equal to or less than the mean, the cumulative value associated with this measure of center is 0.50. However, to determine cumulative normal proportions for other values of Z, both the probability (p) from the normal table (appendix, table A) and the sign of the Z-value (positive or negative) must be known. For example, if $Z = +1.0$, a value greater than the mean, then equation 10.3 is used. For this Z-value, the probability from the normal table is .3413 and the corresponding CNV is 0.8413 (.3413 + .5000). This implies that 84.13 percent of the values in a normal distribution are less than (or equal to) one standard deviation *above* the mean. For negative Z-values, such as $Z = -1.35$, equation 10.5 provides a CNV of 0.0885 (1.0 − (.5000 + .4115) = 0.0885). According to this calculation, slightly less than 9 percent of the values in a normal distribution lie equal to or less than 1.35 standard deviations *below* the mean (figure 10.2).

In addition to calculating the cumulative relative frequencies for both the observed and expected (normal) distributions, the same information can be plotted as ogives. The K-S procedure allows differences between the two cumulative distributions to be determined in one of two ways. Differences can be calculated directly from the data using the two sets of cumulative values. Alternatively, differences can be determined

graphically as vertical deviations for positions along the horizontal axis of the ogive (figure 10.3). The K-S test statistic, D, is the *maximum* (absolute) difference between the two sets of cumulative values:

$$D = \text{maximum} \left| CRF_o(X) - CRF_e(X) \right| \quad (10.6)$$

where: $CRF_o(X)$ = cumulative relative frequencies (observed) for variable X

$CRF_e(X)$ = cumulative relative frequencies (expected) for variable X

When the expected frequency distribution is normal, the cumulative relative frequency [$CRF_e(X)$] corresponds to the cumulative normal value [CNV].

If D is large, the deviation between the actual and theoretical (normal) distribution is also large. As a result, the null hypothesis of no difference between the two distributions is likely to be incorrect, and the probability that the actual distribution is normally distributed is less. Conversely, if D is small, the differences between the actual and expected distributions are all small. The null hypothesis of no significant

Figure 10.2   Cumulative Normal Values (CNV) for Selected Z-value Positions

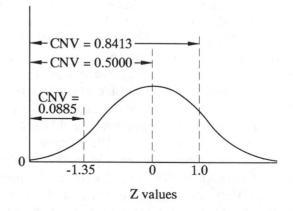

Figure 10.3 Ogive Showing Observed and Expected (Normal) Cumulative Relative Frequency and Differences *(D)*

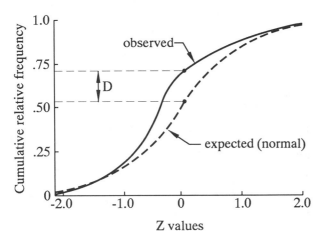

D = maximum vertical difference

Kolmogorov-Smirnov Goodness-of-fit Test

Primary Objective: Compare random sample frequency counts of a single variable with expected frequency counts (goodness-of-fit)

Requirements and Assumptions:
1. Single random sample
2. Population is continuously distributed (test less valid with discrete distribution)
3. Variable measured at ordinal scale or downgraded from interval/ratio scale to ordinal

Hypotheses:

H_o: Population from which sample has been drawn fits an expected frequency distribution (uniform or equal, proportional or unequal, random, etc.); no difference between observed and expected frequencies

H_A: Population does not fit an expected frequency distribution; there is a significant difference between observed and expected frequencies

Test Statistic:
$$D = \text{maximum } |\,CRF_o(X) - CRF_e(X)\,|$$

---

difference between the two distributions is more likely true, implying that the sample is normally distributed.

The significance of the Kolmogorov-Smirnov *D* can be determined in two ways. In the classical or traditional approach to hypothesis testing, the calculated (or graphically measured) *D*-value can be compared with a value taken from a table of critical *D*-values. The null hypothesis can be rejected if the value of *D* exceeds the value found in the table. A *p*-value can be roughly estimated by interpolating within the table of *D*-values.

In the alternative approach, the *D*-value is converted to *Z*, and the table of normal values is used to determine the statistical level of significance or *p*-value. Most computer software packages for the Kolmogorov-Smirnov method calculate both a *D*-value and its corresponding *Z*-value. However, the "manual" computation procedure is rather lengthy and statistically complex. In the application presented later in the chapter, all relevant statistics (*D*, *Z*, and *p*-value) are provided, just as obtained from computer output.

Application to Geographic Problems

Goodness-of-fit tests are demonstrated with three different applications: (1) chi-square is used to determine whether a sample is uniformly or equally distributed, (2) chi-square is again applied to test whether a sample fits an expected proportional or unequal distribution, and (3) Kolmogorov-Smirnov is used to determine whether a sample is normally distributed.

Uniform or Equal Distribution

In the discussion of basic elements of sampling (chapter 6), an air passenger survey was briefly described. Some of the information collected from this survey is used to illustrate the application of chi-square as a goodness-of-fit test for a uniform or equal distribution.

The Salisbury-Wicomico County (MD) Regional Airport serves passengers throughout the Delmarva Peninsula (figure 10.4), an increasingly popular summer resort area. The influx of people from such nearby metropolitan areas as Baltimore, Norfolk, Philadelphia, and Washington, D.C. affects all of the peninsula, especially coastal communities. To plan air transportation more effectively within the region, passengers boarding at the airport were surveyed to determine their travel patterns. One question included in the survey was: "Where in this area did you start your trip today?" A question of concern to planners was whether air travel fluctuated by season for different originating cities.

According to the research hypothesis, the number of travelers coming from coastal communities should demonstrate significant seasonal fluctuation. On the other hand, passengers from noncoastal communities should exhibit a more uniform or equal distribution.

Only those communities with 20 or more surveyed passengers are included in the study (table 10.1): coastal communities appear first, followed by noncoastal communities. The observed passenger counts from the survey are followed (in parentheses) by the number of passengers expected in a uniform or equal distribution with no seasonal fluctuation in airport use.

To calculate the expected frequencies for a uniform or equal distribution, the observed frequency count total is simply divided by the number of categories:

$$E_i = \frac{\sum_{i=1}^{k} O_i}{k} \quad \text{for all categories } (i) \qquad (10.7)$$

where: E_i = expected frequency count for all categories (i)

   O_i = observed or actual frequency count in the ith category

   k = the number of categories

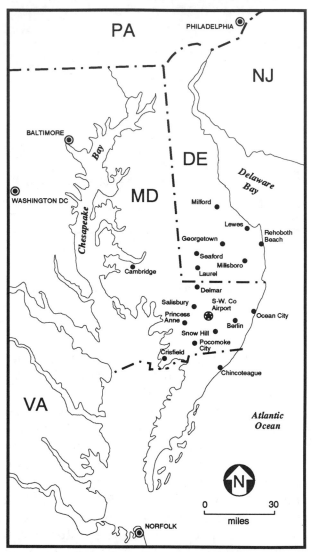

Figure 10.4   Cities Surveyed for the Salisbury/Wicomico County Airport Study

A separate chi-square statistic is calculated for each city, with the procedure illustrated for Pocomoke City, MD (table 10.2). The resulting chi-square value of 5.20 for the Pocomoke City sample has an associated p-value of 0.158. This level of significance suggests, that a chi-square value of this magnitude occurs by chance in 15.8 percent of the random samples drawn from a population with an equal or uniform distribution. Therefore, to conclude that seasonal fluc-

Table 10.1   Number of Passengers Surveyed, by Season: Coastal and Noncoastal Communities

Coastal Communities	Observed (Expected) Number of Passengers Surveyed				
	Summer	Fall	Winter	Spring	Total
Cambridge, MD	5(10.25)	11(10.25)	9(10.25)	16(10.25)	41
Chincoteague, VA	5(8.50)	9(8.50)	12(8.50)	8(8.50)	34
Crisfield, MD	3(7.25)	13(7.25)	5(7.25)	8(7.25)	29
Lewes, DE	7(7.25)	9(7.25)	3(7.25)	10(7.25)	29
Ocean City, MD	89(72.5)	99(72.5)	36(72.5)	66(72.5)	290
Rehoboth Beach, DE	13(9.0)	12(9.0)	4(9.0)	7(9.0)	36

Noncoastal Communities	Observed (Expected) Number of Passengers Surveyed				
	Summer	Fall	Winter	Spring	Total
Berlin, MD	10(11.5)	12(11.5)	14(11.5)	10(11.5)	46
Delmar, MD-DE	6(5.5)	5(5.5)	5(5.5)	6(5.5)	22
Georgetown, DE	2(6.25)	2(6.25)	8(6.25)	13(6.25)	25
Laurel, DE	7(7.5)	5(7.5)	4(7.5)	14(7.5)	30
Milford, DE	4(5.0)	0(5.0)	11(5.0)	5(5.0)	20
Millsboro, DE	14(16.5)	16(16.5)	11(16.5)	25(16.5)	66
Pocomoke City, MD	28(20.0)	20(20.0)	18(20.0)	14(20.0)	80
Princess Anne, MD	10(8.75)	7(8.75)	9(8.75)	9(8.75)	35
Salisbury, MD	344(346.75)	381(346.75)	310(346.75)	352(346.75)	1387
Seaford, DE	25(28.75)	29(28.75)	33(28.75)	28(28.75)	115
Snow Hill, MD	7(6.5)	13(6.5)	2(6.5)	4(6.5)	26

Source: McGrew, J.C. and R.A. Rosing, 1979. *Salisbury/Wicomico County Regional Airport Passenger Survey.* Salisbury, MD: Delmarva Advisory Council.

Table 10.2   Work Table for Chi-square Goodness-of-fit Test for Uniform or Equal Distribution

H_0: population from which sample has been drawn fits a uniform or equal frequency distribution

H_A: population does not fit a uniform or equal frequency distribution

$$\chi^2 = \sum_{i=1}^{k} \frac{(O_i - E_i)^2}{E_i}$$

where: O_i = Observed or actual frequency count in the ith category

E_i = Expected frequency count in the ith category

k = Number of nominal or ordinal categories

From table 10.1 (for Pocomoke City, MD):

Observed Number of Passengers Surveyed

Summer	Fall	Winter	Spring	Total
28	20	18	14	80

If testing goodness-of-fit for uniform or equal distribution, the expected frequency of each category is calculated by dividing the total observed frequency count by the number of categories:

$$E_i = \frac{\Sigma O_i}{k} \quad \text{for all categories } (i)$$

$$E_i = \frac{80}{4} = 20 \quad \text{for Pocomoke City}$$

Expected Number of Passengers (if uniform distribution)

Summer	Fall	Winter	Spring	Total
20	20	20	20	80

$$\chi^2 = \sum_{i=1}^{k} \frac{(O_i - E_i)^2}{E_i} = \frac{(28-20)^2}{20} + \frac{(20-20)^2}{20}$$

$$+ \frac{(18-20)^2}{20} + \frac{(14-20)^2}{20} = 5.20 \qquad p\text{-value} = 0.158$$

tuation exists in airport use from Pocomoke City (i.e., if the null hypothesis is rejected), the chance of a Type I error is 15.8 percent. The logical conclusion would be to support the null hypothesis of no seasonal difference in airport use by the Pocomoke City population.

The chi-square and *p*-value statistics of the other Delmarva communities are presented in table 10.3. Of the six coastal communities, only Ocean City, MD clearly has significant seasonal fluctuation in airport use ($\chi^2 = 32.400; p = 0.000$). Surprisingly, however, the number of passengers sampled from the fall season (99) exceeds the number sampled in summer (89). The Chesapeake Bay coastal community of Crisfield, MD has a calculated chi-square of 7.828 ($p = 0.050$), indicating a relatively significant seasonal fluctuation in airport use. None of the other coastal communities exhibit significant seasonal fluctuation in airport use, although a few have *p*-values in the .10 to .30 range, suggesting a possible seasonal effect.

Table 10.3 Chi-square Statistics and *p*-values for Coastal and Noncoastal Communities

Coastal Communities	Chi-square	*p*-value
Cambridge, MD	6.122	0.106
Chincoteague, VA	2.941	0.401
Crisfield, MD	7.828	0.050
Lewes, DE	3.966	0.265
Ocean City, MD	32.400	0.000
Rehoboth City, DE	6.000	0.112
Noncoastal Communities	**Chi-square**	***p*-value**
Berlin, MD	0.957	0.812
Delmar, DE-MD	0.182	0.980
Georgetown, DE	13.560	0.004
Laurel, DE	8.133	0.043
Milford, DE	4.300	0.116
Millsboro, DE	6.606	0.086
Pocomoke City, MD	5.200	0.158
Princess Anne, MD	0.543	0.909
Salisbury, MD	7.379	0.061
Seaford, DE	1.139	0.768
Snow Hill, MD	10.615	0.014

Among the 11 noncoastal cities, only Georgetown, DE and Snow Hill, MD have calculated chi-square statistics large enough to show significant seasonal fluctuation. In addition, Laurel, DE; Salisbury, MD; and Millsboro, DE, all have *p*-values less than .10, suggesting the possibility of seasonal variation in air passengers. The exploratory research hypothesis is strongly confirmed in several noncoastal cities. Berlin, MD; Delmar, DE-MD; Princess Anne, MD; and Seaford, DE, all have *p*-values exceeding 0.75, implying uniform or even frequency distributions in airport use by season.

Summary Evaluation

The research hypothesis holds to a limited extent. Several noncoastal communities have a more uniform distribution of passengers across the seasons than do any of the coastal communities. The coastal town of Ocean City, MD, in particular, has a decidedly unequal frequency distribution, with many more passengers in the summer and fall than during the winter months. Overall, no definitive general conclusion can be reached from evaluating the chi-square statistics and associated *p*-values. Coastal communities do not generally have more seasonal fluctuation in airport use than do noncoastal communities.

Two potential difficulties associated with the samples should be pointed out. Sample sizes are relatively small to evaluate airport use by season effectively. For more than one half of the communities, the number of passengers surveyed is less than 40, and in several cases less than 25 people were surveyed. From a statistical perspective, the sample sizes are sufficiently large to ensure valid results. However, the small sample sizes also suggest that even though the results are statistically valid, the practical implications may be limited.

The effect of small sample size should always be considered when applying a statistical technique in geographic problem solving. Results from an analysis of small samples can be fully valid from a purely statistical viewpoint, with no violation of assumptions regarding minimum sample size. However, the results may be of limited use for making spatial decisions or reaching realistic conclusions.

The character of the sampling units or values is also a concern in this example. A sampling unit is a passenger boarding the plane at the airport. However, given the sampling design used, it is possible that more than one member of a family or other group traveling together completed the survey questionnaire. In fact, multiple surveys from a number of single families or travel groups were sometimes submitted. Suppose, for example, that a group of several friends from Seaford, DE, enplaned together, and each completed a separate survey form. This situation violates the independent random sample requirement for inferential testing. The frequency and severity of this violation is not known, making it difficult to assess the degree of inaccuracy in this problem.

Proportional or Unequal Distribution

The chi-square statistic is now applied as a goodness-of-fit test to determine whether a sample fits a proportional or unequal distribution. For this application, the χ^2 procedure uses different expected frequency counts in each category. No generally applicable formula is available to calculate these expected frequency counts. Instead, the procedure varies from problem to problem, depending on what is expected from a model, theory, or past experience.

To illustrate this application of the chi-square statistic, an example is presented using information from the General Social Survey (GSS). Conducted by the National Opinion Research Center (NORC) at the University of Chicago, the GSS is administered annually in the United States to a nationwide sample of adults. The sample does not include members of the military or adults residing in institutions such as college dormitories or mental hospitals. The GSS sample selection method is rather complex, involving first a

random sample of cities and counties, followed by a random sample of neighborhoods, a random sample of households, and finally the random selection of an adult from within a household. This sample design procedure may be considered equivalent to a random sample of the population.

A series of questions in the GSS focuses on attitudes toward the role and responsibility of the government in providing various services or performing various tasks. For example, interviewers from NORC ask the following question of respondents: "On the whole, do you think it should or should not be the government's responsibility to provide a job for everyone who wants one?" Four meaningful responses can be made to this question: (1) "definitely should be" [the government's responsibility]; (2) "probably should be"; (3) "probably should not be"; and (4) "definitely should not be." Respondents were asked similarly constructed questions concerning the government's responsibility in the following areas:

- keeping prices under control,
- providing health care for the sick,
- providing a decent standard of living for the old,
- providing industry with the help it needs to grow,
- providing a decent standard of living for the unemployed, and
- reducing income differences between the rich and poor.

Each survey respondent may be categorized by the size and type of place in which he or she lives, which is of interest in geographic research. NORC identifies a respondent's residence by a "size of place" code with 10 categories. To ensure expected frequencies of adequate size for this problem, the NORC classification scheme has been simplified into the following three categories: (1) central city resident (located in the central city of a Metropolitan Statistical Area—city population at least 50,000); (2) suburban resident (located in a suburb of a

Metropolitan Statistical Area); or (3) nonmetropolitan resident (located outside a Metropolitan Statistical Area—either in a small city or town, small incorporated or unincorporated area, or in open country).

The observed "attitude profiles" of central city, suburban, and nonmetropolitan residents are compared with the expected nationwide attitude profiles on each issue. A comparative analysis across the various issues indicates where the most significant differences exist.

Suburbanization has been a dominant trend in the United States since World War II. As the suburban fringe around many American cities has grown over the last few decades, the political, social, and economic viewpoints of these suburban residents can be postulated to reflect the typical national opinions on these issues more closely. A research hypothesis can be formed that the "attitude profile" of suburbanites more closely matches (shows a better goodness-of-fit) the national view than those of center city or nonurban residents.

Both the actual frequency counts by type and size of place and the nationwide "expected" frequency counts for all issues are shown in table 10.4. Note that because some of the original NORC categories were deleted, the frequency counts for the three resident groups cannot be added to obtain the nationwide total.

Central city responses to the question of whether the government should provide a job for everyone who wants one have been used to illustrate the calculation. The categorical frequency distribution of central city responses is compared with the expected frequency distribution of nationwide responses (table 10.5). The nationwide total expected frequency counts have been reduced proportionally to match the 142 responses collected from central city residents. For example, 88 of 638 (or 13.79 percent) of all respondents nationwide indicate that it should definitely be the government's responsibility to provide jobs for all. If central city attitudes perfectly match the national attitude, then 13.79 percent of 142 (or 19.58 respondents) will agree

with this opinion. Thus, the expected number of responses in that category is 19.58.

According to the results of the analysis, the probability that a chi-square value as large as 9.95 could occur by chance if no difference in attitude between central city residents and the national view existed is only .019 (or 1.9 percent). Therefore, concerning views on the government providing jobs for everyone, it appears that the attitude profile of central city residents differs significantly from that found in the national sample.

In a similar manner, chi-square values are calculated for each of the three groups for the series of issues listed earlier (table 10.6). For only two of the seven issues—health care and aid to older Americans—do suburban attitudes more closely reflect the nationwide sample than those of central city or nonmetropolitan residents. That is, for only these two issues is the suburban χ^2 test statistic value lower and p-value higher than central city and nonmetropolitan values. In general, these results do not support the hypothesis that suburban attitudes more closely match the nationwide attitude profile than those found in the central city or nonmetropolitan area.

The differences in viewpoint on health care by type and size of place are particularly significant. Central city residents are much more likely to indicate that the government should provide health care for the sick than would be expected from the nationwide frequency distribution. In table 10.4, the observed number of such responses is 70, but only 51.98 are expected, indicating a rather large difference between observed and expected of $(70 - 51.98) = 18.02$. The opinions of nonmetropolitan area residents on health care $(\chi^2 = 65.636; p = .000)$ show even greater differences from the nationwide view. Very few nonurban respondents support a strong government role in providing health care. This conservatism is reflected by the fact that only five sampled persons in nonmetropolitan locations say the government definitely should be providing health care for the sick, whereas 48.39 such

Table 10.4 Attitude Regarding Government Responsibility to Provide Various Services: *General Social Survey*, 1985

	Attitude Regarding Government Responsibility				
Type or Size of Location	**Definitely Should Be**	**Probably Should Be**	**Probably Should Not Be**	**Definitely Should Not Be**	**Total**
Provide a Job for Everyone Who Wants One.					
Central City	28(19.58)	38(29.82)	34(47.63)	42(44.96)	142
Suburbs	26(28.83)	29(43.90)	80(70.10)	74(66.17)	202
Non-Metro	22(25.66)	45(39.07)	68(62.39)	51(58.89)	186
Nationwide	88	134	214	202	638*
Keep Prices Under Control.					
Central City	51(42.64)	63(64.41)	18(23.30)	10(11.65)	142
Suburbs	57(63.37)	97(95.70)	40(34.62)	17(17.31)	211
Non-Metro	54(57.96)	94(87.54)	29(31.67)	16(15.83)	193
Nationwide	194	293	106	53	646
Provide Health Care for the Sick.					
Central City	70(51.98)	57(68.26)	11(18.52)	7(6.25)	145
Suburbs	70(77.07)	113(101.22)	26(27.45)	6(9.26)	215
Non-Metro	5(48.39)	89(63.66)	34(17.24)	7(5.82)	135
Nationwide	233	306	83	28	650
Provide a Decent Standard of Living for the Old.					
Central City	75(59.42)	59(68.77)	9(14.02)	3(3.78)	146
Suburbs	87(86.69)	102(100.33)	21(20.46)	3(5.52)	213
Non-Metro	65(78.96)	100(91.38)	22(18.63)	7(5.03)	194
Nationwide	267	309	63	17	656
Provide Industry With the Help It Needs to Grow.					
Central City	28(23.21)	64(62.69)	29(34.49)	15(15.62)	136
Suburbs	26(34.64)	105(93.57)	54(51.48)	18(23.31)	203
Non-Metro	35(32.08)	76(86.65)	50(47.67)	27(21.59)	188
Nationwide	107	289	159	72	627
Provide a Decent Standard of Living for the Unemployed.					
Central City	28(21.01)	52(46.52)	32(44.17)	22(22.30)	134
Suburbs	27(31.83)	65(70.48)	75(66.91)	36(33.78)	203
Non-metro	27(29.48)	63(65.27)	68(61.96)	30(31.28)	188
Nationwide	98	217	206	104	627
Reduce Income Differences Between the Rich and Poor.					
Central City	22(22.54)	30(26.79)	31(34.02)	45(44.65)	128
Suburbs	29(35.57)	40(42.28)	61(53.69)	72(70.47)	202
Non-Metro	37(31.17)	36(37.05)	48(47.04)	56(61.74)	177
Nationwide	106	126	160	210	602

Central City—located within the central city of an SMSA/MSA (over 50,000)
Suburbs—located within a suburb of an SMSA/MSA
Non-metro—located outside an SMSA/MSA (either in a small city, town or village, small incorporated or unincorporated area, or in open country)
*Nationwide total does not equal the sum of the category totals, since some of the NORC categories were deleted.

Source: Data from *General Social Surveys* (data files), National Opinion Research Center.

Table 10.5   Work Table for Chi-square Goodness-of-fit Test for Proportional or Unequal Distribution

H_o: population from which sample has been drawn fits a proportional or unequal distribution
H_A: population does not fit a proportional or unequal distribution

$$\chi^2 = \sum_{i=1}^{k} \frac{(O_i - E_i)^2}{E_i}$$

where: O_i = the observed or actual frequency count in the ith category
E_i = the expected frequency count in the ith category
k = the number of nominal or ordinal categories

From table 10.4   Central city resident attitudes toward government providing jobs for everyone:

Observed Attitudes Regarding Government's Responsibility

Definitely Should Be	Probably Should Be	Probably Should Not Be	Definitely Should Not Be	Total
28	38	34	42	142

Calculated Expected Frequency Counts:

	Definitely Should Be	Probably Should Be	Probably Should Not Be	Definitely Should Not Be	Total
Nationwide Frequency Counts	88	134	214	202	638
Percent of Nationwide Responses	13.79	21.01	33.54	31.66	100.00
Expected Frequency Count (if 142 People surveyed)	142 ×.1379 ⎯⎯⎯ 19.58	142 ×.2101 ⎯⎯⎯ 29.82	142 ×.3354 ⎯⎯⎯ 47.63	142 ×.3166 ⎯⎯⎯ 44.96	142

$$\chi^2 = \sum_{i=1}^{k} \frac{(O_i - E_i)^2}{E_i} = \frac{(28-19.58)^2}{19.58} = \frac{(38-29.82)^2}{29.82} + \frac{(34-47.63)^2}{47.63} + \frac{(42-44.96)^2}{44.96} = 9.95$$

p-value = .019

Table 10.6 Chi-square Statistics and *p*-values for Differences in Attitude Regarding Government Responsibilities, by Type of Location: *General Social Survey*, 1985

Type or Size of Location	Chi Square	*p*-value
Provide a Job for Everyone Who Wants One.		
Central City	9.95	.019
Suburbs	7.66	.054
Non-metro	2.98	.394
Keep Prices Under Control.		
Central City	3.11	.375
Suburbs	1.50	.683
Non-metro	0.97	.807
Provide Health Care for the Sick.		
Central City	11.25	.010
Suburbs	3.25	.355
Non-metro	65.64	.000
Provide a Decent Standard of Living for the Old.		
Central City	7.43	.059
Suburbs	1.19	.755
Non-metro	4.66	.198
Provide Industry With the Help It Needs to Grow.		
Central City	1.91	.590
Suburbs	4.89	.180
Non-metro	3.05	.385
Provide a Decent Standard of Living for the Unemployed.		
Central City	6.32	.097
Suburbs	2.28	.516
Non-metro	0.93	.819
Reduce Income Differences Between the Rich and Poor.		
Central City	0.67	.881
Suburbs	2.36	.500
Non-metro	1.68	.642

Central City—located within the central city of an SMSA/MSA (over 50,000)

Suburbs—located within a suburb of an SMSA/MSA

Non-metro—located outside an SMSA/MSA (either in a small city, town or village, small incorporated or unincorporated area, or in open country)

responses are expected if the goodness-of-fit match to the nationwide sample is perfect.

Summary Evaluation

Why do opinions regarding health care differ significantly by type or size of place? Why don't suburban attitudes more closely match the nationwide profile? The general absence of a goodness-of-fit pattern across the various issues makes it difficult to answer these questions and reach any meaningful conclusions.

This problem offers an opportunity to discuss two related validity issues of general concern in geographic problem solving. First, as mentioned in chapter 1, if inappropriate operational definitions are used, subsequent statistical analysis may be invalid. In this example, the terms *central city, suburbs,* and *nonmetropolitan* may not be particularly valid categories for distinguishing attitudes about government responsibility, at least in the context of a national sample of less than 700 people. An alternative study could examine differences in attitude by region of the country or by income or education level of the respondent.

A second validity issue concerns the level of spatial aggregation. The geographer should ask if the spatial aggregation level of the data provide insights about the underlying spatial process. Another research problem could explore attitude differences in and around a single metropolitan area, perhaps by taking a random sample of central city, suburban, and nonmetropolitan residents. Unfortunately, given the national scope of the General Social Survey, this potentially more valid alternative is not possible.

Normal Distribution

In many geographic problems using parametric statistics, the researcher needs to know whether the sample data have been drawn from a normally distributed population. The Kolmogorov-Smirnov one sample test is the most common procedure for evaluating a set of interval or ratio level data for normality. If the goodness-of-fit test shows a close match between an observed distribution of sample data and an expected theoretical normal distribution, the investigator can be confident that the population has a normal distribution.

Thirty years of annual precipitation data have been collected from 10 cities in various world climate regimes. The research question is whether the 30-year precipitation record for each of the cities is normally distributed. Some climatologists argue that places with low levels of precipitation have multi-year precipitation records with considerable relative variability. This suggests a precipitation record that deviates from a normal distribution. Therefore, the hypothesis that areas receiving greater precipitation possess a stronger degree of normality in their distribution of annual precipitation is also examined.

The numerical and graphical methods for completing the K-S test of normality are illustrated using precipitation data for Cairo. This Egyptian city is an example of a desert climate with an average annual precipitation of only 0.81 inches and a maximum rainfall of only 2.87 inches over the sample period. Table 10.7 presents the record of precipitation over the 30-year period and the steps needed to calculate the observed and expected cumulative relative frequencies for the lowest annual precipitation of this period (0.08 inches). Table 10.8 displays the cumulative relative frequencies for all values of precipitation in the sample. These observed and expected cumulative relative frequencies are also shown graphically in an ogive (figure 10.5).

To test whether the two sets of frequencies are significantly different, the difference between the cumulative observed and expected values for each precipitation level (X or Z) is calculated (table 10.8). The same information is shown graphically as the vertical deviation in the ogive (figure 10.5). For the 30-year sample of data in Cairo, the greatest absolute difference between the cumulative relative frequencies

Table 10.7 Work Table for Observed and Expected Cumulative Relative Frequency Values Using the Lowest Precipitation Level for Cairo

H_o: population from which sample is drawn is normally distributed
H_A: population from which sample is drawn is not normally distributed

Sample Data: Thirty Years of Annual Precipitation

0.51	0.31	0.59	0.79	2.87	0.75
0.28	1.06	0.39	1.02	0.67	2.36
0.55	0.75	1.18	1.10	0.47	0.12
0.79	0.83	1.69	0.79	0.59	0.87
0.08	0.31	0.83	0.39	0.39	0.59

Step 1: Rank precipitation data from lowest to highest

0.08	0.39	0.55	0.75	0.83	1.10
0.12	0.39	0.59	0.75	0.83	1.18
0.28	0.39	0.59	0.79	0.87	1.69
0.31	0.47	0.59	0.79	1.02	2.36
0.31	0.51	0.67	0.79	1.06	2.87

Step 2: Calculate mean and standard deviation

$$\overline{X} = 0.81 \text{ inches}; \ s = 0.60 \text{ inches}$$

Step 3: Calculate Z-score for first (lowest) precipitation level ($X = 0.08$ inches)

$$Z = \frac{X - \overline{X}}{s} = \frac{.08 - .81}{.60} = -1.22$$

Step 4: Calculate cumulative relative frequency for observed value ($X = 0.08$)

"What proportion of the *actual* precipitation values are equal to or less than .08 ($Z = -1.22$)?"

Since only one value is less than or equal to 0.08, $CRF_o = 1/30 = 0.033$

Step 5: Calculate cumulative relative frequency for expected (normal) value $Z = -1.22$)

"If the distribution is normal, what proportion of the precipitation values are equal to or less than .08 ($Z = -1.22$)?"

Since $Z < 0$: $CRF_e = 1.00 - (p + .50)$
For $Z = -1.22$, $p = .3880$ and $CRF_e = 1.00 - (.3888 + .50) = .111$

Source: Sample data from National Climatic Data Center, U.S. Dept. of Commerce.

Table 10.8 Kolmogorov-Smirnov Results for Normality Test Using 30 Years of Annual Precipitation Levels for Cairo

X	Z	f	CRF_o	CRF_e	$CRF_o - CRF_e$
.08*	-1.22	1	.033	.111	$-.078$
.12	-1.15	1	.067	.125	$-.058$
.28	-0.88	1	.100	.189	$-.089$
.31	-0.83	1	.133	.203	$-.070$
.39	-0.70	3	.233	.243	$-.010$
.47	-0.56	1	.267	.287	$-.020$
.51	-0.50	1	.300	.310	$-.010$
.55	-0.43	2	.367	.334	.033
.59	-0.36	2	.467	.359	.108
.67	-0.23	1	.500	.410	.090
.75	-0.09	2	.567	.463	.104
.79	-0.03	3	.667	.490	.177
.83	0.04	2	.733	.516	.217
.87	0.11	1	.767	.543	.224**
1.02	0.36	1	.800	.641	$-.159$
1.06	0.43	1	.833	.665	$-.168$
1.10	0.49	1	.867	.689	.178
1.18	0.63	1	.900	.735	.165
1.69	1.48	1	.933	.931	.002
2.36	2.60	1	.967	.995	$-.028$
2.87	3.46	1	1.000	.999	.001

where: X = annual precipitation, in inches
Z = precipitation, in standardized value
f = frequency
CRF_o = cumulative relative frequency (observed)
CRF_e = cumulative relative frequency (expected)

*Calculations for this precipitation level are shown in table 10.7
**D = Maximum | $CRF_o - CRF_e$ | = .224

Figure 10.5 Observed and Expected (Normal) Cumulative Relative Frequency and Maximum Differences (D) for Cairo Precipitation Data

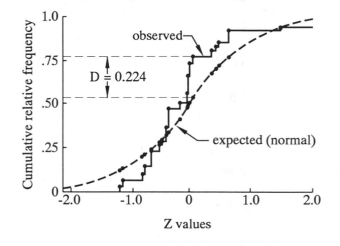

occurs at $Z = +0.11$, where $D = .224$. When this K-S D-value is converted to Z, the computer result ($Z = 1.224$; $p = 0.100$) suggests only a 10 percent chance that the observed sample distribution has been taken from a normal population. Thus, the data from Cairo support the hypothesis that the distribution of annual precipitation has a low degree of normality in an area receiving a low average level of precipitation.

Results from K-S normality tests are shown with the 10 cities ranked from the lowest mean annual precipitation to the highest (table 10.9). As expected, the standard deviation of the 30-year precipitation record generally increases as the mean annual precipitation rises. However, when the dispersion in precipitation is adjusted for the size of the mean using the coefficient of variation, cities with the lowest precipitation generally show the greatest *relative* variability.

Does the degree of normality, as measured by either the K-S D-value or by the corresponding Z-value, also relate to the level of precipitation? In general, this relationship appears to be supported by the sample data (table 10.9). The least normal distribution among the 10 cities occurs in the driest location—Cairo. The wettest location, Rangoon (Yangon), produces a level of normality exceeded only by Vancouver.

Some cities, however, do not fit the hypothesized relationship between degree of normality and level of precipitation. Teheran, for example, the second driest city with an annual average of only nine inches of precipitation, shows a high

Table 10.9 Precipitation Statistics, Kolmogorov-Smirnov Results, and *p*-values for 10 Selected World Cities

City	Mean	Standard Deviation	Coefficient of Variation	K-S *D*	K-S *Z*	*p*-value*
Cairo, Egypt	0.81	0.60	0.741	.224	1.224	0.100
Teheran, Iran	9.01	2.95	0.327	.099	0.545	0.928
Fairbanks, USA	11.28	2.93	0.260	.158	0.867	0.440
Moscow, Russia	23.24	4.44	0.191	.110	0.602	0.862
Rome, Italy	28.37	6.83	0.241	.109	0.599	0.865
Chicago, USA	32.72	6.09	0.186	.113	0.618	0.840
Vancouver, Canada	43.81	6.17	0.141	.083	0.452	0.987
Charleston, USA	52.16	11.66	0.224	.109	0.596	0.870
Singapore	88.03	16.25	0.185	.114	0.627	0.827
Rangoon, Burma**	104.80	13.10	0.125	.083	0.455	0.986

*Two-tailed probability of *Z*
** Yangon, Myanmar

degree of normality. The *p*-value for Teheran (*p* = .928) is exceeded only by those for Vancouver and Rangoon, two locales with much higher levels of precipitation.

Summary Evaluation

Aside from Cairo and Fairbanks, all cities in the study show a very strong probability that the observed precipitation record is normal. In fact, according to the traditional hypothesis testing procedure, Cairo is the only city where the null hypothesis of normality could be rejected. The decision to reject H_0 is made at the relatively weak confidence level of .10, however. In this example, the results suggest that even large deviations between an observed and expected (normal) distribution can meet the test of normality.

Why is the relationship between level of normality and mean annual precipitation not stronger? Perhaps a 30-year sample of data is too small or not representative of precipitation patterns at these locations. The number of cities selected for the example may also be too small to examine the relationship between normality and level of annual precipitation adequately.

10.2 Contingency Analysis

The notion of testing an actual frequency distribution against an expected frequency distribution can be expanded to research problems that examine the relationship between two variables, an extension of the single sample or single variable goodness-of-fit tests. In this instance, however, categorical frequency counts of a single variable are not tested against an expected distribution (such as uniform, proportional, or normal), but rather the frequency distributions of two variables are compared directly with one another.

For example, suppose a city planning agency is developing a five-year capital improvements program (CIP). This program involves the scheduling of public physical improvements based on estimates of future fiscal resources and project costs. As part of the CIP development process, decision makers want citizen input in determining the financing method for several proposed cultural facilities: (1) improving the grounds of a local historical site; (2) constructing a new wing on the art museum; (3) expanding the city aquarium; (4) building a new science center with science education facilities; and (5) constructing a new branch library. Several financing methods are available: general tax revenue, user fees, administrative fees, and revenue loans (bonds). From a survey of residents, a contingency table can be organized that shows for each project the number of people who prefer each funding source. A contingency analysis could then be applied to the frequency counts in this table to determine if the preferred financing methods differ significantly from project to project.

Data values from both variables are assigned to nominal or ordinal categories, and the frequency count of each category from one variable is cross-tabulated with the frequency count of each category from the second variable. The frequencies are summarized in a two-dimensional contingency or **cross-tabulation** table. Each variable must have at least two categories to which data values can be allocated.

The actual frequency count in each cell of the contingency table is compared with the frequency count expected if no relationship exists between the two variables. If at least one difference between actual and expected frequency counts is large, then the variables are not likely to be statistically independent, but rather related in some nonrandom (systematic) fashion. The null hypothesis of no relationship is then rejected. Conversely, if the differences between frequency counts are small, then the variables are concluded to be statistically independent, with only a random association, and the null hypothesis is not rejected.

The chi-square statistic evaluates contingency table frequency counts for differences and is a simple extension of the χ^2 goodness-of-fit test described in the previous section. The entire contingency table is evaluated statistically as a single entity. That is, the analysis determines whether significantly large differences exist between actual and expected frequency counts of the two variables, but does not indicate which cells in the table contain these differences. When both variables have exactly two categories, this analysis poses no problem. With tables having three or more categories for at least one variable, however, further analysis is sometimes needed to learn exactly where the significantly different frequency counts are located.

Certain restrictions apply to the use of chi-square in contingency analysis. Data must be in absolute frequencies rather than in relative frequencies such as percentages or proportions. In addition, no more than one-fifth of the expected frequencies should be less than five and none less than two. As discussed with goodness-of-fit procedures, combining categories is sometimes an option if too many expected frequency counts are too low.

The chi-square statistic used in contingency analysis is:

$$\chi^2 = \sum_{i=1}^{r} \sum_{j=1}^{k} \frac{\left(O_{ij} - E_{ij}\right)^2}{E_{ij}} \qquad (10.8)$$

where: O_{ij} = observed frequency count in the ith row and j^{th} column

E_{ij} = expected frequency count in the ith row and j^{th} column

r = number of rows in the contingency table

k = number of columns in the contingency table

If the observed and expected frequency counts in each cell of the contingency table are similar in value, then the differences $(O_{ij} - E_{ij})$ in all of the cells will be small, the χ^2 statistic will be low, and the null hypothesis of no relationship between the two variables will not be rejected. Conversely, if at least one cell in the contingency table has a large difference between observed and expected frequency counts, χ^2 will also be large, and the null hypothesis will more likely be rejected.

To calculate the expected frequency of a cell in the contingency table, the observed frequency count totals from the row and column for that cell are used:

$$E_{ij} = \frac{R_i C_j}{N}$$ (10.9)

where: R_i = sum of all observed frequency counts in row i

C_j = sum of all observed frequency counts in row j

N = grand total of all observed frequencies

The calculation of expected frequencies is illustrated in the example problem that follows.

Two items regarding these expected frequencies should be noted. First, the frequency count totals of a row or column are often called **marginal totals**. Thus, R_1 is the marginal total for row 1, C_3 is the marginal total for column 3, and so on. Second, even though all observed frequency counts are integers, the calculated expected frequency values

---

Contingency Analysis (χ^2)

Primary Objective: Compare the random sample frequency counts of two variables for statistical independence

Requirements and Assumptions:
1. Single random sample
2. Variables organized by nominal or ordinal categories; frequency counts by category are input to statistical test
3. No more than one-fifth of the frequency counts should be less than five and none of the expected frequencies should be less than two

Hypotheses:
H_o: There is no relationship between two variables in population from which sample has been drawn (variables are statistically independent)
H_A: There is a relationship between two variables in population from which sample has been drawn (variables are not statistically independent)

Test Statistic:

$$\chi^2 = \sum_{i=1}^{r} \sum_{j=1}^{k} \frac{\left(O_{ij} - E_{ij}\right)^2}{E_{ij}}$$

---

will generally not be. Rounding the expected frequency values to integers is not necessary. Doing so would only result in a less precise test.

Application to Geographic Problem

Data from the General Social Survey can also be used to illustrate contingency analysis. Information has been collected on attitudes toward government spending in various areas. Contingency analysis is applied to determine whether these attitudes differ by region of the country and whether attitudes have changed over time.

In the GSS, the following statement is provided to those surveyed: "We are faced with many problems in this country, none of which can be solved easily or inexpensively. I'm going to name some of these problems, and for each one I'd like you to tell me whether you think we're spending too much money on it, too little money, or about the right amount." People were asked to respond to these problems:

- exploring space
- improving and protecting the environment
- improving and protecting the nation's health
- solving the problems of the big cities
- halting the rising crime rate
- dealing with drug addiction
- improving the nation's education system
- improving the conditions of blacks
- meeting the needs of military, armaments, and defense
- aiding other countries
- providing welfare

Certain geographic questions are worth exploring. For example, the individuals surveyed can be categorized by census division. Do significant differences in attitudes toward government spending exist between census divisions? Contingency tables are designed that cross-tabulate the census division of residence with the attitude toward the amount of government spending. By constructing a separate contingency table

for each issue, the issues with the greatest "interdivisional" attitude differences can be identified.

General Social Surveys have been conducted annually for about 20 years. Since many of the same questions are retained from year to year, an analysis of trends in attitudes over time can also be very informative. For example, attitudes about government spending on the space exploration program can be tracked over a number of years to see if nationwide or regional views are converging or diverging. Over which issues has convergence of opinion been most pronounced (perhaps moving toward a national consensus)? Where have regional disparities widened? To explore these questions, the "attitude profile" of 1975 respondents is compared with that of respondents in 1985.

To illustrate the procedure for contingency analysis, those surveyed in 1985 are assigned to their census division of residence (rows of the contingency table). This variable is cross-tabulated with the variable categorizing attitudes about the amount of government spending to solve the problems of big cities (columns of the contingency table). In each cell of the contingency table (table 10.10), the observed frequency

Table 10.10 Contingency Table—Census Division of Respondent Cross-tabulated with Attitude Regarding Government Spending on Big City Problems: *General Social Survey,* 1985

Census Division	The Amount of Government Spending Is—			
	Too Little	About Right	Too Much	Row Total
New England	16 (19.57)	21 (16.24)	7 (8.19)	44
Middle Atlantic	48 (39.59)	29 (32.85)	12 (16.56)	89
E. North Central	59 (57.82)	52 (47.99)	19 (24.19)	130
W. North Central	19 (25.80)	27 (21.41)	12 (10.79)	58
South Atlantic	45 (49.82)	39 (41.34)	28 (20.84)	112
E. South Central	20 (19.57)	12 (16.24)	12 (8.19)	44
W. South Central	28 (28.02)	18 (23.26)	17 (11.72)	63
Mountain	16 (17.35)	19 (14.40)	4 (7.26)	39
Pacific	43 (36.47)	27 (30.27)	12 (15.26)	82
Column Total	294	244	123	661

Note: In each cell, the observed number of respondents is given first, followed by the expected number of respondents in parentheses.

Source: Data from *General Social Surveys* (data files), National Opinion Research Center.

count is given, followed (in parentheses) by the expected frequency count.

The work table for chi-square contingency analysis is shown in table 10.11. To illustrate the calculation procedure for expected frequencies, the expected frequency of row 1 (New England) and column 1 (too little government spending) is also shown.

Chi-square test statistics are calculated for all of the issues for both 1975 and 1985 using a similar procedure. The results are summarized in table 10.12. In 1975, regional differences in attitude about government spending on military, armaments, and defense ($p = .0000$), welfare ($p = .0041$), space exploration ($p = .0043$), and drug rehabilitation ($p = .0070$) were highly significant. In 1985, the most significant regional differences in attitude concerned improving the conditions of blacks ($p = .0000$), the space exploration program ($p = .0001$), military, armaments, and defense ($p = .0027$), and the environment ($p = .0237$).

An increase in the p-value for an issue from 1975 to 1985 indicates a regional convergence of attitudes about government spending on that issue and movement toward a national consensus. Such a trend seems to have occurred for attitudes about halting the rising crime rate. The p-value for government spending to halt crime has moved from $p = .3933$ in 1975 to $p = .9051$ in 1985, reflecting a probable convergence of opinion across census divisions. The contingency table for crime in 1985 shows nearly two-thirds (65.5 percent) of all those surveyed nationwide express the belief that the government spends too little on crime, whereas very few people (5.7 percent) feel that too much money is being allocated to this issue. The contingency table for crime in 1985 looks like table 10.10, but is not shown here. Crime appears to be an issue of national concern, with a similar proportion of people in all divisions of the country desiring increased government involvement.

Table 10.11 Work Table for Chi-square Contingency Analysis

H_o: there is no relationship between two variables (variables are statistically independent, with only a random association)

H_A: there is a relationship between two variables (variables are not statistically independent, but related to one another in some nonrandom fashion)

$$\chi^2 = \sum_{i=1}^{r} \sum_{j=1}^{k} \frac{\left(O_{ij} - E_{ij}\right)^2}{E_{ij}}$$

where: O_{ij} = the observed frequency count in the ith row and j^{th} column

E_{ij} = the expected frequency count in the ith row and j^{th} column

r = the number of rows in the contingency table

k = the number of columns in the contingency table

The row variable is census division and the column variable is attitude regarding the amount of government spending. Example calculation of expected frequency count—New England (row 1) and attitude that amount of government spending on big city problems is too little (column 1)—see table 10.10 for data:

$$E_{ij} = \frac{R_i C_j}{N}$$

$$E_{11} = \frac{R_1 C_1}{N} = \frac{(44)(294)}{661} = 19.57$$

$$\chi^2 = \sum_{i=1}^{r} \sum_{j=1}^{k} \frac{\left(O_{ij} - E_{ij}\right)^2}{E_{ij}} = \frac{(16 - 19.57)^2}{19.57} + \frac{(21 - 16.24)^2}{16.24}$$

$$+ \frac{(7 - 8.19)^2}{8.19} + \ldots + \frac{(27 - 30.27)^2}{30.27} + \frac{(12 - 15.26)^2}{15.26}$$

$$= 25.34$$

$$p\text{-value} = .0640$$

Table 10.12 Chi-square Statistics and *p*-values for Contingency Analyses of Differences in Attitude Between Census Divisions Regarding the Amount of Government Spending on Various Issues: *General Social Surveys*

Issues With Increasing *p*-values from 1975 to 1985 (moving toward national consensus)	1975		1985	
	Chi-square	*p*-value	Chi-square	*p*-value
Halting Rising Crime Rate	16.88	.3933	9.20	.9051
Dealing With Drug Addiction	33.16	.0070	19.53	.2423
Welfare	34.91	.0041	20.69	.1905
Improving the Nation's Education System	22.62	.1244	21.15	.1728
Military, Armaments and Defense	72.15	.0000	36.24	.0027
Issues With Decreasing *p*-values from 1975 to 1985 (moving away from national consensus)				
Improving the Conditions of Blacks	24.56	.0779	62.33	.0000
Space Exploration Program	34.74	.0043	48.13	.0001
Improving and Protecting the Environment	25.42	.0628	29.03	.0237
Improving and Protecting the Nation's Health	12.12	.7358	25.51	.0614
Solving the Problems of Big Cities	17.93	.3278	25.34	.0640
Foreign Aid	13.50	.5357	22.58	.1254

Conversely, if the *p*-value for an issue has decreased from 1975 to 1985, attitudes about government spending are moving away from a national consensus. This trend seems to have occurred with attitudes about improving and protecting the nation's health ($p = .7358$ in 1975 and $p = .0614$ in 1985). Attitudes about government spending on health vary more from region to region in 1985 than 1975, and further analysis is needed to determine which regions are most supportive and least supportive of government spending in health care. National policies or political approaches to health care become more difficult to establish as opinions diverge among regions of the country. With no clear national consensus, national leaders are less likely to suggest solutions.

On the issue of improving and protecting the environment, opinion by census division from 1975 ($\chi^2 = 25.42$; $p = .0628$) to 1985 ($\chi^2 = 29.03$; $p = .0237$) has diverged somewhat. Contingency analysis documents the general divergence of attitude statistically, but cannot determine which specific census divisions have attitude profiles that differ most from the expected frequency counts. A follow-up analysis is needed to learn which pair(s) of census divisions have truly divergent attitudes on the environment.

Summary Evaluation

This contingency analysis raises the same validity issues discussed in the chi-square goodness-of-fit analysis on attitudes about government responsibilities among central city, suburban, and nonmetropolitan residents. Classifying respondents by census division of residence may not be a particularly valid method for distinguishing attitude differences about the level of government spending in various programs. The categories "too little," "about right," and "too much" may not be valid either.

In this analysis, the level of spatial aggregation may not be appropriate to provide insights about the true spatial pattern and processes underlying that pattern. It is quite likely that attitude differences could be explored at a finer spatial level (e.g., around a single metropolitan area) and reveal important spatial trends.

10.3 Difference Test for Ordinal Categories

In section 10.1, the Kolmogorov-Smirnov procedure was used to test differences between a sample distribution and a theoretical normal distribution. Other geographic problems are concerned with examining differences between two distributions drawn from separate independent random samples. If sample observations from the variable being analyzed can be placed in ordinal categories with frequency counts by category, then a two sample version of the K-S procedure can be applied to test for significant differences between samples. In this application, cumulative relative frequencies are compared category by category. As with goodness-of-fit, the maximum difference between the two sample cumulative frequency distributions represents the value of the test statistic.

Suppose a coastal geomorphologist studying the characteristics of beaches wishes to test for differences in sand size between two beaches having different orientations. Beach A is within a sheltered area away from direct wave action, whereas Beach B faces direct and constant impact from the sea. The investigator has collected independent random samples of sand from both beach areas and wants to know whether significant size differences exist. Particles of sand are put through a series of sieves, categorized by particle size, and then counted. Sand particle categories are very coarse, coarse, medium, fine, and very fine. The distribution of sand sizes from the two beaches can be tested for significant difference using the Kolmogorov-Smirnov procedure.

In another example, suppose a health planner is interested in analyzing data on premature births in two hospitals that serve patient populations from varying socioeconomic backgrounds. Information from admissions suggests that the county-funded hospital in the central city serves many people of lower socioeconomic status (including many on public assistance), whereas the privately run, suburban hospital admits more middle- and upper-income patients with medical insurance. The planner suspects a greater incidence of premature births occur at the public hospital. To test this hypothesis, a random sample of 200 births is drawn from each hospital during the study period. Both samples are organized into four classes: births occurring more than 9 weeks prematurely, those 6-8 weeks early, those 2-5 weeks early, and those less than 2 weeks before the expected date. A Kolmogorov-Smirnov two sample test could be applied to test for significant differences in the cumulative relative frequencies for the two hospitals.

When applied to a problem with two samples, two observed cumulative relative frequencies need to be calculated. As in the one sample problem, these results can be shown graphically as an ogive. The procedure examines the maximum difference, D, between the two cumulative distributions, determined either from the table of calculated values or from the ogive:

$$D = \text{maximum} \left| \text{CRF}_o(X_1) - \text{CRF}_o(X_2) \right| \quad (10.10)$$

where: $\text{CRF}_o(X_1)$ = cumulative relative
frequencies (observed)
for sample distribution X_1

$\text{CRF}_o(X_2)$ = cumulative relative
frequencies (observed)
for sample distribution X_2

If D is large, the two sample distributions are likely to have come from different populations. If D is small, differences between the two sample distributions are minor, suggesting that the samples have been drawn from the same population. Like other inferential tests, the p-value associated with the test statistic indicates the probability of such a difference occurring by chance if indeed no difference actually exists in the populations from which the samples have been taken.

Kolmogorov-Smirnov Categorical Difference Test

Primary Objective: Compare the random sample frequency counts of two samples or groups for differences

Requirements and Assumptions:
1.   Two independent random samples or groups
2.   Variables organized by ordinal category for each sample or group; frequency counts by category are input to statistical test
3.   Variables measured at ordinal scale or downgraded from interval/ratio scale to ordinal

Hypotheses:
  H_o: There is no difference between the two population distributions from which the two samples or groups have been drawn
  H_A: There is a difference between the two population distributions from which the two samples or groups have been drawn

Test Statistic:
  $D = \text{maximum} \mid \text{CRF}_o(X_1) - \text{CRF}_o(X_2) \mid$

Application to Geographic Problem

In section 10.2, the chi-square test was used to examine attitude differences among census divisions with respect to government spending in various programs. Efforts to improve and protect the environment are of particular interest to geographers. Results from the chi-square contingency analysis show that regional differences for the environmental issue became more apparent from 1975 to 1985, suggesting a divergence of opinion from a national point of view. However, results from the contingency analysis will not show which parts of the country have the most pronounced differences. Regional differences in environmental attitudes will be explored more fully in this example.

For which pairs of census divisions are attitude differences most pronounced? Where are they most similar? Because the original survey

data categorized the attitudes of respondents concerning spending into three ordinal classes—too little, about right, and too much—the Kolmogorov-Smirnov two sample test for differences in ordinal categories is appropriate. Relative cumulative frequencies are calculated for each census division, and the distributions compared with those for other census divisions. A test statistic, D, is determined for each of the 36 unique pairs of census divisions, resulting in an exploratory, comparative analysis on the issue of environmental protection.

The computations for the New England and Middle Atlantic census division pair are shown in table 10.13. The maximum difference in proportion between the two cumulative frequency distributions is .0799, with a corresponding p-value of .990. This result suggests that virtually no difference exists between the attitudes of New

Table 10.13   Work Table for Kolmogorov-Smirnov Two Sample Test—Government Spending for Improving and Protecting the Environment: New England vs. Middle Atlantic

H_0:   there is no difference in the two population distributions from which the two samples are drawn

H_A: there is a difference in the two population distributions from which the two samples are drawn.

The observed frequency counts for the two samples:

	Too Little	About Right	Too Much	Total
New England	28	15	2	45
Middle Atlantic	66	24	4	94

The frequency counts converted to proportions:

	Too Little	About Right	Too Much
New England	.6222	.3333	.0445
Middle Atlantic	.7021	.2553	.0426

The proportions converted to cumulative proportions:

	Too Little	About Right	Too Much
New England	.6222	.9555	1.0000
Middle Atlantic	.7021	.9574	1.0000
Difference	.0799*	.0019	.0000

K-S (D) = .0799 (p = .990)

*Maximum Difference

Source: Data from *General Social Surveys* (data files), National Opinion Research Center.

England and Middle Atlantic respondents regarding government spending on improving and protecting the environment.

Kolmogorov-Smirnov D values are calculated for all 36 pairs of census divisions (table 10.14). Most of the differences between census divisions are small; in only one of the 36 cells—West North Central vs. Middle Atlantic—is D large enough to produce a p-value significant at the .05 level (D = .2406; p = .023). Over 70 percent of the respondents in the Middle Atlantic division believe the government is spending too little on environmental protection, whereas fewer than 50 percent of those in the West North Central division share this view. In the latter division more than in any other, the attitude seems to be that the government is spending too much on environmental concerns.

Summary Evaluation

From this exploratory analysis, attitudes toward government spending for the environment show few significant differences among census divisions. The sample sizes are probably too small for meaningful interdivisional comparisons, even though statistical conclusions can be reached. P-value sizes are directly influenced by the number of ordinal categories. Because this problem contains only three categories, differences in frequency counts must be very large to produce statistically significant results.

A simple categorization of respondents into census divisions hides key dimensions that may produce strong variation in attitudes. For example, significant differences within divisions likely exist between the young and old, urban and rural residents, as well as those in different socioeconomic groups. If so, this would be an example of using a somewhat invalid level of spatial aggregation to approach the problem.

Table 10.14   Kolmogorov-Smirnov D Statistics and p-values of Differences by Census Division Regarding Government Spending to Improve and Protect the Environment: *General Social Survey*, 1985

Census Division	MA	ENC	WNC	SA	ESC	WSC	M	P
New England (NE)	.0799* (.990)	.0370 (1.000)	.1607 (.498)	.0531 (1.000)	.0499 (1.000)	.0365 (1.000)	.1329 (.812)	.0850 (.984)
Middle Atlantic (MA)		.1169 (.435)	.2406 (.023)	.1330 (.302)	.0983 (.898)	.1164 (.648)	.2128 (.117)	.0904 (.859)
East North Central (ENC)			.1237 (.514)	.0327 (1.000)	.0186 (1.000)	.0386 (1.000)	.0958 (.906)	.0479 (1.000)
West North Central (WNC)				.1512 (.285)	.1422 (.596)	.1571 (.376)	.0936 (.971)	.1502 (.377)
South Atlantic (SA)					.0456 (1.000)	.0166 (1.000)	.0797 (.982)	.0806 (.899)
East South Central (ESC)						.0515 (1.000)	.1144 (.900)	.0351 (1.000)
West South Central (WSC)							.0964 (.957)	.0865 (.936)
Mountain (M)								.1224 (.755)
Pacific (P)								

*Top value in each cell = K-S D
Bottom value in parentheses = p-value.

Key Terms and Concepts

References and Additional Reading

The statistical techniques covered in this chapter are discussed in most introductory textbooks. Refer to the list of general references which cover statistical methods in geography, located at the end of the text prior to the tables in the Appendix.

CHAPTER 11

INFERENTIAL SPATIAL STATISTICS

11.1 Point Pattern Analysis

11.2 Area Pattern Analysis

Geographers often examine spatial patterns on the earth's surface produced by physical or cultural processes. These patterns represent the spatial distribution of a variable across a study area. Sometimes geographic variables are displayed as point patterns with dot maps. In chapter 4, descriptive spatial statistics (or geostatistics) such as the mean center and standard distance were introduced to summarize point patterns. In other instances, explicit spatial patterns representing data summarized for a series of subareas within a larger study region can be displayed effectively using choropleth maps.

Whether data are presented as points or areas, geographers often want to describe and explain an existing pattern. In this chapter, the focus is on inferential spatial statistics that analyze a sample point pattern or sample area pattern to determine if the arrangement is random or nonrandom. When testing a point or area pattern for randomness, the question is whether the population pattern from which the sample has been created was generated by a spatially random process.

Geographers may want to compare an existing spatial pattern to a particular theoretical pattern. Spatial patterns may appear clustered, dispersed, or random (figure 11.1). In case 1, both the point and area patterns have a **clustered** appearance. On the point pattern map, the density of points appears to vary significantly from one part of the study area to another, with many points concentrated in the northwest portion of the area. Perhaps the points represent sites of tertiary economic activity (retail and service functions), which often cluster around a location with high accessibility and high profit potential, such as a highway interchange. On the clustered area pattern map, the shaded subareas could represent political precincts where a majority of registered voters are Democrat, with Republican majority precincts unshaded. Such a clustered pattern would likely occur if there is a distinctly nonrandom spatial distribution of voters by income, race, or ethnicity in the region.

Other spatial patterns seem evenly **dispersed** or regular. The set of points in case 2 (figure 11.1)

appears uniformly distributed across the study area, suggesting that a systematic spatial process produced the locational pattern. The hypothesis in classical central place theory, for example, is that settlements are uniformly distributed across the landscape to best serve the needs of a dispersed rural population. The area pattern in case 2 exhibits a regular or alternating type of spatial arrangement. This pattern could represent county populations in the same region where the central place distribution of settlements is hypothesized. Shaded counties could have above average populations while unshaded counties are below average.

The spatial patterns in case 3 (figure 11.1) appear **random** in nature, with no dominant trend toward either clustering or dispersion. A random

Figure 11.1 Types of Point and Area Patterns

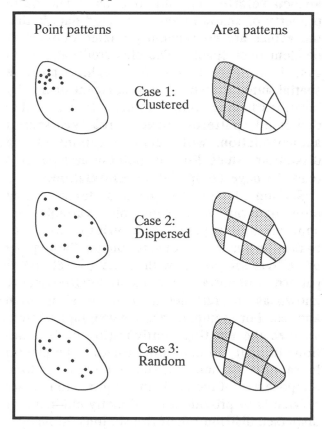

213

point or area pattern logically suggests that a spatially random (Poisson) process is operating to produce the pattern. The Illinois tornado "touch down" locations discussed in section 5.4 illustrate a process that produces a random point pattern. If a choropleth map were designed that showed the number of tornado touchdowns per square mile or kilometer in each Illinois county, a random area pattern would likely result.

In most geographic problems, a point or area pattern will not provide a totally clear indication that the pattern is clustered, dispersed, or random. Rather, many real-world patterns show a combination of these arrangements with tendencies from "purely" random toward either clustered or dispersed.

If a significant nonrandom arrangement is identified in a point or area pattern, **spatial autocorrelation** is said to exist. The basic property of spatially autocorrelated data is that the values are nonrandomly related or interdependent over space. The clustered patterns in case 1 of figure 11.1 seem to exhibit **positive** spatial autocorrelation, with adjacent or nearby locations having similar values. In case 2, the dispersed patterns have **negative** spatial autocorrelation, with nearby locations having dissimilar values. Random point or area patterns (case 3) have no spatial autocorrelation.

Section 11.1 presents methods for analyzing point patterns where geographic variables are appropriately displayed as a series of locations or dots on a map. In some problems, the *spacing of individual points* within the overall point pattern is important, and a statistical procedure known as "nearest neighbor analysis" is often applied. For example, an urban geographer might analyze the existing configuration of fire stations in a city to determine whether the pattern is random or more dispersed than random. Suppose one of the goals in the provision of fire service is to provide a locationally equitable or dispersed distribution of fire stations throughout the city. The geographer might suggest the siting of new stations or relocation of existing

facilities to meet this goal. The proposed new configuration of fire stations could then be analyzed to determine if it were more dispersed than the existing pattern.

In other instances, the *nature of the overall point pattern* is important, and a statistical test known as "quadrat analysis" is often appropriate. For example, from a random sample of trees in a national forest, a biogeographer might determine which trees are diseased, plot the location of each diseased tree as a point on a map, and conduct an analysis of the points. The study would determine whether the trees were randomly distributed throughout the study area or clustered in certain portions of the forest. The result of this analysis would provide some guidance concerning the most appropriate type of treatment (e.g., widespread aerial spraying versus concentrated treatment from the ground).

Section 11.2 presents methods for analyzing area patterns. For example, an economic development planner might be interested in evaluating changing spatial patterns of poverty in Appalachia. Through area pattern analysis of a chronological series of county-level choropleth maps showing poverty in the Appalachian region, the spatial distribution of poverty can be evaluated as becoming more concentrated or clustered over time. An urban geographer could analyze a map pattern depicting the number of existing home sales by census tract to determine the degree to which such sales are spatially concentrated in certain portions of the city. If a nonrandom clustering of high turnover rates is found in certain segments of the city, attention could be focused on why the sales exhibit such spatial patterns.

11.1 Point Pattern Analysis

The spatial patterns of many geographic variables can be portrayed as dot maps. For example, urban geographers plot the location of settlements as points on a map. Economic geographers study the spatial pattern of retail activities by mapping store locations as dots on the map. Physical geographers show glacial features like drumlins as a series of dots.

A primary objective of many geographic studies that begin with locations of a variable on a dot map is to determine the form of the pattern of points. The nature of the point pattern can reveal information about the process that produced the geographic result. In addition, a series of point patterns of the same variable recorded at different times can help to determine temporal changes in the locational process. In short, point pattern analysis offers the researcher quantitative tools for examining a spatial arrangement of point locations on the landscape as represented by a conventional dot map.

Nearest Neighbor Analysis

Nearest neighbor analysis is a common procedure for determining the spatial arrangement of a pattern of points within a study area. The distance of each point to its "nearest neighbor" is measured and the average nearest neighbor distance for all points is determined. The spacing within a point pattern can be analyzed by comparing this observed average distance to some expected average distance, such as that for a random (Poisson) distribution.

The nearest neighbor technique was developed originally by biologists who were interested in studying the spacing of plant species within a region. They measured the distance separating each plant from its nearest neighbor of the same species and determined whether this arrangement was organized in some manner or was the result of a random process. Geographers have applied the technique in numerous research problems, including the study of settlements in central place theory, economic functions within an urban region, and the distribution of earthquake epicenters in an active seismic region. In all applications, the objective of nearest neighbor analysis is to describe the pattern of points within a study

area and make inferences about the underlying process.

The nearest neighbor methodology is illustrated using the example of seven points shown in figure 11.2. A coordinate system is created and the horizontal (X) and vertical (Y) positions of the points recorded (table 11.1). For each of the points, the nearest neighbor (NN) is determined as the point closest in straight-line (Euclidean) distance. The distances to each nearest neighbor (NND) are then calculated from the coordinates or measured from the map. From the set of nearest neighbor distances, the *average* nearest neighbor distance ($\overline{\text{NND}}$) is determined using the basic formula for the mean:

$$\overline{\text{NND}} = \frac{\Sigma \text{NND}}{n} \qquad (11.1)$$

where: n = number of points

Using the seven nearest neighbor distances in table 11.1, the average nearest neighbor distance

---

Figure 11.2   Location of Points for Nearest Neighbor Problem

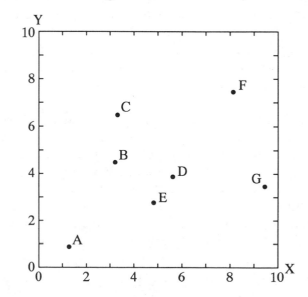

Table 11.1 Coordinates and Nearest Neighbor Information for Example[*]

Point	X	Y	NN	NND
A	1.3	0.9	B	3.98
B	3.2	4.4	C	2.00
C	3.3	6.4	B	2.00
D	5.6	3.8	E	1.36
E	4.8	2.7	D	1.36
F	8.1	7.4	G	4.21
G	9.4	3.4	F	4.21
Sum				19.12

where: NN = Nearest Neighbor
NND = Nearest Neighbor Distance

$$\overline{NND} = \frac{\Sigma NND}{n} = \frac{19.12}{7} = 2.73$$

[*]See figure 11.2 for graph of point locations

is 2.73, indicating that an average distance of 2.73 units separates a point from its nearest neighbor.

The average nearest neighbor distance provides an index of spacing for a set of points. However, the usefulness of this descriptive index comes from comparing the index value for an observed pattern to the results produced from certain distinct point distributions. This objective is analogous to the situation discussed in section 10.1, where an observed frequency distribution was compared to a perfectly normal frequency distribution.

In point pattern analysis, average nearest neighbor distances can be calculated for three distinct point arrangements: random, perfectly dispersed, and perfectly clustered. In each case, the spacing index is determined for an area containing a certain density of points. For example, if points are arranged in a random spatial pattern, the average nearest neighbor distance ($\overline{NND}_R$) would be determined as follows:

$$\overline{NND}_R = \frac{1}{2\sqrt{Density}} \qquad (11.2)$$

where: $\overline{NND}_R$ = average nearest neighbor distance in a random pattern

Density = number of points (n)/Area

Since the study area displayed in figure 11.2 has a dimension of 10 by 10 units, the area represented is 100 square units. The corresponding density of points is therefore 7/100 or .07. For an area with this point density, a random arrangement of points within the study area should produce an average nearest neighbor distance:

$$\overline{NND}_R = \frac{1}{2\sqrt{.07}} = 1.89$$

If the arrangement of points shows maximum dispersion or a perfectly uniform pattern, the average nearest neighbor distance ($\overline{NND}_D$) would be determined as:

$$\overline{NND}_D = \frac{1.07453}{\sqrt{Density}} \qquad (11.3)$$

Since this pattern represents the most dispersed or separated arrangement of points, it also serves as the maximum value for the nearest neighbor index. The average nearest neighbor distance for a dispersed pattern whose point density matches that in figure 11.2 is:

$$\overline{NND}_D = \frac{1.07453}{\sqrt{.07}} = 4.06$$

The pattern of spacing most distinct from dispersion or uniformity is clustering. When all points lie at the same position, the pattern shows maximum or total clustering of points, and each nearest neighbor distance is zero. Therefore, in a perfectly clustered pattern, the average nearest neighbor distance ($\overline{NND}_C$) is also zero, representing the lowest possible value for the index:

$$\overline{NND}_C = 0 \qquad (11.4)$$

Since perfectly clustered and perfectly dispersed patterns provide the extreme spacing arrangements for a set of points, the nearest neighbor index offers a useful method to measure the spacing of locations within an observed point pattern. However, the average nearest neighbor distance is an absolute (as opposed to relative) index, which depends on the units for measuring distance. Therefore, direct comparison of results from different problems or different regions is difficult. Although the minimum value of the nearest neighbor index is always 0 (a perfectly clustered pattern), the maximum value corresponding to a perfectly dispersed pattern is not constant, but a function of the point density. To overcome this, a **standardized nearest neighbor index (R)** is often used. This index is found by dividing the average nearest neighbor distance ($\overline{NND}$) by the corresponding value for a random distribution with the same point density:

$$R = \frac{\overline{NND}}{\overline{NND}_R} \qquad (11.5)$$

With the standardized index, a perfectly clustered pattern produces an R value of 0.0, a random distribution of 1.0, and a perfectly dispersed arrangement generates the maximum R value of 2.149 (figure 11.3). Thus, an actual point pattern can be measured for relative spacing along a continuous scale from perfectly clustered to perfectly dispersed. For the set of points from figure 11.2, the standardized nearest neighbor index is:

$$R = \frac{2.73}{1.89} = 1.44$$

This spacing pattern is moderately dispersed, since it lies between a perfectly dispersed distribution and a random one.

In addition to its use as a descriptive index of point spacing, the nearest neighbor methodology

Figure 11.3 Continuum of R Values in Nearest Neighbor Analysis

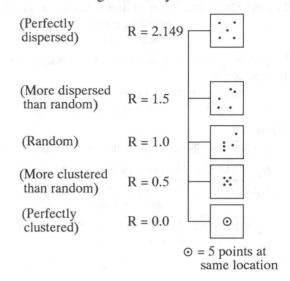

(Perfectly dispersed) $R = 2.149$

(More dispersed than random) $R = 1.5$

(Random) $R = 1.0$

(More clustered than random) $R = 0.5$

(Perfectly clustered) $R = 0.0$

⊙ = 5 points at same location

Source: Modified from Taylor, P. J. 1977. *Quantitative Methods in Geography.* Boston: Houghton Mifflin

can also be used to infer results from a sample of points to the population from which the sample was drawn. A difference test can be used to determine if the observed nearest neighbor index ($\overline{NND}$) differs significantly from the theoretical norm ($\overline{NND}_R$), which would occur if the points were randomly arranged. The expectation is that a Poisson process operates over space to produce a random pattern of points. The null hypothesis is that no difference exists between observed and random nearest neighbor values. The test statistic (Z_n) follows a format similar to other difference tests discussed in chapters 8 and 9:

$$Z_n = \frac{\overline{NND} - \overline{NND}_R}{\sigma_{\overline{NND}}} \qquad (11.6)$$

where: $\sigma_{\overline{NND}}$ = standard error of the mean nearest neighbor distances

The standard error for the nearest neighbor test can be estimated with the following formula:

$$\sigma_{\overline{\text{NND}}} = \frac{0.26136}{\sqrt{n(\text{Density})}} \qquad (11.7)$$

where: n = number of points
        Density = density of points

Either a directional (one-tailed) or nondirectional (two-tailed) approach can be applied, depending on the form of the alternate hypothesis (H_A). If the problem suggests a clear rationale for the actual point pattern being either dispersed or clustered (as opposed to the null hypothesis of randomness), a one-tailed approach is warranted. However, without an underlying reason or theoretical expectation that the null hypothesis should be rejected on one side as opposed to the other, a two-tailed, nondirectional approach is best.

Application to Geographic Problem

The nearest neighbor methodology is applied to point patterns representing community service sites within a city. Some public services are best located in a highly dispersed pattern to provide relatively equal service distance for all parts of the region. Such a concern for spatial equity is especially important for emergency services, such as police and fire, as well as for neighborhood schools. Other community activities may be sited to offer the region a high degree of efficiency, with less concern for equal spacing of services within the region. Many nonemergency services exhibit more clustered arrangements in their point patterns, possibly reflecting financial rather than spatial constraints.

Four community services in Baltimore, Maryland, are selected for a nearest neighbor analysis of their spacing. Two of the services—police and fire—provide emergency protection from their facility sites to various locations within the region. The other two services—elementary education and recreation—provide nonemergency services for persons who travel to these locations. The facility sites for the four services are shown in figure 11.4.

A nearest neighbor analysis is applied to each pattern to measure the spacing of service sites. The calculation procedure for analyzing the spacing of police facilities is summarized in table 11.2. The nearest neighbor distances (in miles) for each of the 11 sites are shown in figure 11.5. The average nearest neighbor distance for police stations in Baltimore is 1.63 miles. If the stations had been distributed randomly across the city, the average spacing would have been 1.36 miles (table 11.2). To hold the influence of point density constant and allow more useful comparisons of nearest neighbor results, a standardized nearest neighbor index (R) is calculated. This ratio of $\overline{\text{NND}}$ to $\overline{\text{NND}}_R$ for police stations in Baltimore ($R = 1.20$) suggests that the actual spacing of points is more dispersed than random. Although dispersal of sites is expected for an emergency public service, does the R value

Figure 11.4 Location of Selected Public Facilities in Baltimore, Maryland

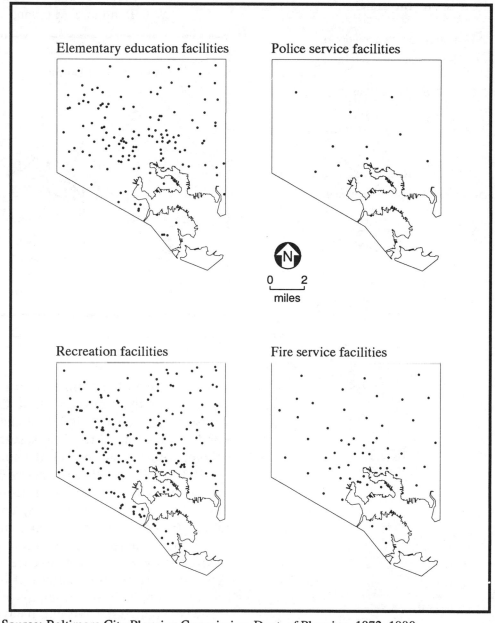

Source: Baltimore City Planning Commission, Dept. of Planning, 1972, 1990.

of 1.20 differ significantly from 1.0, the result for a random pattern of points?

Equations 11.6 and 11.7 are used to test for a significant difference between an observed nearest neighbor index and the corresponding index value for a random spacing of points. The null hypothesis is that the observed and random nearest neighbor distances are equal. In this problem, a one-tailed (directional) test is logical, since police facilities are hypothesized to

Table 11.2 Work Table for Nearest Neighbor
Analysis: Police Service Facilities
in Baltimore, Maryland

H_0: $\overline{NND} = \overline{NND}_R$ (point pattern is random)
H_A: $\overline{NND} \neq \overline{NND}_R$ (point pattern is not random)

Calculate Mean Nearest Neighbor Distance:

$$\overline{NND} = \frac{\Sigma\,NND}{n} = \frac{17.97}{11} = 1.63$$

where: n = number of points

Calculate Random Nearest Neighbor Distance:

$$\overline{NND}_R = \frac{1}{2\sqrt{Density}}$$

where: Density = n/Area

$$\overline{NND}_R = \frac{1}{2\sqrt{11/80.86}} = \frac{1}{2\sqrt{.1360}} = 1.36$$

Calculate Standardized Nearest Neighbor Index:

$$R = \frac{\overline{NND}}{\overline{NND}_R} = \frac{1.63}{1.36} = 1.20$$

Calculate Test Statistic:

$$Z_n = \frac{\overline{NND} - \overline{NND}_R}{\sigma_{\overline{NND}}}$$

where:

$$\sigma_{\overline{NND}} = \frac{0.26136}{\sqrt{n(Density)}}$$

$$\sigma_{\overline{NND}} = \frac{0.26136}{\sqrt{11(1.36)}} = \frac{.26136}{1.223} = .214$$

$$Z_n = \frac{1.63 - 1.36}{.214} = \frac{.27}{.214} = 1.26$$

$$p = 0.1038$$

Figure 11.5 Nearest Neighbor Distances
for Police Service Facilities
in Baltimore, Maryland

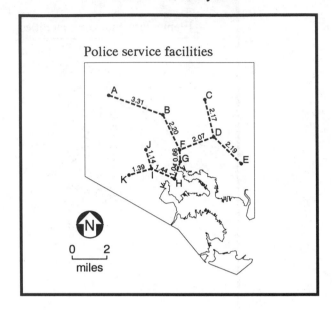

have a dispersed locational pattern. Therefore, the alternate hypothesis is that the observed nearest neighbor index is larger than the corresponding random value. The resulting test statistic for police locations in Baltimore ($Z_n = 1.26$; $p = .1038$) indicates that the researcher has less than 90 percent certainty in rejecting the null hypothesis and concluding the spacing of police stations is more dispersed than random.

Inferential interpretations in this problem can be made only with extreme care. It could be argued that the existing pattern of locations is a "natural" sample from the many possible locations where police stations could have been sited. If this argument is not accepted, the inferential assumptions of a random sample of points, independently selected, are not met. The researcher may want to focus on a descriptive comparison of nearest neighbor indices (R values) for the four community services.

The corresponding nearest neighbor values and test statistic results for each of the four service patterns are shown in table 11.3. The public services are ranked in decreasing size of the R index (from more dispersed to more clustered). The pattern of fire stations in Baltimore is the most dispersed ($R = 1.84$), supporting the assumption of an even distribution of emergency services to meet an equity requirement. The second most dispersed service is elementary education facilities ($R = 1.62$). This result also matches the general perception of location strategy, since primary schools are traditionally sited as neighborhood facilities to minimize the travel distance for younger children.

The most clustered spacing pattern among the four services is recreational facilities ($R = 0.85$), the only service where the random nearest neighbor distance exceeds the observed average nearest neighbor distance. Unlike emergency and educational services, recreation does not require an even or dispersed pattern of facilities, which is reflected in its lower R index.

Summary Evaluation

Spacing of facilities for the four selected services in Baltimore confirms the expected arrangements of emergency and nonemergency services (figure 11.6). As hypothesized, the pattern of emergency fire services exhibits the most

dispersed spacing, followed closely by the pattern of elementary schools. Also as expected, recreation facilities in Baltimore are more clustered than random. Although police services appear more dispersed than random, statistical evidence suggests that the spacing does not differ significantly from a random pattern.

These results, however, are influenced by the way in which the problem was structured. A critical issue when using the nearest neighbor procedure is the delimitation of the study area boundary. In many research problems (like the

Figure 11.6 Continuum of R Values in Nearest Neighbor Analysis: Baltimore Services Displayed

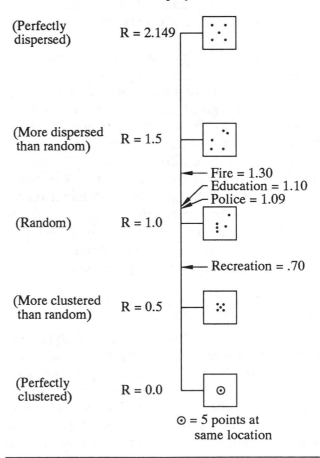

Table 11.3 Nearest Neighbor Values for Selected Public Facilities in Baltimore, Maryland

Public Facility	$\overline{NND}$	Density	$\overline{NND}_R$	R	Z_n	p
Fire	1.12	.680	.61	1.84	11.86	0.0000
Elementary Education	0.63	1.633	.39	1.62	13.33	0.0000
Police	1.63	.136	1.36	1.20	1.26	0.1038
Recreation	0.28	2.325	.33	0.85	–3.85	0.0001

analysis of public services in Baltimore), a political boundary defines the study area. In other problems, where no formal or logical boundary exists for enclosing the point locations, the researcher must designate an operational boundary or spatial limit for the region under study. The position of the boundary does not directly affect the distance between nearest neighbors or the average nearest neighbor index. However, boundary position does affect both the area of the study region and the point density, factors that determine the random nearest neighbor distance. Therefore, specification of the study area boundary influences the outcome of a point pattern analysis. Ideally, a functional study area boundary should be defined that is consistent with the type of points being analyzed. Some researchers suggest that the boundary should be defined just beyond the outermost points within the study area.

Another issue related to study area boundary is the specification of nearest neighbors at the edge of a region. In some problems, the nearest neighbor for a point located near the boundary may actually lie outside the study area. Perhaps the nearest fire station or school lies across the border at a shorter distance than the nearest neighbor lying inside the study area. Therefore, how nearest neighbor distances are handled near the study area boundary influences the results of a study. Administrative boundaries do not necessarily offer the best approach to delimiting the study area in point pattern problems.

Quadrat Analysis

An alternate methodology for studying the spatial arrangement of point locations is **quadrat analysis.** Rather than focusing on the spacing of points within a study area, quadrat analysis examines the frequency of points occurring in various parts of the area. A set of quadrats or square cells is superimposed on the study area, and the number of points in each cell is determined. By analyzing the distribution of cell frequencies, the point pattern arrangement within the study area can be described.

Whereas nearest neighbor analysis concerns the average spacing of the closest points, quadrat analysis considers the variability in the number of points per cell (figure 11.7). If each of the quadrats contains the same number of points (case 1), the pattern would show no variability in frequencies from cell to cell and would be perfectly dispersed. By contrast, if a wide disparity exists in the number of points per cell for the set of quadrats examined (case 2), the variability of the cell frequencies would be large, and the pattern would display a clustered arrangement. In a third alternative (case 3), the variability of cell frequencies is moderate, and the pattern of points would reflect a random or near random spatial arrangement.

However, the absolute variability of the cell frequencies cannot be used as an effective descriptive measure of the point pattern because it is influenced by the density of points, the mean number of points per cell. This relationship is directly analogous to the influence of the mean on the standard deviation of a variable. Recall from section 3.2 that the coefficient of variation is used for meaningful comparisons of relative variability between distributions. In quadrat analysis, an index known as the **variance-mean ratio (VMR)** standardizes the degree of variability in cell frequencies relative to the mean cell frequency:

$$VMR = \frac{VAR}{MEAN} \qquad (11.8)$$

where: VMR = variance-mean ratio
         VAR = variance of the cell frequencies
         MEAN = mean cell frequency

The mean and variance are the basic descriptive statistics that summarize the central tendency and variability of a variable. Since the data from quadrat analysis are usually summarized as frequency counts (number of points in each quadrat), formulas for

Figure 11.7   Quadrat Analysis: The Relationship of Cell Frequency Variability and Point Pattern Form

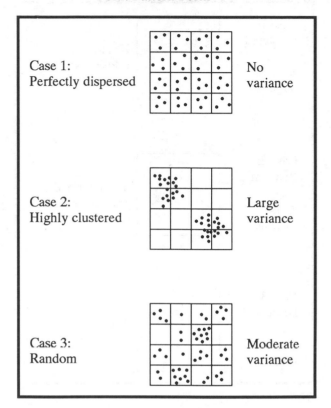

Case 1: Perfectly dispersed — No variance

Case 2: Highly clustered — Large variance

Case 3: Random — Moderate variance

the grouped mean and grouped variance are usually applied (see tables 3.2 and 3.5).

Interpretation of the variance-mean ratio offers insight into the spatial pattern of points within the study area. In a dispersed distribution of points, for example, the cell frequencies will be similar, and variance of cell frequency counts will be very low. In an extreme case (such as case 1 in figure 11.7), if each quadrat contains exactly the same number of points, the variance will be zero, making the variance-mean ratio zero. Conversely, if a point pattern is highly clustered with most cells containing no points and a few cells having many points, the variance of cell frequencies will be large relative to the mean cell frequency. This type of spatial pattern will produce a large value for the variance-mean ratio.

If the set of points is randomly arranged across the cells of the study area, an intermediate value of variance will occur. For a perfectly random point pattern, the variance of the cell frequency is equal to the mean cell frequency. To understand this result, recall from section 5.4 that the Poisson distribution is used to describe the frequency of values for a randomly generated spatial or temporal pattern. In the Poisson distribution, the mean frequency equals the variance of the frequencies. Therefore, when using the variance-mean ratio to describe spatial point patterns, a result close to one (variance equals mean) suggests that the distribution has a random arrangement.

In addition to its use as a descriptive index, the variance-mean ratio can also be applied inferentially to test a distribution for randomness. The test statistic is chi-square, defined as a function of both the VMR and number of cells (m):

$$\chi^2 = \text{VMR} \ (m - 1) \qquad (11.9)$$

The null hypothesis for this test is expressed as no difference between the observed distribution of points and a distribution of points resulting from a random process (that is, VMR = 1). Either a directional (one-tailed) or nondirectional (two-tailed) approach can be used, depending on the test objective of the alternate hypothesis.

Rejection of the null hypothesis can occur if a point pattern is more clustered than random (VMR >1) or more dispersed than random (VMR < 1). The difference test determines whether an observed VMR value differs significantly from one, the theoretical random result. The interpretation of the chi-square test statistic and corresponding p-value reflects these two possibilities. A large VMR generally produces a larger chi-square value (with a small p-value), suggesting greater variability of the cell frequencies and a clustered arrangement of points. Conversely, a small VMR produces a smaller chi-square value, indicating lower variability of cell frequencies and a more dispersed distribution of points. In such

Quadrat Analysis

Primary Objective: Determine whether a point
                    pattern has been generated by a
                    random (Poisson) process

Requirements and Assumptions (when using the test
inferentially):
    1. Random sample of points from a population
    2. Sample points are independently selected

Hypotheses:
    H_o: VMR = 1 (point pattern is random)
    H_A: VMR ≠ 1 (point pattern is not random)
    H_A: VMR > 1 (point pattern is more clustered
            than random)
    H_A: VMR < 1 (point pattern is more dispersed
            than random)

Test Statistic:

$$\chi^2 = \text{VMR } (m-1)$$

Figure 11.8   Quadrat Analysis: The Relationship
of Chi-square Value and *p*-value

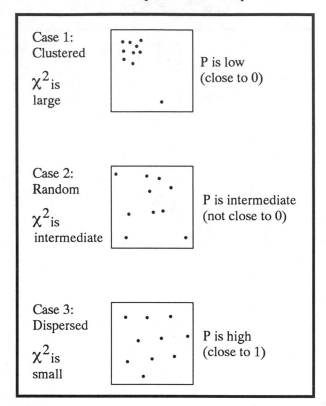

instances, the corresponding *p*-value is larger. Intermediate chi-square values result from variance-
mean ratios closer to one, suggesting a lower
probability of rejecting the null hypothesis and
therefore more likelihood that the distribution of
points is actually random. Thus, a continuum of
p-values from 0 to 1 indicates transition from clustered to random to dispersed (figure 11.8). Either
very high or very low *p*-values suggest rejection of
the null hypothesis, while intermediate values do
not support rejection.

Application to Geographic Problem

The example of Illinois tornado touch down
locations (section 5.4) is reexamined here using
quadrat analysis. The objective now is to describe the pattern of points within the study area
by analyzing the distribution of cell frequencies.
The resulting frequency pattern indicates whether
the points have a dispersed, random, or clustered
arrangement. The chi-square difference test is
then applied to determine whether the observed
pattern differs significantly from a theoretical
random pattern.

In this analysis, the tornado touch down points
are assumed to be an independent random sample.
This assumption is justified if the sample is
considered a "natural" sample over one time
period. This particular set of tornado touch down
sites is one pattern from an infinite number of
patterns that could have occurred from 1916 to
1969.

As discussed in section 5.4, 63 of the 85 cells
covering the state have at least half of their area
in Illinois (figure 11.9). This set of 63 quadrats
excludes only 30 of the 480 recorded tornadoes
for the state. The mean frequency of 7.14 tornadoes per cell represents the density of points
across the study area relative to the quadrat size
(table 11.4). The variability in cell frequencies
around this mean largely determines the nature

Figure 11.9 Illinois Tornado Touch Down Pattern with Quadrats Superimposed

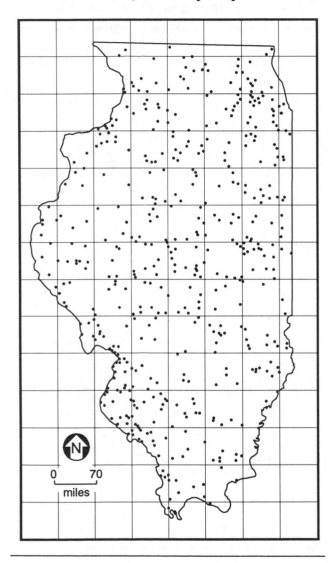

Table 11.4 Work Table for Quadrat Analysis: Illinois Tornado Example

H_o: VMR = 1 (point pattern is random)
H_A: VMR ≠ 1 (point pattern is not random)

(See table 5.6 for frequency data)

Calculate mean cell frequency:

$$\text{Mean Cell Frequency} = \frac{n}{m}$$

where: n = number of points
m = number of cells

$$\text{Mean} = \frac{450}{63} = 7.14$$

Calculate variance of cell frequencies:

$$\text{VAR} = \frac{\sum f_i X_i^2 - \left(\sum f_i X_i\right)^2 / m}{m-1}$$

where: f_i = frequency of cells with i tornadoes
X_i = number of tornadoes per cell

$$\text{VAR} = \frac{3986 - 450^2/63}{62} = \frac{3986 - 3214.29}{62} = 12.45$$

Calculate variance-mean ratio:

$$\text{VMR} = \frac{\text{VAR}}{\text{Mean}} = \frac{12.45}{7.14} = 1.74$$

Test of significance:

$$\chi^2 = \text{VMR}\,(m-1) = 1.74\,(62) = 107.88$$

$$p = .0003$$

of the point pattern. For this set of quadrats, the variance of the tornado frequencies is 12.45 per cell. The two summary statistics produce a variance-mean ratio of 1.74, suggesting a frequency pattern for tornadoes that is on the clustered side of random.

A difference test is applied to determine whether the VMR of 1.74 differs significantly from that of a random pattern (VMR = 1). The null hypothesis

assumes no difference between the observed and expected VMRs. Selection of a directional (one-tailed) or nondirectional (two-tailed) test depends on whether *a priori* evidence exists to suggest a point pattern that is more clustered than random or more dispersed than random. If no evidence suggests the pattern should be clustered (or dispersed), a nondirectional, two-tailed difference test is appropriate. Conversely, an alternate hypothesis that assumes, *a priori,* a clustered or dispersed point pattern requires a directional, one-tailed difference test.

Since no climatological evidence suggests that the spatial pattern of tornadoes in Illinois is anything but random, a two-tailed difference test of the null hypothesis seems the logical choice. The resulting chi-square test statistic of 107.88 produces a corresponding *p*-value of .0003 (table 11.4). Thus, the researcher has very strong confidence that the null hypothesis should be rejected and that the resultant point pattern is more clustered than random.

Does the size of quadrats used to cover the study area influence the results from a quadrat analysis? To examine this question, the same pattern of tornadoes in Illinois is analyzed using a cell design with quadrats one-fourth as large. This cell structure produces 263 smaller cells covering 469 of the state's 480 tornadoes (figure 11.10). To be consistent, boundary cells whose area lies more than half outside Illinois are eliminated.

The denser cell network produces more quadrats with low frequencies and a smaller mean cell frequency of 1.78 tornadoes per quadrat. The variance of the cell frequencies about this mean is also lower at 2.31; since the variance remains larger than the mean, the pattern again displays a clustered arrangement. Standardizing this variance to the mean produces a variance-mean ratio of 1.30, much lower than the value found with the larger quadrat size (VMR = 1.74). However, when the ratio is tested for significant difference from VMR = 1, the chi-square value of 340.6 produces a very significant result (*p* = .0000). Thus, using the 263 smaller quadrats suggests a very small probability of error when rejecting the null hypothesis of no difference between the observed pattern of points and a random distribution. The researcher can be very

Figure 11.10 Illinois Tornado Touch Down Pattern with Smaller Quadrats Superimposed

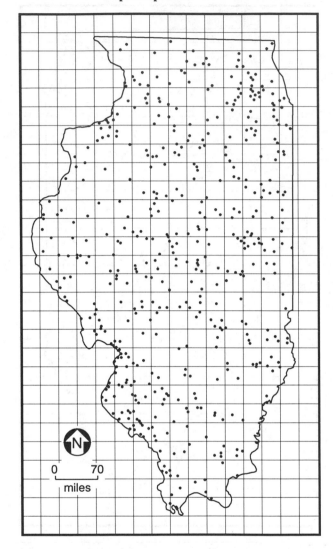

confident that the pattern of tornado touch downs is more clustered than random.

Summary Evaluation

Although quadrat analysis offers a useful approach to studying the spatial arrangement of points, several procedural issues must be considered. Different cell sizes produce varying levels of the mean point frequency and variance per cell. Moreover, studies have shown that if the point pattern is held constant but the quadrat size is decreased (i.e., the number of quadrats is increased), the variance of the point frequencies usually declines faster than the mean. This situation suggests that the smaller-sized quadrats often produce smaller VMR values. Thus, different cell sizes in a single problem can generally be expected to produce different values of the variance-mean ratio.

Although more research needs to be conducted on optimal quadrat size, two guidelines can be offered. When visual examination of a point pattern reveals distinct clusters, the size of quadrats should match that of the clusters. This cell size allows the results from quadrat analysis to reflect more accurately the visual impression of the pattern. Other researchers have suggested a general "rule of thumb" that cell size should be equal to twice the average area per point. In other words, the mean frequency of points should be close to 2.0 per quadrat. This guideline approximates the cell structure used with the 263 cells covering the tornado pattern in Illinois. Taylor (1977) presents a more detailed discussion of cell size in quadrat problems.

Another issue in the use of quadrat analysis also requires some attention. Like nearest neighbor analysis, delimitation of the study area can influence the results of a study. In the use of quadrat analysis, a way to handle cells around the study area boundary needs to be determined. Cells on the boundary will inevitably contain area both inside and outside the study region. In the Illinois tornado problem, cells with more than half of their area outside the study region were eliminated. Alternative decision rules can be used, and these may change the results obtained from the analysis. No matter what decision rule is applied in a particular problem, the geographer needs to be consistent in handling the boundary issue.

11.2 Area Pattern Analysis

The goals and objectives in **area pattern analysis** are similar to those in point pattern analysis. Just as nearest neighbor and quadrat analysis examine the random or nonrandom nature of point patterns, descriptive and inferential procedures are available to analyze area patterns. A choropleth or area pattern map is considered clustered if adjacent contiguous areas tend to have highly similar values or scores. Alternatively, if the values of adjacent areas tend to be dissimilar, then the spatial pattern is considered more dispersed than random (figure 11.1).

Area pattern analysis is appropriate for studying many practical problems in geography. It is particularly valuable to examine the way in which area patterns of a specific variable change over time. For example, suppose a medical geographer is concerned with the diffusion of influenza in a metropolitan area. If the number of cases is reported by census tract for several time periods, the morbidity rate for influenza could be displayed on a series of choropleth maps and the degree of nonrandomness in the patterns analyzed. If the incidence of the disease becomes more dispersed over successive time periods, researchers would be very concerned, because the dispersion indicates that the influenza is spreading more evenly throughout the metro area. It could also mean a significant number of cases were appearing in other areas not close to existing areas of high incidence. Conversely, if the morbidity rate patterns were becoming more clustered over successive time periods, then a closer examination of those areas with higher incidence of influenza would be appropriate. Such knowledge could lead to the implementation of effective strategies for disease control.

Suppose a political geographer wants to study the spatial pattern of voter registration rates across a number of precincts within a community. If registration rates have been mapped by precinct before each of the last several elections, a useful comparative analysis of area patterns is possible. If the pattern is becoming more clustered over successive time periods, a different registration strategy may be necessary. An analysis of area pattern trends might also provide insights into demographic or economic variables that seem to be related to the voter registration rate.

Many methods of area pattern analysis are available, depending on the level of data measurement and the way in which the variable under analysis has been organized. Spatial autocorrelation measures are available for variables organized in nominal, ordinal, and interval/ratio form. However, this chapter focuses exclusively on area patterns shown in binary form. Most of the important concepts and theoretical issues dealing with area pattern analysis can be illustrated with binary maps. In addition, variables are frequently converted into binary form for area pattern analysis. Each census tract in the influenza morbidity rate example could be classified as having either an above average or below average incidence of influenza. Similarly, each of the community precincts in the voter registration rate example could be classified as above or below average. Those interested in area pattern analysis of spatial data not organized in simple binary form should consult the references cited at the end of the chapter.

The Join Count Statistic

The **join count** is the basic organizational statistic for the analysis of area patterns. A "join" is operationally defined as two areas sharing a common edge or boundary. The procedures involved in calculating a join count statistic are relatively straightforward. The fundamental building block is the number of joins in the pattern and the nature of the join structure in the study area. Figure 11.11 illustrates the simplest situation: a binary classification of data and a

Figure 11.11 Join Structure and Examples of Area Patterns Appearing Clustered, Dispersed, and Random

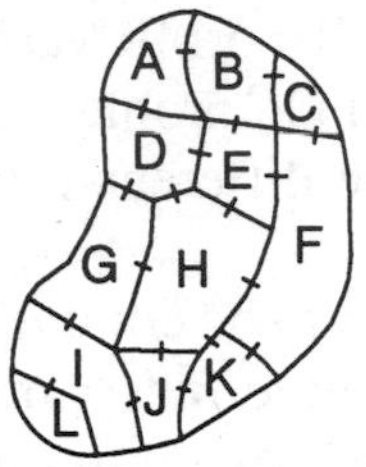

Area	Number of joins
A	2
B	3
C	2
D	4
E	4
F	4
G	3
H	6
I	3
J	3
K	3
L	1

two-category choropleth map. Alternative join count procedures are available for groupings of more than two categories. Many variables in geography are binary in nature, or can be downgraded effectively to binary without sacrificing important locational information.

Figure 11.11 shows the join structure for the areas in the study region. For example, area A is joined to two other areas—B and D. The number of joins associated with each area is listed adjacent to the join structure map.

To comply with accepted practice, each area can be identified as either "black" or "white." It is then possible to refer to the *observed number* of "black-white," "black-black," and "white-

white" joins in the pattern. In case 1 of figure 11.11, the area pattern appears clustered, with relatively few black-white (dissimilar) joins and a rather large number of black-black and white-white (similar) joins. In case 2, a dispersed (alternating checkerboard) pattern has more black-white joins and fewer joins of similar-category areas. With random patterning (case 3), the result is an intermediate number of both similar and dissimilar joins. These situations are all reflected in table 11.5.

To determine if a certain join count distribution has been generated by a random process, the number of black-white joins that would be expected from a theoretical random arrangement of binary observations must be calculated. That is, the *expected number* of black-white joins in a purely random pattern is compared with that actually observed. If the observed join count is significantly different from the expected join count, the observed pattern can be described as nonrandom.

How is the expected number of black-white joins determined? The locational context or problem setting in which the area pattern is being evaluated suggests two possible approaches for generating expected join counts. Some area pattern analyses may be conducted under the

Table 11.5 Join Counts Associated with Area Patterns

Case	Total* Joins	Dissimilar Areas Joined Black-White Joins	Similar Areas Joined Black-Black Joins	White-White Joins	Total
1 (Clustered)	19	5	7	7	14
2 (Dispersed)	19	15	0	4	4
3 (Random)	19	12	4	3	7

*See figure 11.11 for map of join structure and area patterns.

hypothesis of **free sampling**. Free sampling should be used if the researcher can determine the probability of an area being either black or white based on some theoretical notion or with reference to a larger study area. Once the probability of an area being black or white has been determined, the expected number of black-white joins in a random pattern having those probabilities can also be determined. For example, in the influenza morbidity rate problem, suppose each census tract throughout the metropolitan area is classified as having either above average incidence of the disease (black) or below average incidence (white). If the pattern of disease is being analyzed in only one portion of the metropolitan area, logic would suggest that the probability of above- or below-average incidence of influenza should be typical or representative of that found throughout the metro area, and the free sampling hypothesis would be appropriate.

The **nonfree sampling** hypothesis is used when no appropriate reference to a larger study area or general theory is logical or possible. In this case, only the study region itself is considered in the analysis, and the expected number of black-white joins is estimated from a random patterning of only those subareas in the study region. For instance, a political geographer studying the spatial pattern of community voter registration rates by precinct may not have any reason to believe this pattern is typical or representative of any larger area (such as the state in which the community is located). Therefore, if it is hypothesized that the area pattern has resulted from the unique characteristics found within the study area, then a nonfree sampling hypothesis is appropriate.

When the researcher cannot refer to a larger study area to determine the expected number of black-white joins, the nonfree sampling context should be used. In many geographic problems, it is difficult to conclude with any degree of confidence that a subarea pattern is typical of some larger area. Therefore, use of the expected number of black-white joins from a nonfree sampling hypothesis frees the researcher from having to make a restrictive assumption.

Free Sampling Test Procedure

To determine if an area pattern has been generated by a random process, the observed number of black-white joins must be compared with the number of such joins expected in a theoretical random arrangement. In this framework, the null hypothesis is that the observed number of black-white joins is equal to the expected number of black-white joins in a random area pattern. If reason exists to hypothesize that the observed area pattern is nonrandom in a certain direction, then a one-tailed alternate hypothesis is appropriate. For instance, an area pattern could be hypothesized either as more clustered than random or more dispersed than random. Without an *a priori* reason to hypothesize a clustered or dispersed pattern, a two-tailed alternate hypothesis is selected, and the conclusion will be that an area pattern is random or nonrandom.

The test statistic (Z_b) for free sampling is:

$$Z_b = \frac{O_{BW} - E_{BW}}{\sigma_{BW}} \qquad (11.10)$$

where: O_{BW} = observed number of black-white joins
E_{BW} = expected number of black-white joins
σ_{BW} = standard error of the expected number of black-white joins

The observed number of black-white joins is easily determined by counting the actual number of dissimilar joins on the map. The expected number of black-white joins is calculated as follows:

$$E_{BW} = 2Jpq \qquad (11.11)$$

where: J = total number of observed joins
p = actual relative probability that an area will be black
q = actual relative probability that an area will be white

Since free sampling assumes the researcher is working from a specific theory or in reference to a larger study region, the study area probabilities (p and q) are determined by theoretical expectations or derived from the larger region. The largest expected number of black-white (dissimilar) joins will occur in patterns having an equal number of black and white areas. If $p = q = .50$, then the pattern contains the same number of black and white areas, and the product (pq) is maximized [$pq = (.50)(.50) = .25$]. All other combinations of p and q will result in lower-magnitude products and a lower expected number of black-white joins. For example, in an area pattern with 90 percent black areas and 10 percent white areas, it makes sense that most of the joins in the study region would be black-black (similar) joins, and the expected number of black-white joins would be lower than in the 50 percent-50 percent case.

The standard error of expected black-white joins (the denominator of the test statistic) is calculated as follows:

$$\sigma_{BW} = \sqrt{[2J + \Sigma L(L-1)]pq - 4[J + \Sigma L(L-1)]p^2 q^2}$$

$$(11.12)$$

where: σ_{BW} = standard error of the expected number of black-white joins
J = total number of observed joins
ΣL = total number of links in the area pattern
p = actual relative probability that an area will be black
q = actual relative probability that an area will be white

Note: $\Sigma L = 2J$: If areas A and B are joined, then A is "linked" to B and B is "linked" to A, making the sum of all links twice the number of joins

If the observed number of black-white joins is larger than the expected number of black-white joins, then Z_b will be a positive value, indicating a pattern more dispersed than random. A relatively large number of black-white (dissimilar)

joins will occur in a dispersed pattern (figure 11.11). Conversely, if the observed number of black-white joins is smaller than the expected number of such joins, then Z_b will be negative, and the pattern will appear more clustered than random. The greater the magnitude of Z_b (either positive or negative), the greater the likelihood that the area pattern being analyzed is not random. Just like other Z-score difference tests, this Z_b test statistic for area pattern analysis indicates the number of standard deviations separating the expected random black-white join count from the observed value.

Nonfree Sampling Test Procedure

The nonfree sampling procedure is similar to free sampling, with one important difference. Nonfree sampling is used when the area under study does not refer to a larger study area or to general theory. In nonfree sampling, only the study region itself is analyzed. As suggested earlier, nonfree sampling should be chosen when any doubt concerning the tests is present. With this procedure, the researcher is able to avoid making an incorrect assumption regarding the nature of the study area in some larger region.

The null hypothesis in nonfree sampling is the same as in free sampling: the observed number of black-white joins is hypothesized to be no different from the expected number of black-white joins in a random area pattern. If it is appropriate to hypothesize either a clustered or dispersed pattern, the one-tailed alternative should be used.

The test statistic for nonfree sampling appears identical to that used in free sampling. However, the equations for calculating expected black-white joins and the standard error of those joins are somewhat different. The test statistic (Z_b) for nonfree sampling is:

$$Z_b = \frac{O_{BW} - E_{BW}}{\sigma_{BW}} \qquad (11.13)$$

where: O_{BW} = observed number of black-white joins
E_{BW} = expected number of black-white joins
σ_{BW} = standard error of the expected number of black-white joins

In nonfree sampling, the expected number of black-white joins is calculated by incorporating the observed number of black areas and white areas directly into the equation:

$$E_{BW} = \frac{2JBW}{N(N-1)} \qquad (11.14)$$

where: E_{BW} = expected number of black-white joins
J = total number of joins
B = number of black areas
W = number of white areas
N = total number of areas (black plus white)
$\quad = B + W$

In nonfree sampling, the expected number of black-white joins in the study region is determined by asking how many black-white joins would occur from a theoretical random pattern containing the same number of black and white areas as the observed area pattern. The question becomes how many black-white joins would be generated from a random arrangement of the observed number of black and white areas in the area pattern.

Certain generalizations can be made concerning the expected number of black-white joins. Given a particular total number of areas ($N = B + W$), the value of the denominator in equation 11.14 is constant. If most of the areas in the observed pattern are either black or white, then their product (BW) in the numerator will be a relatively smaller number, producing a smaller expected number of black-white joins. Conversely, if about the same number of black and white areas are found in the observed pattern, then their product will be a relatively larger number, producing a relatively larger expected number of black-white joins. These results seem

---

Area Pattern Analysis (Binary Categories)

Primary Objective: Determine whether a binary (two category) area pattern has been generated by a random process

Requirements and Assumptions:
1.   Each area assigned to one of two nominal categories
2.   Each pair of areas must be defined as either adjacent ("joined") or nonadjacent (not "joined") in a consistent manner

Hypotheses:

H_o: $O_{BW} = E_{BW}$ (area pattern is random)

H_A: $O_{BW} \neq E_{BW}$ (area pattern is not random)

H_A: $O_{BW} > E_{BW}$ (area pattern is more dispersed than random)

H_A: $O_{BW} < E_{BW}$ (area pattern is more clustered than random)

Test Statistic:

$$Z_b = \frac{O_{BW} - E_{BW}}{\sigma_{BW}}$$

The area pattern may now be analyzed with the nonfree sampling hypothesis (equation 11.13 with supplementary equations 11.14 and 11.15). As with free sampling, a positive Z_b statistic will result if the observed number of black-white joins is larger than the expected number of black-white joins, indicating a pattern more dispersed than random. Conversely, a negative Z_b statistic occurs if the observed number of black-white joins is smaller than expected, reflecting an area pattern more clustered than random. The greater the absolute magnitude of Z_b, the greater the likelihood that the area pattern under analysis is not random.

Application to Geographic Problems

Free Sampling Test Procedure

Suppose a political geographer wants to determine if the area pattern of political party affiliation of governors in the eastern portion of the United States is random (figure 11.12). Since this study area is part of a larger region (the United States), it can be logically hypothesized that the proportion of eastern states having Democratic and Republican governors should be consistent with the proportion of Democratic and Republican governors nationwide. With this assumption, the free sampling hypothesis is appropriate.

The null hypothesis is that the observed gubernatorial pattern is random. An examination of the pattern in figure 11.12 does not clearly suggest either significant clustering or dispersion of governors by political party affiliation. Evidence of clustering can be seen in the upper Mississippi River basin and in northern New

logical, for a pattern with a large number of black areas and very few white areas (or vice versa) would not be expected to have a large number of black-white joins. The expected number of black-white joins is maximized when an equal number of black and white areas occur in the overall pattern.

Not all randomly generated patterns will contain exactly the same number of black-white joins, so a measure of the amount of variability expected due to sampling must be included. The standard error of expected black-white joins is:

$$\sigma_{BW} = \sqrt{E_{BW} + \frac{\Sigma L(L-1)BW}{N(N-1)} + \frac{4\left[J(J-1) - \Sigma L(L-1)\right]B(B-1)W(W-1)}{N(N-1)(N-2)(N-3)} - E_{BW}^2} \qquad (11.15)$$

where: σ_{BW} = standard error of the expected number of black-white joins
         J = total number of joins
         ΣL = total number of links = $2J$

B = number of black areas
W = number of white areas
N = total number of areas (black plus white)
    = $B + W$

Figure 11.12   Area Pattern Map and Join Counts: Political Party Affiliation
of Governors in Eastern United States, 1990

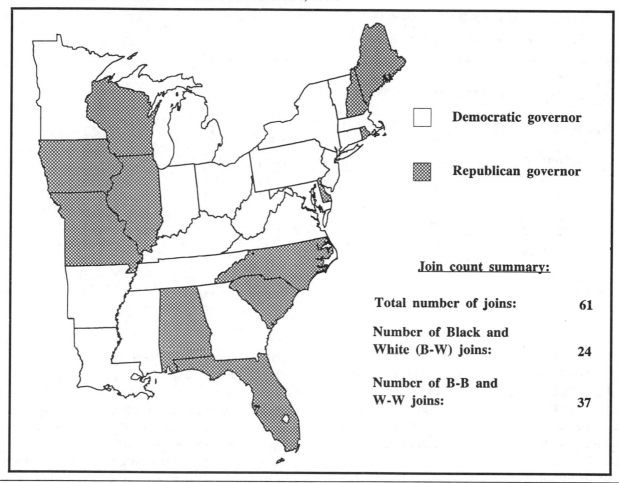

☐ **Democratic governor**

▦ **Republican governor**

Join count summary:

Total number of joins:	**61**
Number of Black and White (B-W) joins:	**24**
Number of B-B and W-W joins:	**37**

England, but dispersal can also be noted as the Republican governors of Delaware and Rhode Island are surrounded by states with Democratic governors. The most appropriate alternate hypothesis would therefore seem to be nondirectional or two-tailed, and the pattern is tested as either random or nonrandom.

The map pattern has 61 total joins, of which 24 are black-white (dissimilar) joins and 37 are either black-black or white-white (similar) joins. A join count summary is provided in figure 11.12. The research question is whether this observed number of black-white joins (24) is significantly different from the number of dissimilar joins expected in a random pattern of Republican and Democratic governors across the United States.

Table 11.6 summarizes the calculation procedure for area pattern analysis used with the free

sampling hypothesis. The observed number of black-white joins is 24. To determine the expected number of dissimilar joins, the nationwide proportions of Democratic and Republican governors are calculated. In the 48 contiguous states, 27 states have Democratic governors and 21 states have Republican governors, for respective probabilities of .5625 and .4375. With the free sampling hypothesis, these national probabilities are incorporated directly into the equation for the expected number of black-white joins in a random pattern (table 11.6). In contrast with the 24 observed black-white joins, this result indicates that approximately 30 dissimilar joins are expected in a random pattern having respective probabilities of .5625 and .4375.

To determine whether this contrast between observed and expected black-white joins is sta-

Table 11.6 Work Table for Area Pattern Analysis (Free Sampling): Political Party Affiliation of State Governors in Eastern United States, 1990

H_0: $O_{BW} = E_{BW}$ (area pattern is random)

H_A: $O_{BW} \neq E_{BW}$ (area pattern is not random)

$O_{BW} = 24$

$E_{BW} = 2Jpq = 2(61)(.5625)(.4375) = 30.02$

$$\sigma_{BW} = \sqrt{[2J + \Sigma L(L-1)]pq - 4[J + \Sigma L(L-1)]p^2 q^2}$$

$$= \sqrt{[2(61) + 430](.5625)(.4375) - 4[61 + 430](.5625)^2 (.4375)^2}$$

$$= 4.11$$

$$Z_b = \frac{O_{BW} - E_{BW}}{\sigma_{BW}} = \frac{24 - 30.02}{4.11} = -1.465$$

p-value $= .1430$

tistically significant, the standard error of expected black-white joins (σ_{BW}) must be incorporated into the test statistic. This requires a tabulation of all linkages between states in the study area pattern (table 11.7). The standard error of expected black-white joins is calculated as 4.11 (table 11.6).

The resultant test statistic value (Z_b) is –1.465. The negative value indicates a tendency toward clustering of Democratic and Republican governors in the eastern United States. However, if evaluated using the two-tailed approach, the result is not highly significant ($p = .1430$). If it had been specifically hypothesized that the pattern were more clustered than random, the resultant p-value would be .0715.

As with the point pattern analyses presented earlier in the chapter, caution must be taken when making inferential (p-value) interpretations in this problem. The "natural" sampling argument may again be valid. However, more emphasis should be placed on the descriptive result (tendency toward clustering) than on the magnitude of the resultant p-value.

Nonfree Sampling Procedure

Now suppose a political geographer wants to evaluate the area pattern of governors by political party affiliation across the entire continental United States (figure 11.13). Alaska and Hawaii are excluded from this analysis, since they are not adjacent or "joined" to any other states. The pattern appears to be fairly complex, with neither strong clustering nor dispersal, making a nondirectional or two-tailed test appropriate. The null hypothesis is that the observed gubernatorial pattern is random, while the alternate hypothesis is that the pattern is nonrandom.

In this example, a nonfree sampling procedure should be chosen. There is no logical reference to a larger study area pattern or to a general theory when determining the expected number of black-white joins: only the pattern of governors in the United States is being analyzed.

Across the continental United States, there are 105 total joins—45 dissimilar (Democratic-Republican), and 60 similar (either Democratic-Democratic or Republican-Republican). The join count summary is shown in figure 11.13. The

Table 11.7 State Linkage Pattern: Political Party Affiliation of State Governors in Eastern United States, 1990

State	Number of Links (L)	L − 1	L(L − 1)
Alabama	4	3	12
Arkansas	4	3	12
Connecticut	3	2	6
Delaware	3	2	6
Florida	2	1	2
Georgia	5	4	20
Illinois	5	4	20
Indiana	4	3	12
Iowa	4	3	12
Kentucky	7	6	42
Louisiana	2	1	2
Maine	1	0	0
Maryland	4	3	12
Massachusetts	5	4	20
Michigan	3	2	6
Minnesota	2	1	2
Mississippi	4	3	12
Missouri	5	4	20
New Hampshire	3	2	6
New Jersey	3	2	6
New York	5	4	20
North Carolina	4	3	12
Ohio	5	4	20
Pennsylvania	6	5	30
Rhode Island	2	1	2
South Carolina	2	1	2
Tennessee	8	7	56
Vermont	3	2	6
Virginia	5	4	20
West Virginia	5	4	20
Wisconsin	4	3	12

$\Sigma L = 122$ $\Sigma L(L-1) = 430$

research task is to determine if the observed number of dissimilar joins (45) is significantly different from the number of dissimilar joins expected in a random gubernatorial pattern.

The worktable for area pattern analysis with the nonfree sampling hypothesis is presented in table 11.8. The observed number of dissimilar joins ($O_{BW} = 45$) has been counted from the map. The expected number of dissimilar joins in a purely random pattern is calculated by incorporating the observed number of states with Democratic and Republican governors directly into the equation. The result (as summarized in table 11.8) indicates that nearly 53 dissimilar joins are expected in a random pattern.

It is worth reemphasizing that many random arrangements of black and white areas are possible in a study region containing 48 total areas, with 21 in one category and 27 in the other. The standard error of expected dissimilar joins (σ_{BW}) must therefore be included in the test statistic. Just as required in free sampling, the linkage pattern must be analyzed, and summaries of the linkage pattern are then incorporated into the standard error calculation. The state linkage pattern is shown in table 11.9, and the resulting standard error of expected dissimilar joins is shown in the summary worktable (table 11.8).

The test statistic value ($Z_b = -0.466$) shows a slight tendency toward clustering, but the probability of the pattern being significantly nonrandom is not high ($p = .3228$). The state pattern of political party affiliation of governors across the continental United States appears to be random, and concluding that the pattern is significantly nonrandom would not be advisable. Even given a one-tailed hypothesis that the pattern was more clustered than random, the resultant p-value ($p = .1614$) suggests that the pattern is not significantly clustered. Again, extreme caution is advised when reaching inferential conclusions, since the gubernatorial pattern is not an artificial sample.

Summary Evaluation

Perhaps the area pattern evaluation would be more informative if comparative data were analyzed. For example, this 1990 gubernatorial pattern could be contrasted with a series of earlier patterns, such as each decade since 1900. A de-

Figure 11.13   Area Pattern Map and Join Counts: Political Party Affiliation of Governors
for the Conterminous United States, 1990

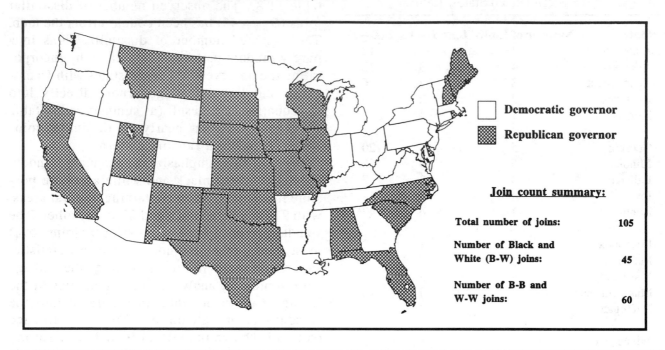

Table 11.8   Work Table for Area Pattern Analysis (Nonfree Sampling): Political Party Affiliation of State
Governors for the Conterminous United States, 1990

H_0: $O_{BW} = E_{BW}$ (area pattern is random)
H_A: $O_{BW} \neq E_{BW}$ (area pattern is not random)

$$O_{BW} = 45$$

$$E_{BW} = \frac{2JBW}{N(N-1)} = \frac{2(105)(21)(27)}{48(47)} = 52.78$$

$$\sigma_{BW} = \sqrt{E_{BW} + \frac{\Sigma L(L-1)BW}{N(N-1)} + \frac{4[J(J-1)-\Sigma L(L-1)]B(B-1)W(W-1)}{N(N-1)(N-2)(N-3)} - E_{BW}^2}$$

$$= \sqrt{52.78 + \frac{(822)(21)(27)}{48(47)} + \frac{4[105(104)-822]21(20)27(26)}{48(47)(46)(45)} - 52.78^2}$$

$$= 16.70$$

$$Z_b = \frac{O_{BW} - E_{BW}}{\sigma_{BW}} = \frac{45 - 52.78}{16.70} = -0.466$$

p-value $= .3228$

Table 11.9 State Linkage Pattern: Political Party Affiliation of State Governors for the Conterminous United States, 1990

State	Number of Links (L)	$L-1$	$L(L-1)$	State	Number of Links (L)	$L-1$	$L(L-1)$
Alabama	4	3	12	Nevada	5	4	20
Arizona	4	3	12	New Hampshire	3	2	6
Arkansas	6	5	30	New Jersey	3	2	6
California	3	2	6	New Mexico	4	3	12
Colorado	6	5	30	New York	5	4	20
Connecticut	3	2	6	North Carolina	4	3	12
Delaware	3	2	6	North Dakota	3	2	6
Florida	2	1	2	Ohio	5	4	20
Georgia	5	4	20	Oklahoma	6	5	30
Idaho	6	5	30	Oregon	4	3	12
Illinois	5	4	20	Pennsylvania	6	5	30
Indiana	4	3	12	Rhode Island	2	1	2
Iowa	6	5	30	South Carolina	2	1	2
Kansas	4	3	12	South Dakota	6	5	30
Kentucky	7	6	42	Tennessee	8	7	56
Louisiana	3	2	6	Texas	4	3	12
Maine	1	0	0	Utah	5	4	20
Maryland	4	3	12	Vermont	3	2	6
Massachusetts	5	4	20	Virginia	5	4	20
Michigan	3	2	6	Washington	2	1	2
Minnesota	4	3	12	West Virginia	5	4	20
Mississippi	4	3	12	Wisconsin	4	3	12
Missouri	8	7	56	Wyoming	6	5	30
Montana	4	3	12				
Nebraska	6	5	30				

$$\Sigma L = 210 \qquad \Sigma L(L-1) = 822$$

scriptive comparison of 9 or 10 test statistics and p-values would indicate if trends toward either clustering or dispersal have occurred during the past century. Speculation could then be made about why the area pattern changes have occurred.

The area pattern map (figure 11.13) is a valid representation of political party affiliation for governors in the continental United States. However,

it would probably be misleading to use the same map as an indication of the spatial patterning in political philosophy (such as conservative versus liberal). For example, the Democratic governors of a New England state and a Deep South state likely have more diverse political views than the Democratic and Republican governors of adjacent New England states.

Additional Measures of Area Patterns

Although the primary focus of section 11.2 is binary area pattern analysis, spatial data may be presented in a variety of nonbinary forms. Areas on a choropleth map may be assigned to more than two nominal (qualitative) categories. For example, counties in a state could be classified by dominant religious affiliation, or census tracts in a metropolitan area could be classified by predominant ethnic group.

Other choropleth maps may depict a weakly-ordered ordinal variable. In these cases, the value of each area on the map is placed in a category and the categories are rank ordered. The state-level choropleth map of the United States percentage population change from 1980 to 1990 (figure 1.2) is a good illustration of a weakly-ordered area pattern. The population change value of each state is allocated to one of five ordinal categories. Does the spatial pattern of population change have a clustered appearance? Is regionalization significant in the pattern of population change in the 1980s? If Sunbelt growth and Snowbelt stagnation or decline produce regional patterns, an area pattern analysis should reveal a clustered rather than random result.

If data are available showing the ranking of states from highest percentage population change (assigned rank 1) to lowest percentage population change (rank 48, for the conterminous U.S.), then a strongly-ordered ordinal variable map could be analyzed. As yet another alternative, the actual percentage growth of each state might be available, creating a ratio scale variable.

Area pattern analyses are possible for variables on all of these measurement scales, thus providing a very flexible set of statistical tests. With an ordinal, interval, or ratio variable on a map, a join cannot simply be considered as "similar" or "dissimilar," as can be done with nominally scaled area patterns. Rather, a join is assigned a numerical value representing the magnitude of difference in value of the two areas it connects.

Contemporary research in geography concerning spatial autocorrelation measures and issues is extensive. In fact, "The problem of spatial interdependence or spatial autocorrelation is central in most applications of statistical methods to geographic data, either as a focus of investigation in itself, or as a factor which complicates investigation of other hypotheses" (Odland, J. *et al.*, 1989). The interested reader should examine the appropriate references listed with this chapter.

Key Terms and Concepts

area pattern analysis	227
clustered, dispersed and random patterns	213
free and nonfree sampling	229
join and join count statistic	228
nearest neighbor analysis	215
quadrat analysis	222
spatial autocorrelation (positive and negative)	214
standardized nearest neighbor index (R)	217
variance-mean ratio	222

References and Additional Reading

Cliff, A. and J. Ord. 1973. *Spatial Autocorrelation.* London: Pion.

Ebdon, D. 1985. *Statistics in Geography: A Practical Approach.* Oxford: Basil Blackwell.

Getis, A. 1964. "Temporal analysis of land use patterns with nearest neighbor and quadrat methods" *Annals, Assoc. of Amer. Geog.,* 54, 391-399.

Goodchild, M. 1988. *Spatial Autocorrelation,* CATMOG series, No. 47. Norwich, England: Geo Books.

Gregory, S. 1978. *Statistical Methods and the Geographer.* London: Longman.

Griffith, D. A. 1988. *Spatial Autocorrelation: A Primer*. Washington, D.C.: Assoc. of Amer. Geog.

Griffith, D. and C. Amrhein. 1991. *Statistical Analysis for Geographers*. Englewood Cliffs, NJ: Prentice Hall.

Haggett, P., A. Cliff, and A. Frey. 1977. *Locational Models*. London: Edward Arnold.

Odland, J. 1988. *Spatial Autocorrelation*. Volume 9 in Scientific Geography Series. Beverly Hills, CA: Sage.

Odland, J., R. G. Golledge, and P. A. Rogerson. 1989. "Mathematical and Statistical Analysis in Human Geography" in G. L. Gaile and C. L. Willmott (editors). *Geography in America*. Columbus, OH: Merrill.

Silk, J. 1979. *Statistical Concepts in Geography*. London: George Allen & Unwin.

Taylor, P. 1977. *Quantitative Methods in Geography: An Introduction to Spatial Analysis*. Boston: Houghton-Mifflin.

Thomas, R. 1977. *An Introduction to Quadrat Analysis*, CATMOG series, No. 12. Norwich, England: Geo Books.

PART V

STATISTICAL RELATIONSHIPS BETWEEN VARIABLES

CHAPTER 12

CORRELATION

One of the more important concerns in geographic analysis is the study of relationships between spatial variables. Many geographic studies involve mapping variables and determining the degree of relationship between two or more map patterns. Using visual comparison of maps to measure correspondence or association is subjective because only a general impression of the relationship is gained. Two persons can view the same maps and interpret their association very differently.

Suppose a geographer is investigating urban crime problems within a city. Research and empirical studies on crime patterns from numerous urban areas suggest that a strong underlying association has developed between murders and the presence of drug activity. Although urban researchers from many disciplines are working to understand the complex relationship between these two social problems, the focus of geographic inquirics is often to establish the **spatial association** between the two variables. This general relationship can be seen by comparing a map of homicides to one of known drug sale points in Washington, D.C. (figure 12.1). Although visual examination appears to support the link between murders and drug activity, comparison of the two maps is extremely subjective. An accurate, unbiased estimate of the degree of association between the patterns on the two maps is difficult to obtain.

Correlation analysis provides a more objective, quantitative means to measure the association between a pair of spatial variables. Both the direction and strength of association between the two variables can be determined statistically. In the previous example, the degree of spatial association between homicides and drug activity could be established more precisely using correlation. Moreover, the technique can be applied either directly to numerical data or to mapped information converted to numerical form.

In section 12.1, the nature of correlation analysis is discussed. The scattergram is highlighted as an analytic tool used to study both the direction and strength of association. Several indices used to measure correlation or association between variables are defined and explained. Each index is illustrated with geographic examples.

The most widely used index of correlation, Pearson's correlation coefficient, is discussed in section 12.2, and examples of association for interval/ratio level data are included. Other geographic studies involve the use of ordinal or rank-order data. Section 12.3 presents Spearman's correlation index, the most widely used coefficient for measuring association of ordinal data. The use of correlation for measuring map association with examples of dot, choropleth, and isoline maps is explored in section 12.4. The last part of the chapter (section 12.5) presents related correlation issues of particular interest to geographers.

12.1 The Nature of Correlation

A geographic investigation often begins with graphic display of the data. A common tool for portraying the relationship or association between two variables is a two-dimensional graph called a **scattergram** or **scatterplot** (See section 2.5). Three examples of scattergrams are shown in figure 12.2. With one variable plotted on each axis, the pattern of points in a scattergram helps to provide an understanding of the nature of a particular relationship.

Two types of information—direction and strength of association—can be identified from the scattergram. Suppose a line is placed through the pattern of points in the scattergram to summarize the relationship between the two variables. The slope of this line indicates the direction of the relationship, and the amount of scatter of points about the line reveals the strength of association. This line, called the "least squares regression line," will be discussed in more detail in the following chapter.

244

Figure 12.1 Homicides and Drug Sale Points in Washington, D.C.

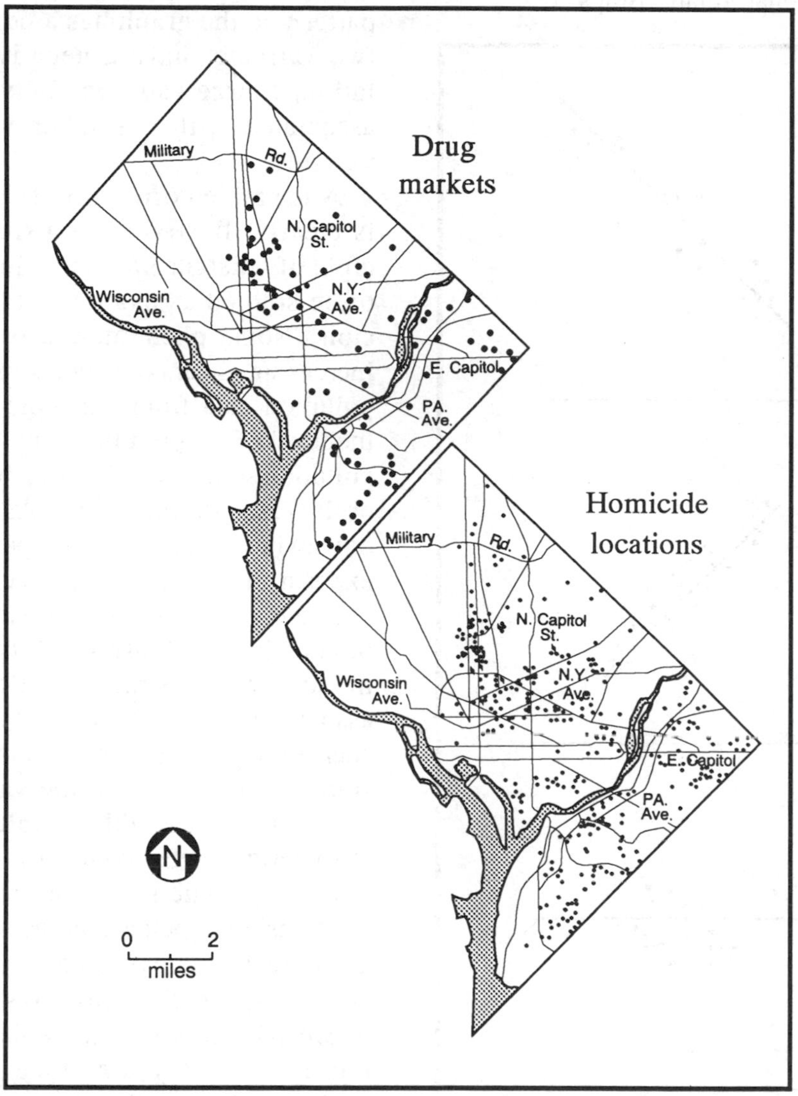

Source: Redrawn from maps in *The Washington Post*, January 13, 1989, p. E1, with permission.

If the general trend of points is from lower left to upper right (figure 12.2, case 1), the **direction of association** is **positive,** and the line summarizing the points has a positive slope. In a positive or direct relationship, a larger value in one variable generally corresponds to a larger value in the second variable; alternatively, a smaller value in the first variable usually coincides with a smaller value in the second variable. Such a correspondence will result in a positive correlation.

In geography, a positive correlation is found with many variables related to population size. For example, the association between popula-

Figure 12.2   Generalized Scattergrams Showing
Directional Relationships

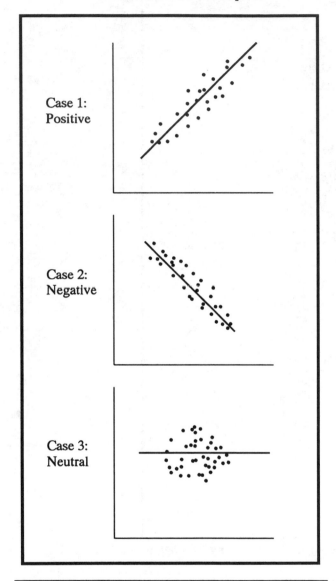

upper left to lower right (figure 12.2, case 2). In this example, the line summarizing the point pattern on the graph has a negative slope. When two variables have a negative or inverse correlation, a large value in one variable is generally associated with a smaller value in the second variable.

A negative correlation between two variables is clearly illustrated by using the general principle of "distance decay." In such relationships, often studied as examples of "contagious diffusion," some phenomenon or idea declines with increasing distance from a source or origin. Using pollution data from a sample of monitoring sites, the level of air pollution and the distance from the pollution source would probably exhibit a negative or inverse relationship. As one moves farther downwind from a site of pollution, the level of exposure to the pollutant declines.

Some variables may not exhibit either a positive or negative relationship. If the pattern of points in the scattergram is **random** (figure 12.2, case 3), no association exists between the two variables. In such examples, the line summarizing the relationship has no slope, and the values of one variable are not associated with the values of the other. This is sometimes referred to as a **neutral** relationship.

In geographic research, variables may show no relationship or pattern on the scattergram. Suppose a geographer studying housing characteristics in a city wishes to determine if an association exists between house size (square feet of living area) and the number of years since construction. It is unlikely that a significant association exists between these two variables. There may be no logical rationale for expecting a correlation between house size and age of unit.

The **strength of association** between two variables can be determined by examining the amount of point spread around the summarizing line of a scattergram. If the point pattern is tightly packed near the line (sometimes described as cigar-shaped), the relationship is said to be strong (figure 12.3, case 2). However, if the points are more widely spread around the line

tion and number of retail functions in a sample of settlements usually exhibits a positive or direct relationship. As demonstrated in central place theory, settlements with more people at higher levels of the urban hierarchy generally contain more retail establishments.

The direction of relationship is **negative** if the general trend of points in the scattergram is from

Figure 12.3 Generalized Scattergrams Showing Strength of Association

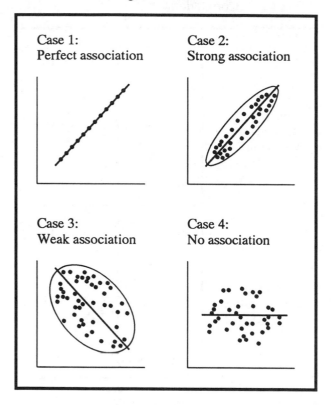

(termed a football-shaped pattern), the association is moderate to weak (figure 12.3, case 3). The strongest association occurs when the points lie exactly along a line—perfect correlation (figure 12.3, case 1); the weakest association occurs when the points are distributed with no pattern—no association (figure 12.3, case 4).

Any two variables can be correlated, and the strength and direction of relationship calculated. However, extreme caution must be used when evaluating or interpreting correlations. A relationship or association between variables does not necessarily imply the existence of a cause-effect relationship. For example, a geographer could correlate annual precipitation and pH level of the soil at a sample of locations. Even if a

nonzero correlation is derived, it is highly unlikely that any type of causal relationship exists between these variables.

Although the visual examination of a scattergram is a useful way to begin a geographic investigation, the direction and strength of association between two variables can be determined only in general, often subjective terms. In geographic studies, a more objective, rigorous method for determining the strength of relationship or degree of association between two variables is needed. Statisticians have defined various indices, usually called "correlation coefficients," to measure the strength of relationships. Most of these coefficients are constructed to have a maximum value of 1.0, which indicates perfect positive or direct correlation between variables. A minimum value of −1.0 represents a perfect negative or inverse correlation, and a value of 0.0 is used to denote no correlation or association between variables. Thus, like a scattergram, correlation coefficients can indicate both the direction of the relationship (positive or negative) as well as the strength of association.

As in other areas of statistical analysis, the level of measurement (nominal, ordinal, interval/ratio) largely determines which index of correlation or association is applied to a problem. The most commonly used correlation coefficients are Pearson's product-moment for interval or ratio data and Spearman's rank-order for ordinal or ranked data. Indices to measure association in categorical data (such as measuring the strength of association in a contingency table) are less frequently used in geographic problem solving and will not be discussed here.

12.2 Association of Interval/Ratio Variables

The most powerful and widely used index to measure the association or correlation between two variables is **Pearson's product-moment correlation coefficient**. In fact, when the term "correlation" appears in geographic literature, it is often assumed that the writer is referring to Pearson's correlation index. To use this measure of association, data must be of interval or ratio scale. It is also assumed that the variables have a linear relationship. In addition, if the index is used in an inferential rather than descriptive manner, both variables should be derived from normally distributed populations.

Pearson's correlation coefficient relates closely to the statistical concept of **covariation**—the degree to which two variables "covary" (vary together or jointly). If the values of the two variables covary in a similar manner, the data contain a large covariation, and the two variables will show strong correlation. Alternatively, if the paired values of the variables show little consistency in how they covary, the correlation will be very weak.

The concept of covariation and its relationship to correlation can be more easily understood by comparing a set of scattergrams, each having four quadrants produced from the mean values of the X (horizontal) and Y (vertical) variables (figure 12.4). In each scattergram, the mean values and total variation in the two variables are held constant. This allows a direct comparison of the relative covariation, which differs in each scattergram.

An understanding of covariation begins first with the deviations of X $(X - \overline{X})$ and Y $(Y - \overline{Y})$ from their respective means. As discussed earlier, these deviations are basic building blocks of standard deviation and standardized scores. The X and Y deviations of each data value (matched pair) are multiplied together and summed for the set of values to produce:

$$CV_{XY} = \Sigma\left(X - \overline{X}\right)\left(Y - \overline{Y}\right) \qquad (12.1)$$

where: CV_{XY} = covariation between X and Y
$(X - \overline{X})$ = deviation of X from its mean $(\overline{X})$
$(Y - \overline{Y})$ = deviation of Y from its mean $(\overline{Y})$

Figure 12.4   Generalized Scattergrams Showing the Relationship of Covariation to Correlation

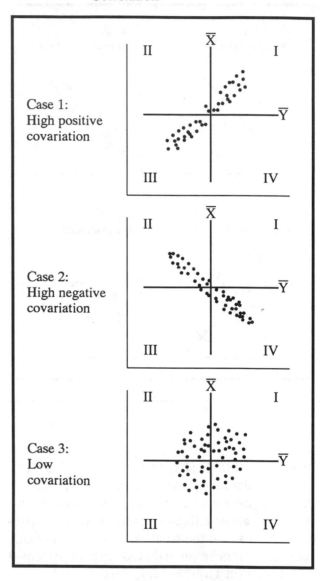

Mathematically, covariation is analogous to the important concept of total variation—the sum of the squared deviations from the mean—an integral component of analysis of variance and regression:

$$TV_X = \Sigma(X - \overline{X})(X - \overline{X}) = \Sigma(X - \overline{X})^2 \qquad (12.2)$$

$$TV_Y = \Sigma(Y - \overline{Y})(Y - \overline{Y}) = \Sigma(Y - \overline{Y})^2 \qquad (12.3)$$

where: TV_X = total variation in X
  TV_Y = total variation in Y

The point patterns in each scattergram of figure 12.4 represent different examples of covariation. In case 1, virtually all points in the scattergram lie in quadrants I and III. In quadrant I, the deviations of X $(X - \overline{X})$ and Y $(Y - \overline{Y})$ are both positive, since each X and Y value is greater than its respective mean. In quadrant III, where most other points in case 1 are located, the X and Y deviations are both negative, since all X and Y values are less than their respective means. Because the X and Y values for each observation in quadrants I and III covary in the same direction from their means, the product of the two deviations will produce a *positive* result for each point. When these individual products are summed according to equation 12.1, the resultant value will be large and positive. Thus, case 1 represents an example of a large positive covariation. As seen earlier, this scattergram also corresponds to a positive or direct correlation between the two variables.

The situation is different for the scattergram in figure 12.4, case 2, where most points lie in quadrants II and IV. In quadrant IV, deviations in X are positive (X values are greater than $\overline{X}$). However, deviations in Y are negative because values of Y are less than $\overline{Y}$. A reversed situation occurs in quadrant II, where deviations in X are negative (X values are less than $\overline{X}$) and deviations in Y are positive. In case 2, the product of the deviations will be *negative* because the values of the X and Y variables in both quadrants covary in opposite directions from their means. When the products from the X and Y deviations are summed, the resultant covariation will be large and negative. The example of covariation in case 2 represents two variables that have a large, inverse correlation.

In the third scattergram (figure 12.4, case 3), points are scattered in a random pattern, with nearly equal dispersal of points in each quadrant. Some of the X and Y deviations will be positive and others negative. When the deviations are multiplied together for each unit of data, both positive and negative products will occur. When these products are summed for all points, the values generally cancel each other out and produce a covariation close to zero. This low covariation corresponds to a very small correlation, suggesting little or no relationship between the two variables.

Pearson's correlation coefficient (r) can be expressed mathematically in several different ways:

1. with deviations from the mean and standard deviations (equation 12.4)
2. with X and Y values transformed to Z-scores (equation 12.5)
3. with the original values of the X and Y variables (equation 12.6)

Any of these formulas produces an equivalent result.

Based on the conceptual definition, Pearson's correlation is expressed as the ratio of the covariance in X and Y to the product of the standard deviations of the two variables:

$$r = \frac{\left[\Sigma(X - \overline{X})(Y - \overline{Y}) \right]/N}{s_X s_Y} \qquad (12.4)$$

where: r = Pearson's correlation coefficient
  N = number of paired data values
  s_X, s_Y = standard deviation of X and Y, respectively

If the data are converted to standardized or Z-score form (section 5.5), an alternative formula exists to calculate the Pearson's correlation coefficient:

$$r = \frac{\Sigma[Z_X Z_Y]}{N} \qquad (12.5)$$

where: $Z_X = \dfrac{(X - \overline{X})}{s_X}$ and $Z_Y = \dfrac{(Y - \overline{Y})}{s_Y}$

$Z_X = X$ variable transformed to Z-score

$Z_Y = Y$ variable transformed to Z-score

N = number of paired data values

Using equation 12.5, the correlation between two variables is equal to the sum of the product of Z-scores for each data value divided by the number of paired values. This formula is valid because Z-scores take into account deviations from the mean and the standard deviation.

If a researcher wishes to use the original values of the X and Y variables directly, a computational formula is used to derive Pearson's correlation coefficient:

$$r = \dfrac{\Sigma XY - (\Sigma X)(\Sigma Y)/N}{\sqrt{\left[\Sigma X^2 - (\Sigma X)^2/N\right]}\sqrt{\left[\Sigma Y^2 - (\Sigma Y)^2/N\right]}} \quad (12.6)$$

Although this formula appears more complex than the previous equations, it uses only the original data, and no prior calculation of means, standard deviations, or deviations from the mean is needed.

In addition to its use as a descriptive index of the strength and direction of association, correlation can also be used to infer results from a sample to a population. The sample correlation coefficient (r) is the best estimator of the population correlation coefficient (ρ). In this application, the null hypothesis is that no correlation exists in the populations of the two variables (H_0: $\rho = 0$). Since the populations of the variables are assumed to have a linear association (see boxed insert), the null hypothesis of no correlation is equivalent to stating that X and Y are independent.

As in most other inferential tests, the alternate hypothesis (H_A) can be stated as directional (one-tailed) or nondirectional (two-tailed). In the one-tailed approach, the researcher has a logical basis for expecting the correlation to be either posi-

---

Pearson's Correlation Analysis

Primary Objective: Determine if an association exists between two variables

Requirements and Assumptions:
 1. Random sample of paired variables
 2. Variables have a linear association
 3. Variables measured at interval or ratio scale
 4. Variables are bivariate normally distributed

Hypotheses:
 H_0: $\rho = 0$
 H_A: $\rho \neq 0$ (two-tailed)
 H_A: $\rho > 0$ (one-tailed) or
 H_A: $\rho < 0$ (one-tailed)

Test Statistic:
$$t = \dfrac{r\sqrt{n-2}}{\sqrt{1-r^2}}$$

---

tive or negative. When no rationale exists for the direction of correlation between two variables, a two-tailed alternate hypothesis should be used.

Although either the t or Z distributions are used to evaluate the significance of a Pearson's correlation coefficient r, the most common approach uses t. Since a Z distribution is appropriate only when sample size exceeds 30, the use of t for testing correlation coefficients offers more flexibility for problems with various sample sizes. The most common test statistic is:

$$t = \dfrac{r}{s_r} \quad (12.7)$$

where: s_r = standard error estimate

$$s_r = \sqrt{\dfrac{1-r^2}{n-2}} \quad (12.8)$$

The test statistic for Pearson's r can therefore be rewritten as:

$$t = \dfrac{r\sqrt{n-2}}{\sqrt{1-r^2}} \quad (12.9)$$

Geographic Example of Pearson's Correlation Coefficient

The use of Pearson's product-moment correlation coefficient in geographic analysis is demonstrated with an example from physical geography. One of the mechanisms influencing precipitation on land areas is the presence of large bodies of water nearby. These bodies of water provide an important source of moisture, which produce sizable amounts of precipitation when brought over a land area by prevailing winds. However, as distance from the body of water increases, the water's influence decreases, and precipitation levels generally decline. Thus, areas adjacent to water tend to have higher amounts of precipitation than do areas farther away.

An interesting example of this land-water relationship occurs in regions adjacent to the southern shores of the Great Lakes in New York, Pennsylvania, Ohio, and Indiana. The presence of the Great Lakes and a strong northwesterly wind flow during the winter combine with other influences to create an important regional climatic phenomenon called "lake effect snow." The spatial pattern of snowfall in northeastern Ohio along the southern shore of Lake Erie is examined using Pearson's correlation.

Data from this region are used to examine the influence of distance from Lake Erie on the levels of average annual snowfall. A systematic sample of 38 locations is taken from an isoline map that shows average snowfall amounts for the region (figure 12.5). For each of the sampled points, two variables are recorded: (1) average snowfall, interpolated from the map of isohyets, and (2) straight-line distance from Lake Erie. If Lake Erie influences snowfall within the region, a negative correlation between snowfall and distance is expected. As distance from the Lake increases, average annual snowfall amounts should decrease.

A scattergram of the 38 points in the study area shows the relationship between snowfall levels (*Y*) and distance from Lake Erie (*X*) (figure 12.6). The pattern of points is clearly not linear. Contrary to the general relationship just hypothesized, snowfall amounts for locations immediately adjacent to the Lake are lower than snowfall levels slightly farther from the Lake. A closer look at the climatological processes operating in this lake-effect area suggest the simple hypothesized relationship between distance and snowfall is not fully accurate. A more detailed explanation is necessary.

During the winter, the water temperature of Lake Erie is warmer than adjacent land areas. As cold air currents from the northwest pass over this warmer water, the air is heated and can hold more moisture. As the moisture-laden air

---

Figure 12.5   Northeastern Ohio Study Area: Average Annual Snowfall, in inches

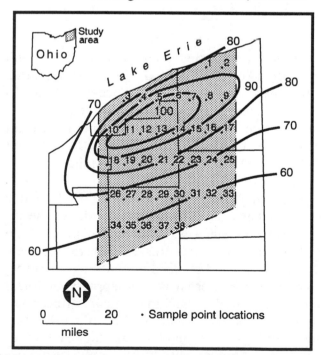

Source: Redrawn from Kent, Robert B. (editor), 1992. *Region in Transition: An Economic and Social Atlas of Northeast Ohio*, (fig. 23.6). Akron: The University of Akron Press.

Figure 12.6   Scattergram Showing Relationship
Between Annual Snowfall and
Distance from Lake Erie

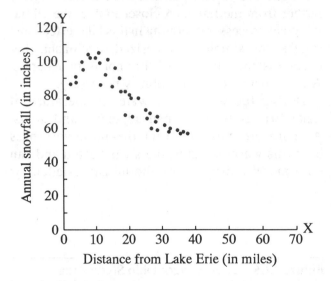

effectively in geographic studies, results are often more difficult to interpret.

Yet another alternative is to eliminate that portion of the study area less than about six miles from the Lake shore (figure 12.7). The remaining 33 sample points in the revised study region generate a scattergram that shows a linear relationship between distance from the Lake and amount of snowfall (figure 12.8). This strategy meets the linearity assumption of the Pearson correlation technique and can be justified on a climatological basis.

The distance (X) and snowfall (Y) values for the revised study area are shown in table 12.1. Since the original X and Y values are directly available, the appropriate computational formula is equation 12.6. The intermediate summation values and the resulting correlation coefficient are shown in table 12.2. The null hypothesis is that no association

continues moving onto cooler land surfaces south of the Lake, it is cooled, the moisture often condenses, and precipitation occurs. Immediately adjacent to the Lake, the air temperature is often warm enough for the precipitation to fall as rain. Somewhat further from the Lake (about 7-10 miles), the air is more frequently cold enough to create snowfall.

Pearson's correlation cannot be used for this data set because the scattergram shows the relationship between snowfall and distance from Lake Erie to be curvi-linear. Several alternate approaches could be used to continue the investigation. One alternative is to apply a nonlinear correlation model matching the curvi-linear pattern on the scattergram. While fully valid, this methodology is beyond the introductory level of this text. Another possibility would be to convert the nonlinear data into linear form by using a logarithmic transformation. Pearson's correlation could then be applied to this transformed data. Although this approach has been used

Figure 12.7   Revised Northeastern Ohio Study
Area: Average Annual Snowfall,
in inches

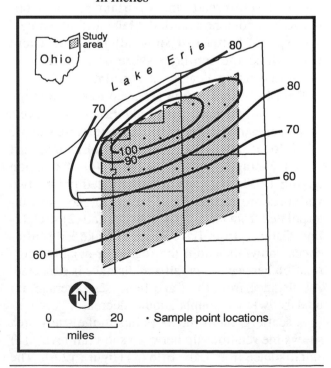

Figure 12.8 Scattergram Showing Relationship Between Annual Snowfall and Distance from Lake Erie for Revised Study Area

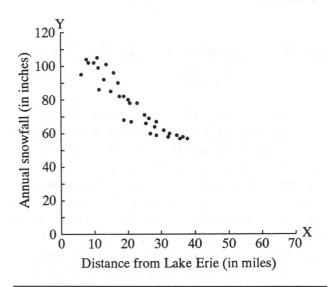

Distance from Lake Erie (in miles)

Table 12.1 Data for Northeastern Ohio: Average Annual Snowfall and Distance from Lake Erie

Observation	Distance (miles) X	Snowfall (inches) Y
1		
2	(Points 1-5 were eliminated	
3	from study area. See text for	
4	explanation.)	
5		
6	7.7	104
7	9.9	102
8	11.2	99
9	12.9	92
10	6.0	95
11	8.3	102
12	10.9	105
13	13.6	101
14	15.8	96
15	17.1	90
16	18.8	82
17	20.6	78
18	11.4	86
19	14.9	85
20	17.5	82
21	20.1	80
22	22.8	78
23	25.0	71
24	26.3	69
25	28.5	67
26	18.8	68
27	21.0	67
28	25.4	66
29	28.0	64
30	30.7	62
31	32.4	60
32	34.6	59
33	36.4	58
34	26.7	60
35	28.5	59
36	32.0	58
37	35.5	57
38	37.7	57

Source: Adapted from Kent, Robert B. (editor), 1992. *Region in Transition: An Economic and Social Atlas of Northeast Ohio*, (fig. 23.6). Akron: The University of Akron Press.

(correlation) exists between distance and snowfall in the study area. The alternate hypothesis (H_A) is that an inverse association exists between distance and snowfall in the study area. Since the direction of correlation is hypothesized as negative or inverse, a one-tailed test is appropriate.

The resulting r value of -0.93 indicates that a high inverse relationship exists between snowfall and distance from Lake Erie. This correlation is statistically significant with a p-value of 0.000. For locations between about 6 and 40 miles from Lake Erie, the hypothesis of decreased snowfall associated with increasing distance from the Lake is confirmed, and the null hypothesis of no correlation can be safely rejected.

Summary Evaluation

The lake effect problem demonstrates the use of correlation analysis in geographic research. Not only was a statistically significant relationship found between distance from the Lake and snowfall, but also a solid climatological basis exists to support the correlation.

Table 12.2   Work Table for Correlation Example:
Northeastern Ohio Snowbelt

Intermediate Summation Values

$\Sigma X = 707.2$ $\Sigma X^2 = 17858.2$ $\Sigma XY = 50288.7$
$\Sigma Y = 2559$ $\Sigma Y^2 = 207181$ $N = 33$

H_o: no association exists between snowfall and
distance
H_A: an inverse association exists between snowfall
and distance

$$r = \frac{\Sigma XY - (\Sigma X)(\Sigma Y)/N}{\sqrt{\left[\Sigma X^2 - (\Sigma X)^2/N\right]}\sqrt{\left[\Sigma Y^2 - (\Sigma Y)^2/N\right]}}$$

$$r = \frac{50288.7 - (707.2)(2559)/33}{\sqrt{\left[17858.2 - (707.2)^2/33\right]}\sqrt{\left[207181 - (2559)^2/33\right]}}$$

$$r = -0.93$$

$$t = \frac{r\sqrt{n-2}}{\sqrt{1-r^2}} = \frac{-.93\sqrt{31}}{\sqrt{1-.93^2}} = 14.08$$

$$p = 0.0000 \ \text{(one-tailed)}$$

Another characteristic of this problem is the complexity of delineating a suitable boundary for the study area. At first, it was thought that an analysis would be appropriate from the Lake shore to the southern boundary some 40 miles away. The subsequent curvi-linear shape of the scattergram indicates that complex climatological processes are operating. Therefore, a simple linear relationship between distance and snowfall does not exist across the entire study area.

Another issue in this problem is the nature of the dependent variable, snowfall. Since the region has very few snowfall monitoring stations, a generalized isoline map represents the only snowfall data available. It would certainly be preferable to measure snowfall amounts directly at monitoring stations throughout the study region. Because this was not possible, the precision of the data may be questionable.

12.3   Association of Ordinal Variables

In geographic problems with data in ranked form, **Spearman's rank correlation coefficient** (r_s) is the most widely used measure of the strength of association between two variables. It is appropriate when (1) variables are measured on an ordinal (ranked) scale, or (2) interval/ratio data are converted to ranks. Spearman's correlation coefficient should be applied when the assumptions of Pearson's correlation are not fully met. For example, Spearman's coefficient may be appropriate if samples are drawn from highly skewed or severely non-normal populations. The statistical power of Spearman's correlation has been shown to be nearly as strong as Pearson's r.

Spearman's rank correlation coefficient is applicable in situations where the X and Y variables have a **monotonic relationship**. Two variables share an association that is monotonically increasing when X increases as Y increases (or remains constant). Conversely, a monotonically decreasing relationship between variables occurs when X increases, but Y decreases (or remains constant). Because of the nature of ordinal data, Spearman's rank correlation coefficient does not distinguish between a **linear relationship** and a monotonic one.

Similar to Pearson's correlation, r_s ranges from a maximum of 1.0 for perfect positive or direct correlation to a minimum of –1.0 for perfect negative or inverse correlation. When no association exists between variables, $r_s = 0.0$. Spearman's correlation coefficient measures the degree of association between two sets of ranks using the following equation:

$$r_s = 1 - \frac{6(\Sigma d^2)}{N^3 - N} \qquad (12.10)$$

where: d = difference in ranks of variables X and Y for each paired data value
Σd^2 = sum of the squared differences in ranks
N = number of paired data values

In a problem where a perfect match occurs in the ranks of all paired data values, each difference value (d) equals 0, and the sum of all squared differences also equals 0. In this situation, equation 12.10 generates a perfect positive correlation of 1.0 (table 12.3, case 1). At the other extreme, if the ranks are exactly reversed between the two variables (top rank in variable X matches bottom rank in variable Y, and so on), the Spearman index equals –1.0 (table 12.3, case 2).

Similar to other statistical tests using ordinal data, the presence of tied rankings influences the Spearman's coefficient. However, the effect of ties on the resultant correlation index will be significant only when the proportion of tied rankings to total number of values sampled is very large. In these instances, a correction factor for ties needs to be applied. As a general rule, the correction factor is not necessary if the number of tied rankings is less than 25 percent of the total number of pairs.

Spearman's correlation coefficient is commonly used as a descriptive index of association between two ordinal variables. However, when the paired variables represent a sample drawn at random from a population of bivariate data values, an r_s value can be tested for significant difference from 0. The sample correlation coefficient (r_s) is the best estimator of the population correlation coefficient (ρ_s). In these applications, the null hypothesis is that no relationship exists between the two variables in the population (H_0: $\rho_s = 0$). Confirmation of the null hypothesis is equivalent to affirming independence between the X and Y variables.

Table 12.3   Perfect Correlations of Two Ranked Variables

Case 1: Positive Correlation

Variable X Ranks	Variable Y Ranks	Difference in Ranks (d)	d^2
1	1	0	0
2	2	0	0
3	3	0	0
4	4	0	0
5	5	0	0
			$\Sigma d^2 = 0$

$$r_s = 1 - \frac{6(\Sigma d^2)}{N^3 - N} = 1 - \frac{6(0)}{5^3 - 5} = 1 - \frac{0}{120} = 1.0$$

Case 2: Negative Correlation

Variable X Ranks	Variable Y Ranks	Difference in Ranks (d)	d^2
1	5	–4	16
2	4	–2	4
3	3	0	0
4	2	2	4
5	1	4	16
			$\Sigma d^2 = 40$

$$r_s = 1 - \frac{6(\Sigma d^2)}{N^3 - N} = 1 - \frac{6(40)}{5^3 - 5} = 1 - \frac{240}{120} = -1.0$$

Either a one- or two-tailed test is used, depending on the form of the alternate hypothesis. Like the corresponding test of the Pearson coefficient, either the t or Z distributions can be applied to test an r_s value for significance. When the Z distribution is used, the test statistic is determined by the Spearman correlation and the sample size:

$$Z_{r_s} = r_s \sqrt{n-1} \qquad (12.11)$$

Geographic Example of Spearman's Rank Correlation

The use of Spearman's rank correlation coefficient is demonstrated by examining recent population changes in the United States. Throughout most of the country's history, the percent of total population living in urban areas has increased, and growth has been particularly strong in densely populated areas. Recently, however, a number of demographers have suggested that this pattern of growth has been altered so that rural areas, small towns, and regions with relatively few people are among the fastest growing locations in the United States. If this hypothesized trend toward nonmetropolitan growth or decentralization is true, then regions with relatively low population densities should have the highest growth rates.

To test this "rural renaissance" or decentralization theory, data on population density and population growth were obtained for U.S. states for 5-year intervals from 1960 to 1985. Population per square mile at the beginning of each period is correlated with the percentage population change during that period. Spearman's correlation values are used here to describe trends in U.S. population growth patterns. If population growth was more pronounced in low-density areas, then Spearman's correlation index will be negative. If a positive correlation coefficient occurs, population growth will be directly related to population density at the state level. In the latter case, the population trend will reflect growth occurring in the high-density areas.

Spearman's correlation coefficient is calculated using the paired rankings of population density and percentage population change for each of the five time periods. The methodology for the problem is illustrated for the 1980-1985 data, with 12 of the 50 states shown in table 12.4.

The rank correlation analysis for the 50 states shows clear changes in growth trends over the 25-year period (table 12.5). Between 1960 and 1970, population growth was directly related to population density, as indicated by the positive correlation indices for the two earliest time periods. These coefficients reflect a long-standing trend of metropolitanization and continuing growth in previously settled areas of the country. However, because the correlations are relatively close to zero, the tendency for growth to occur in the densest regions was relatively weak in the 1960s.

The occurrence of strong negative rank correlation coefficients between density and population growth during the 1970s provides clear evidence of population movement away from dense areas and toward more sparsely inhabited states. During this decade, the states with the greatest percentage growth in population had lower population densities. This negative relationship between population growth and density continued into the first half of the 1980s, but at a slightly weaker level, as seen by the Spearman coefficient of -0.406.

Table 12.4   Spearman Correlation Example: State Population Change and Density, 1980-85

State*	Original Data		Ranked Data			
	Population Percentage Change 1980-85	Density** 1980	Population Percentage Change 1980-85	Density 1980	d	d²
(1) Alabama	3.34	76.64	21	25	−4	16
(2) Alaska	30.25	0.70	50	1	49	2401
(3) Arizona	17.26	23.95	49	11	38	1444
(4) Arkansas	3.24	43.88	20	16	4	16
(5) California	11.39	151.43	42	37	5	25
(6) Colorado	11.80	27.90	44	13	31	961
		...				
		...				
(45) Vermont	4.49	55.21	28	22	6	36
(46) Virginia	6.73	134.65	33	35	−2	4
(47) Washington	6.76	62.10	34	23	11	121
(48) W. Virginia	−0.72	80.85	3	26	−23	529
(49) Wisconsin	1.47	86.47	9	27	−18	324
(50) Wyoming	8.07	4.86	37	2	35	1225
						$\Sigma d^2 = 29278$

*Data listed for only 12 states
**Persons per square mile

$$r_s = 1 - \frac{6\left(\Sigma d^2\right)}{N^3 - N}$$

$$r_s = 1 - \frac{6(29278)}{125000 - 50} = 1 - \frac{175668}{124950} = 1 - 1.406 = -0.406$$

Source: Bureau of the Census, Dept. of Commerce

Summary Evaluation

The Spearman analysis has shown that recent growth occurred in lower-density states. However, without further analysis, it is unclear whether growth within any particular state took place in its metropolitan or nonmetropolitan portions. All that is known is the overall statewide growth rate; not known is the extent to which that growth is urban or rural.

In the eastern and midwestern states, which have a longer settlement history, growth in the

Table 12.5 Spearman Correlation Coefficients for State Population Change and Density

Time Period	Spearman Correlation
1960-1965	+0.166
1965-1970	+0.176
1970-1975	−0.539
1975-1980	−0.561
1980-1985	−0.406

1970s was probably associated with population moving away from metropolitan areas to rural regions. However, in the more recently settled western states, extensive growth may have occurred in a small number of expanding metropolitan areas, like Phoenix, Arizona and Las Vegas, Nevada.

The use of Spearman's correlation coefficient in this example documents an interesting transition in population movement patterns within the United States that occurred around 1970. The analysis offers strong evidence of a major change away from a concentration of growth in densely populated states of the country and the emergence of lower-density states as significant areas of population growth. However, as just suggested, a more complete understanding of the nature of population change during this period requires further geographic analysis.

12.4 Use of Correlation Indices in Map Comparison

Correlation measures the degree of association between variables. In most cases, data are available in numerical form and can be analyzed directly with correlation techniques. Sometimes, however, spatial information is available only in map form. How can a geographer measure the association between two map patterns when the original data are not readily available?

As discussed in the opening of this chapter, visual comparison of maps may be subjective and lead to biased conclusions. More productive geographic research requires an objective analysis of map patterns. With the use of spatial sampling methods, correlation indices can be applied to numerical data acquired from maps. The methodology is illustrated here with three types of maps: dot maps, isoline maps, and choropleth maps.

Dot Maps

The location of items within a geographic area can often be portrayed effectively with a dot map. A dot on the map represents the location of each item or group of items. The resulting pattern of points defines the geographic distribution of the variable under study. Procedures for analyzing a single point pattern were presented in section 11.1. This section focuses on the relationship between two or more point patterns.

Suppose a geographer wants to investigate the relationship between spatial patterns of two variables portrayed by dot maps covering the same area. An index of association can be calculated using information taken directly from the dot maps. To acquire data for correlating the two dot map patterns, a set of equal-sized quadrats (usually square cells) is placed over the study area. The quadrat size should be adequate to depict the spatial complexity of the pattern and allow the calculation of a representative correlation index. The discussion of cell size in quadrat analysis (section 11.1) also applies to this application of correlating dot maps.

The quadrats should be placed over each map to produce identical cell location and orientation patterns on both maps. If the maps have the same scale, the quadrats used for the first dot map will cover exactly the same areas on the second map. However, if the two maps have different scales, one of two procedures should be followed. The maps could be converted to the same scale, and a set of quadrats used in the manner discussed previously. Alternatively, different-sized quadrats could be used for the two maps if each cell represents the same "real-world" area.

Each of the quadrats represents a locational unit of data, and the frequency of points per quadrat from the two maps are the X and Y values. Using this data set created from the dot maps, either Pearson's or Spearman's correlation indices can be calculated.

This procedure is demonstrated using the geographic problem relating crime and drug activity discussed at the beginning of the chapter. The two maps in figure 12.1 appeared in *The Washington Post* illustrating the spatial connection between locations of drug sales and homicides in Washington, D.C. The maps use visual correspondence to demonstrate evidence of a relationship. However, a more objective measure of the degree of association between the two variables can be determined using Pearson's correlation.

The dot maps are constructed at the same scale, which permits the same quadrat pattern to be used for both maps. A set of 16 cells is placed over the dot maps (figure 12.9), and the frequency of drug sale points and homicides is recorded for each cell. Pearson's correlation coefficient is then calculated using the two sets of frequencies collected for the 16 quadrats. The resultant index value ($r = 0.953$) is significantly different from 0 ($p = 0.000$) and strongly confirms the hypothesis suggested by the newspaper that drug activity and murders in Washington, D.C. are spatially related.

Isoline Maps

Isoline maps are useful for presenting data, such as precipitation, temperature, barometric pressure, and elevation, which are distributed continuously across an area. By selecting a suitable isoline interval and connecting locations having equal values with an isoline, the pattern of lines on the map represents the geographic distribution of the variable over the area.

The methodology used to investigate the association between two isoline maps is analogous to that used for dot maps. Instead of placing a set of quadrats over the maps and recording the frequency of dots per cell, a set of sample points is placed systematically on each isoline map. If the maps have equal scales, the same point pattern grid can be used for both maps. If the maps have different scales, the maps must be converted to the same scale, or the grid system must be altered so that an identical set of point locations occurs on each map.

Once the sample points are properly placed on the maps, the value of the continuously distributed variable is recorded for each matching pair of points. Some interpolation of numerical values is needed where points do not fall directly on an isoline. The set of points represents the values, and the recorded values from the two isoline maps provide the corresponding matched X and Y values. Using the data taken from the maps, a correlation coefficient is calculated that measures the strength of association between the two map variables.

This procedure is illustrated with maps of rural population density and annual precipitation for the People's Republic of China. In a now classic study by Robinson and Bryson (1957), these variables were used to correlate map patterns in the state of Nebraska.

Geographical studies have shown that various climatic and geomorphological factors profoundly influence the ability of the land to support population. Because they have greater potential for food production, areas with higher precipitation levels tend to have higher population densities. This is especially true in developing areas where a high proportion of the population is involved in primary activities, such as agriculture. People in these areas more directly depend on favorable environmental conditions to survive and lack the resources to modify adverse environments.

Figure 12.9   Dot Maps of Homicides and Drug Sale Points in Washington, D.C., with Quadrats Superimposed

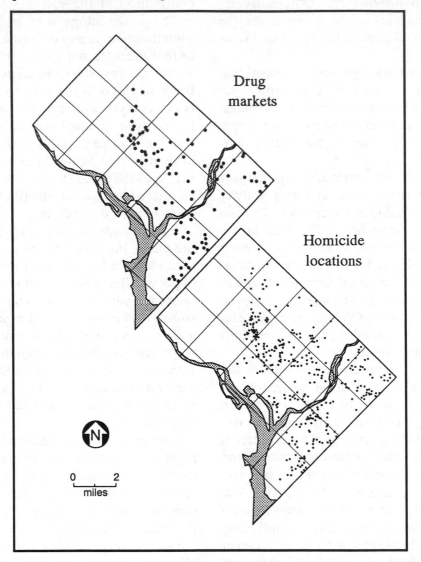

A regular grid pattern of 50 points is super-imposed on the two isoline maps of China (figure 12.10). For both maps, the value of the variable is interpolated at each point. A Pearson's correlation coefficient is then calculated from these data. The resulting value of 0.396 ($p = .002$) shows a moderately positive and statistically significant relationship between precipitation

Figure 12.10 Isoline Maps of Rural Population Density and Annual
Precipitation in China with Sample Point Locations

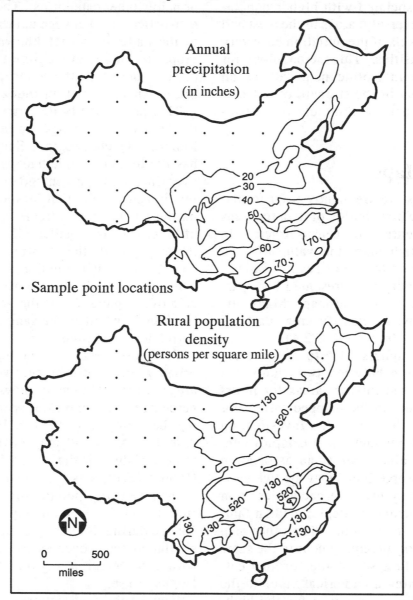

Source: Central Intelligence Agency. 1971. *People's Republic of China Atlas.*
Washington, D.C.: U.S. Government Printing Office.

and population density. Areas in China with higher levels of precipitation (the east and southeast) tend to be associated with higher population densities. Conversely, areas to the west and north with lower levels of precipitation have very low population densities. This methodology for analyzing patterns in isoline maps allows an objective assessment of the strength of association for variables that are continuously distributed.

Choropleth Maps

Choropleth maps are frequently used to display the geographic distribution of data for areas such as states, counties, or census tracts. In creating a choropleth map, the value of each areal unit is allocated to one of several categories, and each category is represented by a particular pattern or color on the map. Many different methods exist to classify areal data for choropleth mapping (see section 2.4).

Sometimes geographers wish to measure the degree of association between two choropleth maps having the same internal areas for information available only in the mapped form. For example, an economic geographer investigating spatial patterns of income and unemployment may want to determine whether an association exists between choropleth maps of the two variables. One map may use five categories from highest to lowest income to depict median family income in a set of regions. The other map may show the rate of unemployment in the same areas, classified into a set of four ordinal categories. By assigning a numerical value to the categories (1 for the lowest to n for the highest), the map information can be transformed into a numerical data set suitable for correlation analysis.

For a problem of this type, Spearman's correlation index is a better choice than Pearson's to show the generalized association between the two variables, because choropleth map patterns are converted to numerical values at the ordinal

level. The researcher knows only that the values of a particular map category can be ranked above or below other categories. The interval between map patterns (or between numeric values assigned to the patterns) is not known. Therefore, the proper technique to measure the degree of association between the choropleth patterns is Spearman's correlation index.

This procedure is illustrated with a descriptive example of changes in language use in the Montreal, Quebec region. Suppose a geographer has a series of choropleth maps that show changes over time in the percentage of population speaking French and English for the 28 municipalities on the island of Montreal. Since the core city of Montreal has considerable language diversity, the maps divide the Montreal municipality into 24 subareas, each containing a generally homogeneous population structure. From this study area of 50 spatial units, the geographer wants to investigate whether the language use maps are correlated over time.

For those persons speaking French as their primary language at home, two choropleth maps are available. One map shows the change in the percentage of population speaking French during the period from 1971 to 1981 (figure 12.11, case 1). A second map covers change in the same variable for the period from 1981 to 1986 (figure 12.11, case 2). Does the change in French-speaking population for the earlier period (1971-81) correlate positively with the change during the later period (1981-86)? If so, similar spatial processes are operating over time within the Montreal region to affect patterns of language use.

Each of the map patterns is converted into a numerical variable with five categories. The lowest map category, those municipalities showing at least a 7.5 percent *decline* in French-speaking population, is assigned a value of 1. The assignment of numerical values to the other map categories continues up to 5, those municipalities having an *increase* in French-speaking population of at least 7.5 percent. The Spearman

Figure 12.11 Choropleth Maps of Percentage Change in Population
Speaking French at Home: Montreal, Quebec

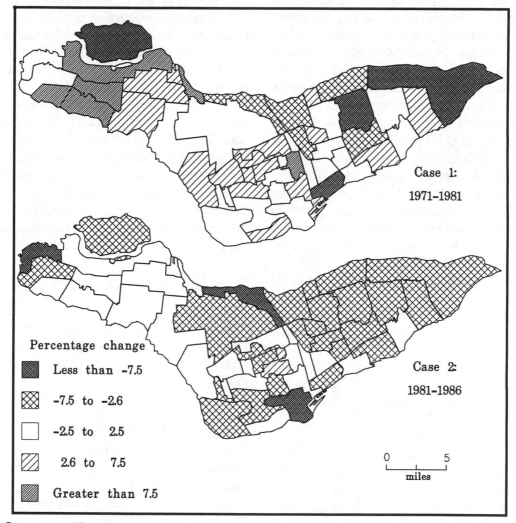

Case 1: 1971–1981

Case 2: 1981–1986

Percentage change

- Less than –7.5
- –7.5 to –2.6
- –2.5 to 2.5
- 2.6 to 7.5
- Greater than 7.5

0   5
miles

Source: modified from Trudel, Daniel. 1992. "A Spatial Analysis of Language-Use Patterns in
Montreal, 1971-1986". Master's Thesis, The University of Akron (Figure 4.4)

correlation is then calculated for the two variables in rank form. The resulting correlation index (r_s = .695) shows a moderately strong positive association between the two time periods. This correlation value indicates a relatively consistent temporal pattern of change for the percentage of French-speaking population within the Montreal region. That is, most of the areas either increased, decreased, or maintained their percentage of French language use during both time periods. If the map patterns had shown greater difference from area to area when the early time period was compared to the later period, the resulting correlation would be weaker. A weaker correlation would suggest that different spatial processes were operating at different times to affect the language use patterns within the Montreal region.

Using a similar methodology, choropleth maps of the Montreal region showing the change in the percentage of the population speaking English are also correlated over the two time periods. The Spearman's correlation value for the English population ($r_s = .368$) also shows a positive association between the two periods, but at a weaker level. This descriptive analysis suggests that the regional patterns of change in English language use are less consistent over time than those for the French language. Thus, the two choropleth maps showing change in the percentage of English-speaking population differ more widely from the early time period to the later period. Further analysis of these patterns is needed to gain a clearer understanding of the spatial processes that affect the dynamics of language in a multilingual region such as Montreal. Nevertheless, the correlation methodology for relating patterns in choropleth maps is a useful, objective measurement of the strength of association between two variables defined for areas.

12.5 Issues Regarding Correlation

When geographers apply statistical analysis to spatial data, the level of aggregation of the observation units may influence the results. This concern is especially important when inferences are drawn from the results of geographic analyses. Significant findings at one level of aggregation may not occur at other levels. For example, correlation results at the individual level are probably not equally significant at higher levels of aggregation. Although it may be observed that the level of income and amount of education are highly correlated for individuals, these same variables may not be similarly associated at the county or state levels. Thus, different scales of analysis may produce different degrees of correlation.

Another critical geographic concern is the so-called **ecological fallacy** concept, a reversal of the problem of aggregation just discussed. Researchers sometimes use highly aggregated data and then attempt to infer these results to lower levels of aggregation or to the individual level. Again, such inferences may not be valid. For example, just because crime rates are statistically correlated with percentage of persons under the poverty level at the state or census tract level, this does not imply that all persons below the poverty level are criminals. The correlation only shows that crime rates are higher in areas with a larger percentage of persons under the poverty level. Any inferences made beyond this statement, such as to individuals, cannot be supported by the data or the statistical analysis.

Key Terms and Concepts

References and Additional Reading

Abler, R., J. Adams, and P. Gould. 1971. *Spatial Organization: The Geographer's View of the World.* Englewood Cliffs, NJ: Prentice-Hall.

Barber, G. 1988. *Elementary Statistics for Geographers.* New York: Guilford.

Ebdon, D. 1985. *Statistics in Geography: A Practical Approach.* Oxford: Basil Blackwell.

Robinson, A. and R. Bryson. 1957. "A Method for Describing Quantitatively the Correspondence of Geographical Distributions." *Annals, Association of American Geographers* 47: 379-391.

Taylor, P. 1977. *Quantitative Methods in Geography: An Introduction to Spatial Analysis.* Boston: Houghton-Mifflin.

REGRESSION

In chapter 12 Pearson's correlation coefficient was discussed as the index for computing the degree of association between variables measured on an interval/ratio scale. In correlation analysis, the researcher does *not* have to assume a functional or causal relationship between the two variables. A correlation can be computed for any two variables, as long as the correlation index is consistent with the level of measurement for the data.

Geographers often work in research areas where relationships between variables need to be explored in more detail. The assumption or hypothesis may be that one variable influences or affects another or that a functional relationship ties one variable to another. For these geographical problems, regression analysis is a more useful statistical procedure than correlation.

A classic problem studied by geographers is the spatial relationship between level of precipitation in an agricultural region and the population density the area can support. It is hypothesized that the amount of moisture available at locations within a region influences or determines the density pattern of farm population in the region. Data could be collected for the two variables at various sites in the region, and regression used to answer questions about how population density relates to precipitation level. The nature or form of this relationship can be explored and the strength of the relationship determined. Assuming that the relationship is not exact or perfect, regression allows sources of error in the relationship to be examined and helps uncover additional variables that may influence the geographic pattern. The problem differs from simple correlation analysis because a functional relationship is expected between the variables, and the nature of that relationship needs to be explored more fully. Simply stated, regression should not be used unless a clear rationale or model links one variable to another, as is the case with population density and rainfall.

Regression can be applied successfully in all areas of geography. For example, a medical geographer could use regression to examine the relationship between the number of physicians located in the counties of a state and the income level of persons residing in these areas. A physical geographer may wish to see if a relationship exists between the acidity level in various locations in a chain of lakes and distance from a point pollution source. A political geographer could use regression to compare the strength of votes for a political party and the educational, financial, or racial composition of voters in the wards of a city. A cultural geographer surveying current attitudes could apply regression to see whether the level of support for a controversial issue like abortion relates to such socioeconomic characteristics as occupation, income, religious belief, or education.

The discussion in sections 13.1 through 13.4 concerns **bivariate regression**, which examines the influence of one variable on another. The variable creating the influence or effect is called the **independent variable**, and the variable receiving the influence or effect is termed the **dependent variable**. In regression terminology, the dependent variable is affected by (or perhaps caused by) the independent variable. The bivariate regression sections include a detailed examination of a geographic problem that illustrates the technique. This discussion provides an understanding of both the concepts and calculation procedures for simple regression analysis.

The most common application of regression is the identification of linear relationships between variables. In linear form, changes in values of the variables are constant across the range of the data. In these instances, the pattern of points from a scattergram approximates a straight line (figure 13.1, case 1). Real-world relationships, however, sometimes show variables that are related in curvi-linear ways where variable changes are not constant across all values

Figure 13.1   Linear and Curvi-linear Relationships

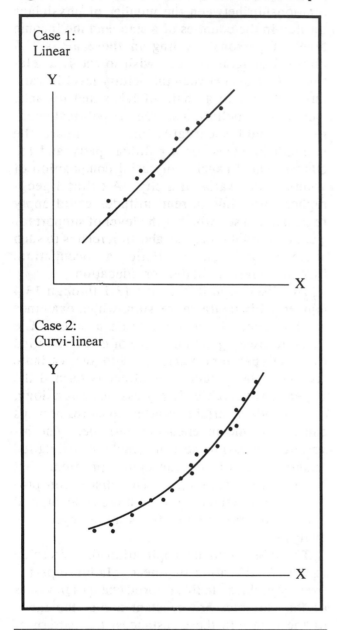

In section 13.5, **multivariate regression** is introduced with discussion limited to a general understanding of the concepts involved. More detailed information, including formulas and computational procedures, can be found by consulting the references cited at the end of the chapter.

13.1 Form of Relationship in Bivariate Regression

Like correlation, bivariate regression attempts to determine how one variable relates to another. The basic question posed by regression is: "What is the *form* or *nature* of the relationship between the variables under study?" As discussed in the correlation chapter, the relationship between two variables is easily visualized by constructing a scatterplot or graph. The point pattern on the graph determines the form of the relationship between variables.

Recall the example in which the average amount of snowfall at locations in northeast Ohio was related to distance from Lake Erie. Data for these interval-level variables were plotted for 33 sites in the study area to produce a scattergram (figure 12.8). The association was negative or inverse because snowfall amounts generally declined at distances further from the Lake. This example is now extended to illustrate the use of regression to analyze geographic data.

Although correlation and regression can both begin with data placed on a scattergram, the rationale for assigning variables to the two axes differs. In correlation, the two variables are assigned arbitrarily to the horizontal and vertical axes, and variables are not identified as being either independent or dependent. Correlation merely determines the degree of association between variables. Thus, when correlation was used to analyze the snowbelt data for association, independent and dependent variables were not specified, and the assignment of the snowfall and distance variables to axes of the scatterplot was arbitrary.

(figure 13.1, case 2). The two variables in the original snowbelt study area (chapter 12) depicted a curvi-linear relationship when plotted as a scattergram. In this chapter, only regression applied to linear relationships is discussed.

In bivariate regression, however, one variable serves as the dependent variable and the other as the independent variable. Since regression examines the influence of the independent variable on the dependent variable, proper specification of the two variables is necessary, and the assignment of variables to axes of the scattergram is not arbitrary. The independent variable is always placed on the horizontal or X axis (abscissa) and the dependent variable on the vertical or Y axis (ordinate). If the assignment of variables to axes is reversed, a different regression result will occur.

In the snowbelt problem, distance from Lake Erie is the independent variable and snowfall the dependent variable. The logical hypothesis is that the level of snowfall is affected or influenced by distance from the Lake (and not vice versa), so the snowfall variable must be placed on the vertical axis and distance on the horizontal axis.

The form of association between two variables can be portrayed graphically by plotting the data values on a scattergram. Bivariate regression describes this pattern of points more objectively by placing a line through the scatter of points. This line, called the "best fitting" or "least-squares" line of regression, summarizes the overall trend in the data and represents the form of the relationship between the independent and dependent variables.

Although an infinite number of lines could be drawn to summarize the points in a scattergram, the **least-squares regression line** is unique. As the name implies, the line minimizes the sum of squared vertical distances between each data point and the line (figure 13.2):

$$\text{minimize } \Sigma \ d_i^2$$

where: d_i = vertical distance separating point i from the regression line

No other line can be generated where the sum of the squared distances between the points and the line (measured vertically) is a smaller value than that calculated for the least-squares line. This line

Figure 13.2   The Objective of Least-squares Regression

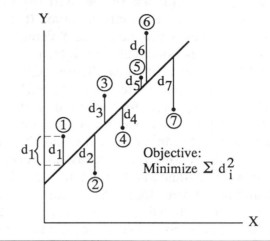

represents the best estimate of the relationship between the independent and dependent variables. It also serves as a predictive model by generating estimates of the dependent variable using both the values of the independent variable and knowledge of the relationship which connects the two variables.

In a bivariate regression with independent variable (X) and dependent variable (Y), the least-squares regression line is denoted by the following linear (straight-line) equation:

$$Y = a + bX \qquad (13.1)$$

In addition to the two variables, the equation contains two constants or parameters (a and b), which are calculated from the actual set of data. These values uniquely define the equation and establish the position of the best fitting line on the scattergram. The equation of the least-squares line for the snowbelt example is:

$$Y = 113.51 - 1.68X$$

where: X = distance from Lake Erie (miles)
Y = annual snowfall (inches)
a = 113.51
b = −1.68

The constant *a*, called the **Y-intercept**, represents the expected value of *Y* when the value of *X* is zero, the point where the regression line crosses the Y axis. It should be noted that *a* is the predicted best estimate of the *Y* value when *X* is zero. In the snowbelt example, the value of *a* is 113.51 inches of snow (figure 13.3). This result indicates that more than 113 inches of snowfall are expected for locations along the lakefront of Lake Erie where distance equals 0.

This interpretation of *a* in the snowbelt example is invalid, however. Because the relationship between distance and snowfall is curvi-linear rather than linear, locations within 6 miles of Lake Erie are excluded from the study area. The regression model uses the 33 locations outside the 6-mile band, and results apply only within the existing bounds of distance (formally termed the "domain" of *X*). Thus, for locations between 0 and 6 miles from Lake Erie (below the lower bound of distance), interpretation is invalid. In the same way, using the model to estimate snowfall for distances outside the 40-mile limit of the study area would also not be valid.

Figure 13.3 Interpreting the Regression Line for the Snowbelt Example

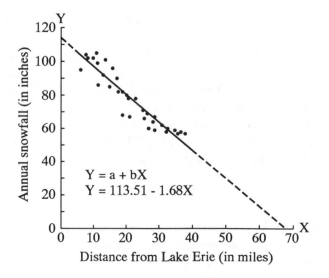

The other constant in the regression equation, *b*, represents the **slope** of the line. This value, also called the **regression coefficient**, shows the *absolute* change of the line in the *Y* (vertical) direction associated with an increase of 1 in the *X* (horizontal) direction. The slope reveals how responsive the dependent variable is to a unit increase in the independent variable.

Both the sign and magnitude of *b* offer useful information about the bivariate relationship. The sign (+ or −) of the slope determines the direction of relationship between the two variables. If the slope is positive (*b* > 0), the line trends upward from low values of *X* to high values of *X* (figure 13.4, case 1). On the other hand, if the slope is negative (*b* < 0), the line moves downward (figure 13.4, case 2). This interpretation of direction is equivalent to that for the correlation coefficient. When the relationship between the independent and dependent variables is direct, *b* will be positive and the line trends up from left to right. However, when the relationship is inverse, the value of *b* is negative, and the line trends down. When no relationship exists, the value of *b* is zero, and the line parallels the X axis (figure 13.4, case 3).

In absolute terms, the magnitude of the parameter *b* indicates the flatness or steepness of the regression line when moving from lower to higher values of *X*. When *b* is large (regardless of sign), the change in *Y*, the dependent variable, is large relative to a unit increase in *X*. In this situation, the slope of the regression line tends to be steep. When *b* is small, the opposite interpretation occurs; the change in the vertical direction is small when compared to a unit increase in the independent variable, and the line has a flatter slope.

In the snowbelt example, the calculated slope or *b* value is −1.68. Since the regression coefficient is negative, the least-squares line declines from left to right on the scattergram, indicating lower snowfall levels at greater distances from Lake Erie (figure 13.3). Although these results agree with those from the correlation analysis, the slope value measures the inverse relation-

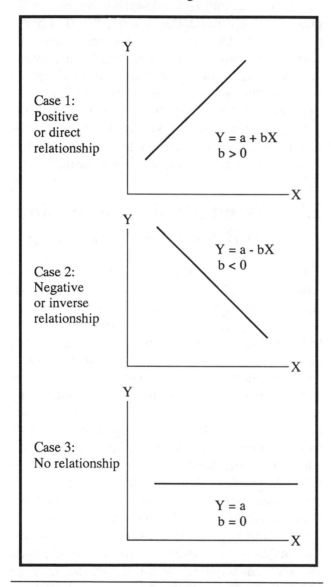

Figure 13.4 Interpretation of Slope in Bivariate Regression

Case 1: Positive or direct relationship

Y = a + bX
b > 0

Case 2: Negative or inverse relationship

Y = a - bX
b < 0

Case 3: No relationship

Y = a
b = 0

one variable are altered, the value of *b* will also change. For example, note the difference in slope or steepness of the regression line in the snow-belt example when distance is measured in kilometers (figure 13.5, case 1) rather than miles (figure 13.5, case 2). Therefore, although the slope accurately relates the *absolute* change in the independent and dependent variables, it cannot be used as a valid index of the *relative* relationship between the two variables because its numerical value is tied to the units of measurement.

The *a* and *b* parameters for the least-squares regression line are determined as follows:

$$b = \frac{n \, \Sigma XY - (\Sigma X)(\Sigma Y)}{n \, \Sigma X^2 - (\Sigma X)^2} \qquad (13.2)$$

$$a = \frac{\Sigma Y - b \, \Sigma X}{n} \qquad (13.3)$$

where: ΣX = sum of the values for variable X
ΣY = sum of the values for variable Y
ΣX^2 = sum of the squared values for variable X
ΣXY = sum of the product of corresponding X and Y values
n = number of observations

Note that the slope or regression coefficient is calculated first, then used to compute the Y intercept or *a* value. Table 13.1 shows the intermediate steps used to calculate *a* and *b* for the snowbelt example.

The two parameters uniquely define the least-squares regression line that best summarizes the relationship between the independent and dependent variables. This line can be placed on the scattergram by calculating and plotting two points that lie on the line and then connecting them with a straight line. The two points can be determined by selecting any two values of X, substituting them into the regression equation, and calculating the corresponding values of Y.

By selecting certain particular points to plot, the task of drawing the regression line is usually

ship in more explicit terms—for each unit increase in X (1 mile of distance further from the Lake), the amount of snowfall is reduced by approximately 1.68 inches.

Such interpretations of flatness or steepness, however, can sometimes be misleading, since the value of *b* depends on the units of measurement used for the two variables. If the units of

Figure 13.5   Influence of Measurement Units on Magnitude of Slope

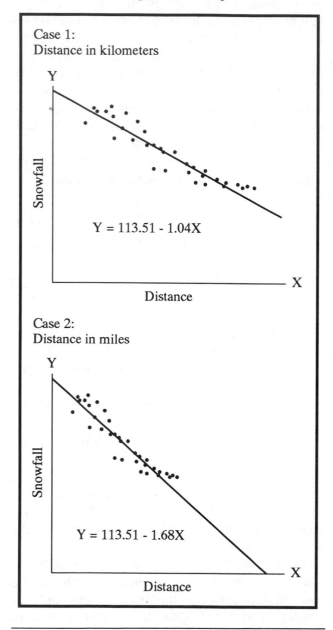

Case 1:
Distance in kilometers

Snowfall

$Y = 113.51 - 1.04X$

Distance

Case 2:
Distance in miles

Snowfall

$Y = 113.51 - 1.68X$

Distance

on the regression line is the X-intercept, where $Y = 0$ and $X = -a/b$. The X-intercept for the snowbelt example is the location where the regression line crosses the X axis, $Y = 0$ and $X = 67.57$ (figure 13.6, point 2). The regression line also passes through the point determined by the means of the X and Y variables. Since the mean values of the two variables are readily available, they represent a useful choice when plotting the regression line. In the snowbelt example, the means of the two variables are $\overline{X} = 21.4$ miles and $\overline{Y} = 77.5$ inches (figure 13.6, point 3).

13.2 Strength of Relationship in Bivariate Regression

For any realistic application, the scatterplot of points will not be depicted perfectly by the least-squares line of regression. All points will not lie exactly on the line, which implies that the independent variable cannot fully explain the dependent variable. This "error" in regression analysis can be traced to several sources. Since geographers study patterns on the earth's surface that are frequently produced by complex processes, expecting one independent variable in a regression model to account fully for the variation in the dependent variable is unreasonable. Even when multiple influences are considered, some portion of most real-world patterns is either attributable to unknown variables or is the result of unpredictable, random occurrences. The inability to measure or operationalize variables in an accurate or valid fashion may be another source of error. As a basis for evaluating the explanatory ability of the regression model, the strength of a relationship must be determined.

The issue of strength in regression analysis can be viewed in both conceptual and practical terms. In bivariate regression, one independent variable (X) is used to explain or account for variation in the dependent variable (Y). The ability of the independent variable to account for the variation in Y provides a measure of strength or level of

made easier. For example, the Y intercept, where $X = 0$ and $Y = a$, usually represents a conveniently graphed point through which the regression line passes. In the snowbelt example, the Y intercept occurs at $X = 0$ and $Y = 113.51$ (figure 13.6, point 1). A second convenient point

Table 13.1 Bivariate Regression Example for Northeastern Ohio Snowbelt:
The Influence of Distance from Lake Erie on Average Annual Snowfall

Site	X	Y	X^2	XY
1				
2				
3	(Data points were eliminated from the study area)			
4				
5				
6	7.7	104	59.8	804.2
7	9.9	102	97.7	1008.1
8	11.2	99	124.8	1106.0
9	12.9	92	166.2	1186.0
10	6.1	95	36.2	571.5
11	8.3	102	69.4	849.5
12	10.9	105	119.6	1148.5
13	13.6	101	185.3	1374.7
14	15.8	96	249.2	1515.5
15	17.1	90	292.1	1538.2
16	18.8	82	354.6	1544.2
17	20.6	78	423.2	1604.6
18	11.4	86	129.4	978.1
19	14.9	85	222.5	1267.9
20	17.5	82	307.2	1437.2
21	20.1	80	405.5	1610.9
22	22.8	78	520.3	1779.1
23	25.0	71	624.2	1773.9
24	26.3	69	691.1	1814.0
25	28.5	67	810.2	1907.1
26	18.8	68	354.6	1280.6
27	21.0	67	441.3	1407.5
28	25.4	66	646.1	1677.7
29	28.0	64	785.7	1793.9
30	30.7	62	942.6	1903.5
31	32.4	60	1052.5	1946.5
32	34.6	59	1198.4	2042.4
33	36.4	58	1321.9	2108.8
34	26.7	60	714.2	1603.5
35	28.5	59	810.2	1679.4
36	32.0	58	1024.5	1856.4
37	35.5	57	1259.4	2022.8
38	37.7	57	1418.5	2146.8
SUM	707.1	2559	17858.2	50288.7

$$b = \frac{N\Sigma XY - (\Sigma X)(\Sigma Y)}{N\Sigma X^2 - (\Sigma X)^2} \qquad b = \frac{(33)(50288.7) - (707.1)(2559)}{(33)(17858.2) - (707.1)^2} = -1.68 \qquad a = \frac{2559 - (-1.68)707.1}{33} = 113.51$$

$$a = \frac{\Sigma Y - b\Sigma X}{N}$$

explanation. To understand this process from another perspective, the strength of relationship in regression is determined by the amount of deviation between the points on the scattergram and the position of the best-fit line. In general, the closer the set of points lie to the regression line, the stronger the linear relationship between the variables. Determining the strength of relationship between two variables in regression is the same as measuring the relative ability of the independent variable to account for variation in the dependent variable.

A useful analogy involving a bucket and sponge was used by Abler, Adams, and Gould (1971) to illustrate the bivariate regression process (figure 13.7). A bucket full of water represents the total variation in the dependent variable, Y. A sponge denotes the independent variable, X, which will be used to explain variation in Y. When the sponge is dipped into the

bucket and removed, some of the water (symbolizing variation in Y) is absorbed. This represents the amount of variation in Y that can be explained by X. Although some of the water is absorbed by the sponge, some of it remains in the bucket. This residual amount is the portion of the variation in Y that cannot be explained by X.

The sponge analogy can be developed further. The central issue in measuring the strength of relationship is the determination of the relative ability of the sponge to remove water from the bucket. If the sponge is "super absorbent," it will remove a large proportion of the water. A "less absorbent" sponge removes less water. Calculating the ratio of volume of water removed by the sponge to the total volume of water originally in the bucket provides a strength index for the sponge.

The bucket and sponge analogy translates directly into regression terminology. The **total variation** in Y, the dependent variable, is represented by the original volume of water in the bucket (figure

---

Figure 13.6 Plotting the Regression Line

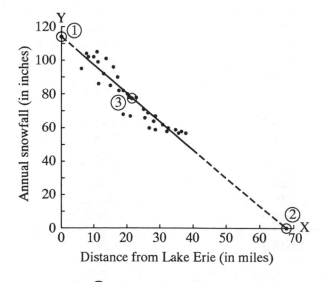

Point ① : Y = a, X = 0 (Y = 113.51, X = 0)
Point ② : Y = 0, X = -a/b (Y = 0, X = 67.57)
Point ③ : Y = $\overline{Y}$, X = $\overline{X}$ ($\overline{Y}$ = 77.5, $\overline{X}$ = 21.4)

Figure 13.7   Bucket and Sponge Analogy
                   in Bivariate Regression

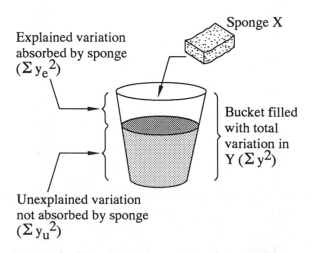

Source: modified from Abler, R., J. Adams, P. Gould. 1971. *Spatial Organization: The Geographers's View of the World.* Englewood Cliffs, NJ: Prentice Hall.

13.7). This provides a measure of the variation available for explanation by the independent variable:

$$\Sigma y^2 = \Sigma(Y - \overline{Y})^2 \qquad (13.4)$$

where: Σy^2 = total variation in Y

Note that Y ("large y") symbolizes values of the dependent variable, whereas y ("small y") denotes the deviation of each value of the dependent variable from its mean $(Y - \overline{Y})$. Because the calculation of total variation involves summing the squared deviations from the mean, the term is often referred to as the **total sum of squares**. An alternative formula for calculating the total sum of squares is:

$$TSS = \Sigma y^2 = \Sigma Y^2 - \frac{(\Sigma Y)^2}{n} \qquad (13.5)$$

where: TSS = total sum of squares

As illustrated in the bucket and sponge analogy, the total variation in the dependent variable can be broken into two parts: (1) the **explained variation** and (2) the **unexplained variation**:

$$\Sigma Y^2 = \Sigma y_e^2 + \Sigma y_u^2 \qquad (13.6)$$

where: Σy_e^2 = explained variation
Σy_u^2 = unexplained variation

The explained variation, also called the "explained sum of squares," is the amount of variation that can be accounted for by the independent variable X; in the analogy, it is the amount of water absorbed by the sponge. The unexplained variation, also called "residual variation," is the portion of total variation in the dependent variable that cannot be accounted for by the independent variable. It is analogous to the water not removed by the sponge. Because unexplained variation relates directly to analysis of residuals or error in regression, this concept will be discussed later in this section.

The explained variation is calculated by taking the ratio of the square of the covariation between X and Y to the variation in X:

$$\Sigma y_e^2 = \frac{(\Sigma xy)^2}{\Sigma x^2} \qquad (13.7)$$

where: Σxy = covariation of X and Y
Σx^2 = total variation in X

As discussed in section 12.2, the covariation between X and Y indicates how the two variables vary together (covary) and is used to help interpret the correlation coefficient. If two variables tend to covary consistently in the same direction, covariation (and the correlation coefficient) is high and positive (See figure 12.4, case 1). If they vary systematically, but in opposite directions, the covariation and resulting correlation is high and negative (See figure 12.4, case 2).

Although the direction of association represented by the sign of the covariation term is important in correlation analysis, it can be ignored when using covariation to determine explained variation. Remember that regression is concerned with the ability of the independent variable to account for variation in the dependent variable. By squaring the covariation in equation 13.7, the influence of a negative covariation is eliminated.

When the squared covariation (numerator of equation 13.7) is large relative to the total variation in X (denominator in equation 13.7), the explained variation is also large. In such cases, the points of the scattergram will tend to lie close to the regression line, and the independent variable X will account for more of the variation in the dependent variable Y. This is equivalent to selecting an absorbent sponge.

On the other hand, if the independent and dependent variables do not covary systematically in a scattergram, the resulting covariation measure (and correlation coefficient) will be low (See figure 12.4, case 3). In this situation, the ratio of the squared covariation to the total variation in X will also be low, producing a smaller amount of explained variation. When the level of covariation of the X and Y variables is weak, the points tend to scatter more widely about the regression line, and the general strength of relationship is weak.

The amount of explained variation is an absolute measure calculated in units of the dependent variable and is comparable to the volume of water removed by the sponge in the analogy. A more useful way to express the strength of a regression relationship is with a relative index that is not tied to the units of measurement. Termed the **coefficient of determination** or r^2, the relative strength index for regression is simply the ratio of the explained variation to the total variation in Y:

$$r^2 = \frac{\Sigma y_e^2}{\Sigma y^2} \qquad (13.8)$$

The index is often multiplied by 100 for ease of interpretation. In this form, the strength index ranges from 0 to 100 and can be interpreted as the *percentage* of variation in the dependent variable that is explained by the independent variable or in the bucket-sponge analogy, the percentage of water removed by the sponge.

Information on the strength of relationship for the snowbelt example is shown in table 13.2. The coefficient of determination ($r^2 = 0.872$) indicates that more than 87 percent of the variation in snowfall levels across this portion of northeast Ohio can be accounted for by the independent variable, distance from Lake Erie. The remaining 12.8 percent of regional snowfall variation is not explained by the distance variable in this regression model.

Although the coefficient of determination (r^2) is closely related to the correlation coefficient (r), the two indices have different purposes and interpretations. The correlation coefficient shows the direction and level of association between

Table 13.2 Work Table for Calculating Strength of Relationship for the Snowbelt Example

Total Variation

$$\text{Total SS} = \Sigma y^2 = \Sigma\left(Y - \overline{Y}\right)^2 = \Sigma Y^2 - \frac{(\Sigma Y)^2}{N}$$

$$= 207181 - \frac{(2559)(2559)}{33} = 207181 - 198438.8$$

$$= 8742.2$$

Explained Variation

$$\Sigma y_e^2 = \frac{(\Sigma xy)^2}{\Sigma x^2}$$

$$\Sigma xy = \Sigma XY - \frac{\Sigma X \Sigma Y}{N} = 50288.7 - \frac{(707.1)(2559)}{33}$$

$$= 50288.7 - 54832.4 = -4543.7$$

$$\Sigma x^2 = \Sigma X^2 - \frac{(\Sigma X)^2}{N} = 17858.2 - \frac{(707.1)(707.1)}{33}$$

$$= 17858.2 - 15151.2 = 2707.0$$

$$\Sigma y_e^2 = \frac{(\Sigma xy)^2}{\Sigma x^2} = \frac{(-4543.7)(-4543.7)}{2707.0} = 7626.6$$

Coefficient of Determination

$$r^2 = \frac{\Sigma y_e^2}{\Sigma y^2} = \frac{7626.6}{8742.2} = 0.872$$

any two variables and does not imply a functional or causal relationship. The coefficient of determination, on the other hand, is used as a regression index to measure the degree of fit of the points to the regression line or the ability of the independent variable to account for variation in the dependent variable. As a result, the use of r^2 requires a logical rationale for the existing relationship and the specification of independent and dependent variables.

13.3 Residual or Error Analysis in Bivariate Regression

Residual analysis provides additional spatial and nonspatial information about the variation in the dependent variable that cannot be explained by the independent variable. Because geographic relationships seldom allow perfect explanations of the dependent variable, points will only rarely lie on the regression line in the scatterplot. The amount of deviation of each point from the regression line is termed the absolute **residual**. It represents the vertical difference between the actual and predicted values of Y:

$$RES = Y - \hat{Y} \qquad (13.9)$$

where: RES = residual
Y = actual value of the dependent variable
$\hat{Y}$ = predicted regression line value of Y

The predicted values of Y are generated from the regression line. Each $\hat{Y}$ value represents a best estimate of the dependent variable produced from the a and b parameters that define the regression line. By substituting any value of X, (for example X_1), into the regression equation, the corresponding predicted value ($\hat{Y}_1$) is calculated:

$$\hat{Y}_1 = a + bX_1 \qquad (13.10)$$

By using the parameters of the line and the values of the independent and dependent variables, residuals can also be calculated without having to compute the predicted Y values directly:

$$RES = Y - (a + bX) \qquad (13.11)$$

Why is so much attention given to calculation of the residuals from regression? Geographers gain two general insights when examining residual values associated with a matched pair of data values. First, the size or magnitude of each residual provides the absolute amount of error associated with that data value. For some values, the residual or error is small, and the independent variable accurately predicts the value of the dependent variable. In other cases, the residual is large, indicating poor prediction of the dependent variable. The smaller the magnitude of a residual, the smaller the vertical distance on the scattergram separating the point from the regression line and the less error associated with the data value.

Second, the direction of residuals from the line is important to researchers. For some data values, the actual value of Y exceeds the predicted value, and the residual is positive (RES > 0). In these instances, the point lies above the regression line on the graph, and the model underestimates the actual value of the dependent variable. For other values, the predicted value of Y exceeds the corresponding actual value, the residuals are negative (RES < 0), and the points lie below the line. In these cases, the regression model has overestimated the actual value of Y.

The insights that can be gained from residual analysis are illustrated by examining two matched pairs of data values from the snowbelt example—sample locations 14 and 18. Using equation 13.11 and the data values listed in table 13.1, residuals for these values are calculated as follows:

$$\begin{aligned} \text{Point 14:} \ RES_{14} &= Y_{14} - (a + bX_{14}) \\ &= 96 - (113.51 + (-1.68)(15.8)) \\ &= 96 - 86.97 = 9.03 \end{aligned}$$

$$\begin{aligned} \text{Point 18:} \ RES_{18} &= Y_{18} - (a + bX_{18}) \\ &= 86 - (113.51 + (-1.68)(11.4)) \\ &= 86 - 94.36 = -8.36 \end{aligned}$$

Point 14, located approximately 16 miles from Lake Erie and near the center of study area, represents an observation with a positive residual (figure 13.8, point 14). Positioned above the regression line on the scattergram, with a predicted Y value of 86.97, the model underestimates snowfall at this site by 9 inches.

On the other hand, sample point 18, located about 11 miles from the Lake near the northwest corner of the study area, produces a negative residual whose magnitude (absolute error) is quite large (figure 13.8, point 18). With an actual snowfall amount of 86 inches, but a predicted value of 94.36, the resulting residual (−8.36) lies well below the regression line. The predicted values of Y and the corresponding residuals for each of the 33 study area sites are shown in table 13.3.

Residuals from regression can be used to examine absolute error associated with any data

value. Residual analysis can also be used to determine additional information concerning the total error in the model as well as the relative error for any observation. Recall that the unexplained or residual variation is that part of the variation in Y that cannot be explained by the independent variable, X. Also called the "residual sum of squares," this component of variation measures the total error in the regression model and can be derived by summing the squared residuals:

$$RSS = \Sigma y_u^2 = \Sigma(RES)^2 = \Sigma(Y - \hat{Y})^2 \quad (13.12)$$

where: RSS = residual sum of squares

Unexplained variation can also be computed by subtracting the explained sum of squares from the total sum of squares:

$$\Sigma y_u^2 = \Sigma y^2 - \Sigma y_e^2 \quad (13.13)$$

Although total error may be a useful index in some regression problems, a measure of the error associated with a typical or average value is often more desirable. An index of relative error, called the **standard error of the estimate**, represents the typical distance separating a point from the regression line on the scattergram. Standard error is analogous to standard deviation, which measures the deviation of a typical value from the mean of a distribution. The standard error of the estimate can be calculated in several ways, using the total error, the residuals, or the actual and predicted values of Y:

Figure 13.8   Interpretation of Residuals from Regression

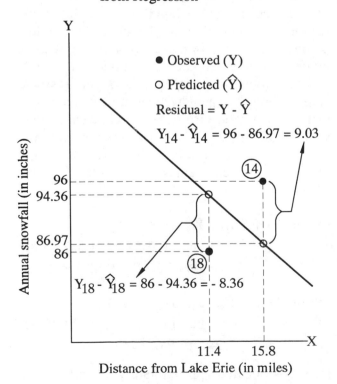

$$SE = \sqrt{\frac{\Sigma y_u^2}{n-2}} = \sqrt{\frac{\Sigma RES^2}{n-2}} = \sqrt{\frac{\Sigma(Y - \hat{Y})^2}{n-2}} \quad (13.14)$$

where: SE = standard error of the estimate
$n - 2$ = degrees of freedom

Table 13.3 Actual and Predicted Values of the Dependent Variable and Absolute and Standardized Residual Values for Snowbelt Example

Site	Y	X	$\hat{Y}$	RES	SRES
1					
2					
3					
4					
5					
6	104	7.7	100.57	3.43	0.57
7	102	9.9	96.88	5.12	0.86
8	99	11.2	94.69	4.31	0.72
9	92	12.9	91.84	0.16	0.03
10	95	6.1	103.26	−8.26	−1.38
11	102	8.3	99.57	2.43	0.41
12	105	10.9	95.20	9.80	1.64
13	101	13.6	90.66	10.34	1.73
14	96	15.8	87.97	9.03	1.51
15	90	17.1	84.93	5.22	0.87
16	82	18.8	81.90	0.07	0.01
17	78	20.6	78.90	−0.90	−0.15
18	86	11.4	94.36	−8.36	−1.40
19	85	14.9	88.48	−3.48	−0.58
20	82	17.5	84.11	−2.11	−0.35
21	80	20.1	79.74	0.26	0.04
22	78	22.8	75.21	2.79	0.47
23	71	25.0	71.51	−0.51	−0.09
24	69	26.3	69.33	−0.33	−0.05
25	67	28.5	65.63	1.37	0.23
26	68	18.8	81.93	−13.93	−2.32
27	67	21.0	78.23	−11.23	−1.87
28	66	25.4	70.84	−4.84	−0.81
29	64	28.0	66.47	−2.47	−0.41
30	62	30.7	61.92	0.08	0.01
31	60	32.4	59.08	0.92	0.15
32	59	34.6	55.38	3.62	0.60
33	58	36.4	52.36	5.64	0.94
34	60	26.7	68.65	−8.65	−1.44
35	59	28.5	65.63	−6.63	−1.11
36	58	32.0	59.75	−1.75	−0.29
37	57	35.5	53.87	3.13	0.52
38	57	37.7	50.17	6.83	1.14

Y = dependent variable (snowfall)
X = independent variable (distance)
$\hat{Y}$ = predicted Y value = $(a + bX)$
RES = residual = $(Y - \hat{Y})$

SE = standard error = $\sqrt{\dfrac{\Sigma RES^2}{n-2}} = 5.99$

SRES = standardized residual = (RES / SE)

Calculation of total and relative error for the snowbelt example illustrates the different uses of these regression indices. The unexplained variation or residual sum of squares can be calculated from the residuals or from the actual and predicted Y values:

$$\Sigma y_u^2 = \Sigma(\text{RES})^2 = \Sigma\left(Y - \hat{Y}\right)^2$$
$$= 1115.6$$

It can also be calculated from the total and explained variation:

$$\Sigma y_u^2 = \Sigma y^2 - \Sigma y_e^2$$
$$= 8742.2 - 7626.6$$
$$= 1115.6$$

These calculations show that of the 8742.2 units of total variation in the dependent variable; 1115.6 units of variation (inches of snowfall, in this case) are left unaccounted for by the single independent variable of distance from Lake Erie. To see how much of this total error is associated with a typical value or location in northeast Ohio, the standard error of the estimate must be computed:

$$\text{SE} = \sqrt{\frac{\Sigma y_u^2}{n-2}} = \sqrt{\frac{1115.6}{31}} = \sqrt{35.98} = 5.99$$

In this example, the resulting standard error of 5.99 suggests that the typical data value differs from its predicted or regression value by about 6 inches of snowfall. In other words, the independent variable, distance from Lake Erie, can be used to estimate snowfall at locations in this region with a general precision level of 6 inches.

The standard error of the estimate is also used to produce **standardized residual values.** This simple procedure, analogous to generating Z-scores for a distribution, converts absolute residuals into relative residuals:

$$\text{SRES} = \frac{\text{RES}}{\text{SE}} \qquad (13.15)$$

where: SRES = standardized residual

In this form, standardized residuals relate the magnitude of each residual to the size of the typical residual, represented by the standard error. They can be interpreted as the typical amount of error associated with a value measured in standard error units.

For example, a value whose residual equals 0 lies on the regression line and has a standardized residual of 0. A residual equal to the standard error value (i.e., a point 1 standard error unit above the regression line) would have a standardized residual of 1. Standardized residual values have the same sign as their corresponding absolute residual values. Table 13.3 displays the absolute and standardized residual values of each observation in the snowbelt example.

Residual analysis may also be used to identify additional factors that may influence the dependent variable. In bivariate regression, the independent variable explains a portion of the total variation in the dependent variable and leaves the remainder unexplained (residual error). If this error is interpreted as a new dependent variable, other variables can be identified to explain more of the remaining variation. Thus, residuals from regression should not be viewed as the end of the research process, but rather as an intermediate step in uncovering further influences on the dependent variable.

Geographic use of regression usually involves independent and dependent variables that can be mapped as spatial distributions. In these instances, residual analysis offers a useful method for determining additional influences on the dependent variable: a map of the absolute or standardized residuals. Such maps represent the spatial pattern of error from the regression and show geographically the inaccuracy of the independent variable in estimating the dependent variable. By analyzing spatial trends or nonrandom patterns on residual maps, new variables can often be discovered which serve as additional explanatory influences on the dependent variable.

The use of residual maps in geography is illustrated using the spatial distribution of the absolute residuals from the snowbelt example (figure 13.9). The map displays the pattern of error that results when the single variable of distance is used to estimate the level of snowfall across the region. The western portion of the study area appears to have large negative residuals. In this area, the model has greatly overpredicted snowfall levels; the actual levels are much lower than predicted. Locations in the northern and northeastern portions of the study area generally have large positive residuals. In these places, the actual amount of snowfall exceeds the estimated level, indicating substantial underprediction by the regression model.

How does a geographer use a residual map to help identify other variables that may influence the spatial pattern of the dependent variable? To answer this question, two important issues should be considered: (1) Is there another variable with a spatial pattern similar to that of the residuals? (2) Is there a logical, rational reason for this variable to influence the dependent variable? The first factor shows a practical strategy for discovering possible influences in the problem; namely, analyzing the residual map for clues to the new variable. The second factor requires the researcher to have a clear understanding of the geographic patterns and processes being analyzed. Here is another instance where the successful use of statistical analysis in geography requires a strong geographic knowledge or background in the subject under investigation.

Figure 13.9   Residual Map for the
Snowbelt Example

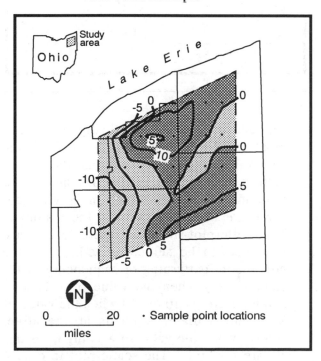

13.4 Inferential Use of Regression

Sometimes the results from bivariate regression can be tested for significance and results from a sample inferred to the population from which the sample was drawn. This can be done in several ways. For example, inferences can be made concerning the slope and the Y-intercept, with the two parameters defining the regression line. Inferential testing can also be applied to the coefficient of determination, with the measure of strength in regression. However, no matter which inferential procedure is chosen, a stringent set of assumptions apply.

Regression has more assumptions than the other inferential procedures discussed in the text. Variables must be measured on an interval or ratio scale, using one of two modeling schemes. In a **fixed-X model**, the investigator preselects certain values of the independent variable (X), perhaps as part of a controlled experiment. Sample values for the dependent variable (Y) are then derived using a component of randomness. In the **random-X model**, sample values for both the independent and dependent variables are chosen at random. A series of assumptions must be met when using regression in either the fixed-X or random-X models.

1. Since regression analysis places a best-fitting line through a scatterplot, the variables are assumed to have a linear relationship (figure 13.10, case 1). If the association between variables cannot be represented by a straight line, the regression line will not accurately depict the true relationship between the variables.

2. For every value of the independent variable (X), the distribution of residual or error values $(Y - \hat{Y})$ should be normal, and the mean of the residuals should be zero. If these assumptions are met, a normal distribution of residuals is "centered" on the regression line for any value of X (figure 13.10, case 2). Meeting these requirements makes it virtually certain that variables X and Y are themselves normally distributed. In practice, however, geographers seldom have an adequate number of Y values for every value of X, making it difficult (if not impossible) to validate these assumptions with sample data.

3. For every value of the independent variable (X), the variance of residual error is assumed to be equal. This is known as the "homoscadasticity" or "equal variance" requirement. The "funnel-shaped" scatterplot (figure 13.10, case 3) is almost certainly heteroscedastic, for the variance of residuals at X_1 appears considerably greater than the variance of residuals at X_2.

4. The value of each residual is independent of all other residual values. This requirement assumes that the residuals are randomly arranged or sequenced along the regression

Figure 13.10   Scatterplots and Residual Patterns in Regression Analysis

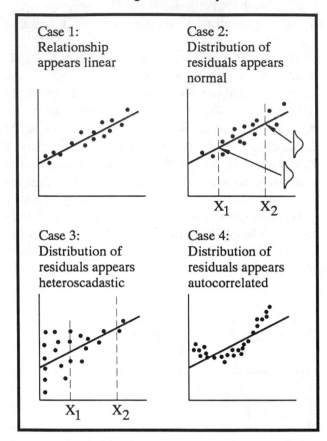

line and are not systematically influenced or affected by the magnitude of X. If the data values represent locations, the spatial pattern of residuals is also assumed to be random. If this assumption is violated, autocorrelation is present in the residuals. The scatterplot shown in figure 13.10, case 4 appears to be autocorrelated. Note the nonrandom patterning or sequencing of residuals: (a) when the value of X is low, residuals are positive; (b) when the value of X is intermediate, residuals are negative; and (c) when the value of X is high, residuals are positive. The scatterplot in case 4 also appears curvi-linear, violating assumption 1.

Inferential testing of the regression parameters considers the Y-intercept (α) and slope (β), which define the population regression line relating the independent and dependent variables:

$$Y = \alpha + \beta X \qquad (13.16)$$

The sample Y-intercept (a) and sample slope (b) are the best estimators of their respective population parameters.

Inferential testing of the population intercept is seldom used in geographic problem solving and regression modeling. For many geographic applications, the predicted or estimated value of Y when X is zero has little practical meaning. If the Y-intercept value lies outside the domain of X, testing its significance is not particularly worthwhile. In the snowbelt example, the relationship between distance from Lake Erie and snowfall is curvi-linear, and as a result, the study area excluded locations within 6 miles of the Lake. Thus, the intercept when $X = 0$ has no valid interpretation.

Inferential testing of the population slope or regression coefficient is often very useful in geographic analysis. The slope indicates the responsiveness of the dependent variable to changes in the independent variable, and this relationship may have direct spatial interpretations. In the snowbelt problem, a geographer can predict the magnitude of decrease in snowfall amount expected at greater distances from Lake Erie. Also, if results from inferential testing show no significant slope (i.e., if $\beta = 0$), then no significant relationship exists between the two variables. In these instances, the regression model has no predictive value.

Inferential significance tests can also be applied to the population coefficient of determination (ρ^2), using the sample coefficient of determination (r^2) as the best estimator. Analysis of variance (or F statistic) is used to evaluate the significance of r^2. In this context, the null hypothesis is that the population coefficient of determination is not significantly greater than

zero (H_0: $\rho^2 = 0$), and the alternate hypothesis is the converse (H_A: $\rho^2 \neq 0$).

Recall that the total variation in the dependent variable is equal to the sum of two components, the explained variation and the unexplained variation (equation 13.6). It was also shown that the coefficient of determination is the ratio of explained variation to total variation in Y (equation 13.8). Thus, returning to the bucket and sponge analogy, testing the significance of r^2 to determine if it is significantly greater than zero is equivalent to testing whether the sponge (explained variation) removes a significant amount of water (total variation) from the bucket.

These components of variation in Y may be expressed in terms of the "sum of squares," illustrating the direct equivalence between regression and analysis of variance:

$\Sigma y^2 = \Sigma y_e^2 + \Sigma y_u^2$ (from equation 13.6)

$\Sigma y^2 = \text{TSS}$ (total sum of squares)

$\Sigma y_e^2 = \text{RSS}$ (regression or explained sum of squares)

$\Sigma y_u^2 = \text{ESS}$ (error, residual, or unexplained sum of squares)

Therefore: $\quad \text{TSS} = \text{RSS} + \text{ESS}$

and

$$r^2 = \frac{\Sigma y_e^2}{\Sigma y^2} = \frac{\text{RSS}}{\text{TSS}}$$

The F statistic from analysis of variance is expressed in terms of the coefficient of determination:

$$F = \frac{r^2(n-2)}{1-r^2} \qquad (13.17)$$

This F statistic for regression is the square of the t statistic used in the previous chapter to test the significance of r, the simple correlation coefficient

Bivariate Regression Analysis

Primary Objective: Determine if an independent variable (X) accounts for a significant portion of the total variation in a dependent variable (Y)

Requirements and Assumptions:
1. Variables measured on interval or ratio scale
2. Fixed-X Model: Values of independent variable (X) chosen by the investigator, and values of dependent variable (Y) randomly selected for each X

   Random-X Model: Values of both X and Y randomly selected
3. Variables have a linear association
4. For every value of X, the distribution of residuals ($Y - \hat{Y}$) should be normal, and the mean of the residuals should equal zero
5. For every value of X, the variance of residual error is equal (homoscedastic)
6. The value of each residual is independent of all other residual values (no autocorrelation)

Hypotheses:
H_0: $\rho^2 = 0$
H_A: $\rho^2 \neq 0$

Test Statistic:

$$F = \frac{r^2(n-2)}{1-r^2}$$

(equation 12.9). In the snowbelt example, the significance testing results in the following:

$$F = \frac{r^2(n-2)}{1-r^2} = \frac{.873(33-2)}{1-.873} = 213.09$$

$$p = 0.0000$$

From this result, it can be concluded that the independent variable (distance from Lake Erie) accounts for a significant amount of the total variation in the dependent variable (snowfall).

13.5 Basic Concepts of Multivariate Regression

Although bivariate regression provides an effective way to understand the use and interpretation of regression in geographic research, most real-world problems require a multivariate approach. **Multiple regression**, where a set of independent variables is used to explain a single dependent variable, offers a logical extension to the simple, two-variable regression procedure. In the bucket and sponge analogy, this approach is equivalent to using additional sponges to absorb a greater volume of water from the bucket.

As in the bivariate case, the selection of independent variables for multiple regression is an important step in applying this statistical method properly. In geographic studies, the focus of investigation is on explaining the spatial pattern of the dependent variable. Independent variables must be obtained that relate to or explain this dependent variable. Since use of regression suggests a functional relationship between the variables, the independent variables need to be evaluated carefully to ensure that they all show a logical relationship to the dependent variable.

Multivariate regression proceeds in a way similar to that used in bivariate analysis. The statistical technique determines the functional relationships linking the independent variables to the dependent variable. The strength of these relationships is measured and the remaining error analyzed. Because multiple regression uses more than one independent variable, the procedure provides information on the absolute and relative ability of each variable to explain the dependent variable.

Multiple regression first examines the functional relationship between the independent variables and the single dependent variable. In bivariate regression, this objective could be explored graphically by observing the pattern of points plotted in a two-dimensional scattergram. However, one dimension (axis on a graph) is required for each variable. Except for problems using only two independent variables, graphical representation is not practical in multivariate regression. Nevertheless, the form of the relationship connecting the independent variables to the dependent variable can be presented and interpreted algebraically.

Using the least-squares objective discussed earlier in this chapter, a best-fitting line is generated to represent the prevailing trend of points in the two variable scattergram:

$$Y = a + bX_1 \qquad (13.18)$$

In those multivariate problems having two independent variables, observations are located in a three dimensional space, and a best-fitting plane (figure 13.11) can be derived:

$$Y = a + b_1X_1 + b_2X_2 \qquad (13.19)$$

For problems having three or more independent variables, the same procedure is applied algebraically:

$$Y = a + b_1X_1 + b_2X_2 + \ldots + b_nX_n \qquad (13.20)$$

where:
$$Y = \text{dependent variable}$$
$$X_1 \ldots X_n = \text{independent variables}$$
$$a = Y\text{-intercept or constant}$$
$$b_1 \ldots b_n = \text{regression coefficients}$$

In both the bivariate and multivariate equations, the constant or a value shows the value of Y when the values of all X variables are zero. The major difference between the equations lies in the regression coefficients. In the two-variable problem, one regression coefficient or b value is produced to show the influence of a single independent variable on the dependent variable. In the multivariate case, however, one regression coefficient (b_i) is calculated for each independent variable, (X_i). Each coefficient indicates the absolute influence of an independent variable on the dependent variable. Like the bivariate case, each regression coefficient shows the change in Y associated with a unit change in the given X variable.

Figure 13.11   Three-dimensional Graph of Multiple Regression Plane

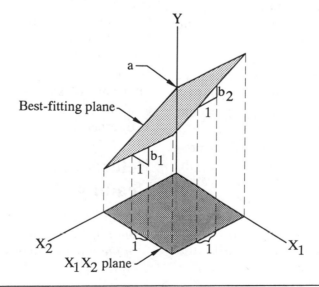

Because multiple regression uses more than one independent variable to explain a dependent variable, the relative importance of each variable is a major concern. At first, it may appear that the regression coefficients could be used directly to determine the importance of independent variables. However, because these parameters are influenced by the measurement units, they cannot be used directly.

This situation is analogous to the problems encountered when using standard deviation to compare relative levels of variation between two or more variables. As discussed in chapter 3, when variables are measured on different scales, the coefficient of variation provides a better measure for comparing the relative variation between variables. In multiple regression, a valid comparison of the *relative* ability of each independent variable to explain the variation in the dependent variable is accomplished by computing **standardized regression coefficients**, also known as "beta values." These coefficients serve as relative indices of strength, allowing a direct comparison of the influence of each independent variable in accounting for the variation in the dependent variable. Among the set of independent variables, the one with the largest beta value produces the strongest relationship to the dependent variable. Like the interpretation of regression coefficients, the sign of each beta value (positive or negative) determines whether this relationship is direct or inverse.

In problems having more than one independent variable, measuring the total amount of variation explained by the set of independent variables is also important. In earlier discussions on bivariate regression, two indices provided absolute and relative measures of strength. Explained variation measured the total level of variation in the dependent variable that could be explained or accounted for by the single independent variable. The explained variation was then converted to a relative index, called the "coefficient of determination" or r^2.

Similarly, in multivariate regression, a **coefficient of multiple determination** (R^2) is calculated:

$$R^2 = \frac{\Sigma y_e^2}{\Sigma y^2} \qquad (13.21)$$

where: R^2 = coefficient of multiple determination
Σy_e^2 = explained variation
Σy^2 = total variation in Y

This index measures the ratio of variation explained by the set of independent variables to the total variation in the dependent variable. When the R^2 index is multiplied by 100, the result is interpreted as the percentage of variation explained.

Key Terms and Concepts

References and Additional Reading

Abler, R. F., J. S. Adams, and P. R. Gould. 1971. *Spatial Organization: The Geographer's View of the World.* Englewood Cliffs, NJ: Prentice-Hall.

Clark, W. A. V. and P. L. Hosking. 1986. *Statistical Methods for Geographers.* New York: Wiley.

Draper, N. and H. Smith. 1981. *Applied Regression Analysis* (2nd edition). New York: Wiley.

Ebdon, D. 1985. *Statistics in Geography: A Practical Approach* (2nd edition). Oxford: Basil Blackwell.

Griffith, D. and C. Amrhein. 1991. *Statistical Analysis for Geographers.* Englewood Cliffs, NJ: Prentice-Hall.

Johnston, R. J. 1978. *Multivariate Statistical Analysis in Geography.* London: Longman.

Neter, J., W. Wasserman, and M. Kutner. 1983. *Applied Linear Regression Models.* Homewood, IL: Irwin.

GENERAL REFERENCES— STATISTICAL METHODS IN GEOGRAPHY

Barber, G. 1988. *Elementary Statistics for Geographers*. New York: Guilford Press.

Clark, W. and P. Hosking. 1986. *Statistical Methods for Geographers*. New York: Wiley.

Ebdon, D. 1985. *Statistics in Geography: A Practical Approach* (2nd edition). Oxford: Basil Blackwell.

Fitzgerald, B. (ed). 1974. *Science in Geography* (Vols. 1-4). Oxford: Oxford University Press.

Gregory, S. 1978. *Statistical Methods and the Geographer* (4th edition). New York: Longman.

Griffith, D. and C. Amrhein. 1991. *Statistical Analysis for Geographers*. Englewood Cliffs, NJ: Prentice-Hall.

Hammond, R. and P. McCullagh. 1974. *Quantitative Techniques in Geography: An Introduction*. Oxford: Oxford University Press.

Johnston, R. 1978. *Multivariate Statistical Methods in Geography*. London: Longman.

Lewis, P. 1977. *Maps and Statistics*. London: Methuen.

Matthews, J. 1981. *Quantitative and Statistical Approaches to Geography*. Oxford: Pergamon.

Norcliffe, G. 1977. *Inferential Statistics for Geographers*. New York: Halstead Press.

Shaw, G. and D. Wheeler. 1985. *Statistical Techniques in Geographical Analysis*. New York: Wiley.

Silk, J. 1979. *Statistical Concepts in Geography*. London: George Allen & Unwin.

Taylor, P. 1977. *Quantitative Methods in Geography: An Introduction to Spatial Analysis*. Boston: Houghton-Mifflin.

Unwin, D. 1981. *Introductory Spatial Analysis*. New York: Methuen.

Wilson, A. and M. Kirby. 1975. *Mathematics for Geographers and Planners*. Oxford: Clarendon Press.

Yeates, M. 1974. *An Introduction to Quantitative Analysis in Human Geography*. New York: McGraw-Hill.

APPENDIX

STATISTICAL TABLES

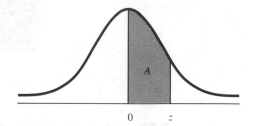

Table A   The Normal Table

Note: To get *A* for a given value of *Z*, insert a decimal point before the four digits. For example, *Z*=1.43 gives *A*=0.4236.

z	0.00	0.01	0.02	0.03	0.04	0.05	0.06	0.07	0.08	0.09
0.0	0000	0040	0080	0120	0160	0199	0239	0279	0319	0359
0.1	0398	0438	0478	0517	0557	0596	0636	0675	0714	0753
0.2	0793	0832	0871	0910	0948	0987	1026	1064	1103	1141
0.3	1179	1217	1255	1293	1331	1368	1406	1443	1480	1517
0.4	1554	1591	1628	1664	1700	1736	1772	1808	1844	1879
0.5	1915	1950	1985	2019	2054	2088	2123	2157	2190	2224
0.6	2257	2291	2324	2357	2389	2422	2454	2486	2517	2549
0.7	2580	2611	2642	2673	2704	2734	2764	2794	2823	2852
0.8	2881	2910	2939	2967	2995	3023	3051	3078	3106	3133
0.9	3159	3186	3212	3238	3264	3289	3315	3340	3365	3389
1.0	3413	3438	3461	3485	3508	3531	3554	3577	3599	3621
1.1	3643	3665	3686	3708	3729	3749	3770	3790	3810	3830
1.2	3849	3869	3888	3907	3925	3944	3962	3980	3997	4015
1.3	4032	4049	4066	4082	4099	4115	4131	4147	4162	4177
1.4	4192	4207	4222	4236	4251	4265	4279	4292	4306	4319
1.5	4332	4345	4357	4370	4382	4394	4406	4418	4429	4441
1.6	4452	4463	4474	4484	4495	4505	4515	4525	4535	4545
1.7	4554	4564	4573	4582	4591	4599	4608	4616	4625	4633
1.8	4641	4649	4656	4664	4671	4678	4686	4692	4699	4706
1.9	4713	4719	4726	4732	4738	4744	4750	4756	4761	4767
2.0	4772	4778	4783	4788	4793	4798	4803	4808	4812	4817
2.1	4821	4826	4830	4834	4838	4842	4846	4850	4854	4857
2.2	4861	4864	4868	4871	4875	4878	4881	4884	4887	4890
2.3	4893	4896	4898	4901	4904	4906	4909	4911	4913	4916
2.4	4918	4920	4922	4925	4927	4929	4931	4932	4934	4936
2.5	4938	4940	4941	4943	4945	4946	4948	4949	4951	4952
2.6	4953	4955	4956	4957	4959	4960	4961	4962	4963	4964
2.7	4965	4966	4967	4968	4969	4970	4971	4972	4973	4974
2.8	4974	4975	4976	4977	4977	4978	4979	4979	4980	4981
2.9	4981	4982	4982	4983	4984	4984	4985	4985	4986	4986
3.0	4987	4987	4987	4988	4988	4989	4989	4989	4990	4990
3.1	4990	4991	4991	4991	4992	4992	4992	4992	4993	4993
3.2	4993	4993	4994	4994	4994	4994	4994	4995	4995	4995
3.3	4995	4995	4996	4996	4996	4996	4996	4996	4996	4997
3.4	4997	4997	4997	4997	4997	4997	4997	4997	4998	4998
3.5	4998	4998	4998	4998	4998	4998	4998	4998	4998	4998

The area, *A*, stays at 0.4998 until *Z*=3.62. From *Z*=3.63 to 3.90 *A*=0.4999. For *Z* > 3.90, *A*=0.5000, to four decimal places.
From Leon F. Marzillier, *Elementary Statistics* ©1990 by Wm. C. Brown Publishers.

Table B   Table of Random Numbers

31871	60770	59235	41702	89372	28600	30013	18266	65044	61045
87134	32839	17850	37359	27221	92409	94778	17902	09467	86757
06728	16314	81076	42172	46446	09226	96262	77674	70205	98137
95646	67486	05167	07819	79918	83949	45605	18915	79458	54009
44085	87246	47378	98338	40368	02240	72593	52823	79002	88190
83967	84810	51612	81501	10440	48553	67919	73678	83149	47096
49990	02051	64575	70323	07863	59220	01746	94213	82977	42384
65332	16488	04433	37990	93517	18395	72848	97025	38894	46611
42309	04063	55291	72165	96921	53350	34173	39908	11634	87145
84715	41808	12085	72525	91171	09779	07223	75577	20934	92047
63919	83977	72416	55450	47642	01013	17560	54189	73523	33681
97595	78300	93502	25847	19520	16896	69282	16917	04194	25797
17116	42649	89252	61052	78332	15102	47707	28369	60400	15908
34037	84573	49914	59688	18584	53498	94905	14914	23261	58133
08813	14453	70437	49093	69880	99944	40482	04254	62842	68089
67115	41050	65453	04510	35518	88843	15801	86163	49913	46849
14596	62802	33009	74095	34549	76634	64270	67491	83941	55154
70258	26948	60863	47666	58512	91404	97357	85710	03414	56591
83369	81179	32429	34781	00006	65951	40254	71102	60416	43296
83811	49358	75171	34768	70070	76550	14252	97378	79500	97123
14924	71607	74638	01939	77044	18277	68229	09310	63258	85064
60102	56587	29842	12031	00794	90638	21862	72154	19880	80895
33393	30109	42005	47977	26453	15333	45390	89862	70351	36953
92592	78232	19328	29645	69836	91169	95180	15046	45679	94500
27421	73356	53897	26916	52015	26854	42833	64257	49423	39440
26528	22550	36692	25262	61419	53986	73898	80237	71387	32532
07664	10752	95021	17030	76784	86861	12780	44379	31261	18424
37954	72029	29624	09119	13444	22645	78345	79876	37582	75549
66495	11333	81101	69328	84838	76395	35997	07259	66254	47451
72506	28524	39595	49356	92733	42951	47774	75462	64409	69116
09713	70270	28077	15634	36525	91204	48443	50561	92394	60636
51852	70782	93498	44669	79647	06321	04020	00111	24737	05521
31460	22222	18801	00675	57562	97923	45974	75158	94918	40144
14328	05024	04333	04135	53143	79207	85863	04962	89549	63308
84002	98073	52998	05749	45538	26164	68672	97486	32341	99419
89541	28345	22887	79269	55620	68269	88765	72464	11586	52211
50502	39890	81465	00449	09931	12667	30278	63963	84192	25266
30862	61996	73216	12554	01200	63234	41277	20477	71899	05347
36735	58841	35287	51112	47322	81354	51080	72771	53653	42108
11561	81204	68175	93037	47967	74085	05905	86471	47671	18456

From Leon F. Marzillier, *Elementary Statistics* ©1990 by Wm. C. Brown Publishers.

Table C Student's *t* Distribution

							Degrees of Freedom									
t	1	2	3	4	5	6	7	8	9	10	11	12	13	14	15	16
0.1	0317	0353	0367	0374	0379	0382	0384	0386	0387	0388	0389	0390	0391	0391	0392	0392
0.2	0628	0700	0729	0744	0753	0760	0764	0768	0770	0773	0774	0776	0777	0778	0779	0780
0.3	0928	1038	1081	1104	1119	1129	1136	1141	1145	1148	1151	1153	1155	1157	1159	1160
0.4	1211	1361	1420	1452	1472	1485	1495	1502	1508	1512	1516	1519	1522	1524	1526	1528
0.5	1476	1667	1743	1783	1809	1826	1838	1847	1855	1861	1865	1869	1873	1876	1878	1881
0.6	1720	1953	2046	2096	2127	2148	2163	2174	2183	2191	2197	2202	2206	2210	2213	2215
0.7	1944	2218	2328	2387	2424	2449	2467	2481	2492	2501	2508	2514	2519	2523	2527	2530
0.8	2148	2462	2589	2657	2700	2729	2750	2766	2778	2788	2797	2804	2810	2815	2819	2823
0.9	2333	2684	2828	2905	2953	2986	3010	3028	3042	3054	3063	3071	3078	3083	3088	3093
1.0	2500	2887	3045	3130	3184	3220	3247	3267	3283	3296	3306	3315	3322	3329	3334	3339
1.1	2651	3070	3242	3335	3393	3433	3461	3483	3501	3514	3526	3535	3544	3551	3557	3562
1.2	2789	3235	3419	3518	3581	3623	3654	3678	3696	3711	3723	3734	3742	3750	3756	3762
1.3	2913	3384	3578	3683	3748	3793	3826	3851	3870	3886	3899	3910	3919	3927	3934	3940
1.4	3026	3518	3720	3829	3898	3945	3979	4005	4025	4041	4055	4066	4075	4084	4091	4097
1.5	3128	3638	3847	3960	4030	4079	4114	4140	4161	4177	4191	4203	4212	4221	4228	4235
1.6	3222	3746	3960	4075	4148	4196	4232	4259	4280	4297	4310	4322	4332	4340	4348	4354
1.7	3307	3844	4062	4178	4251	4300	4335	4362	4383	4400	4414	4426	4435	4444	4451	4458
1.8	3386	3932	4152	4269	4341	4390	4426	4452	4473	4490	4503	4515	4525	4533	4540	4546
1.9	3458	4026	4232	4349	4421	4469	4504	4530	4551	4567	4580	4591	4601	4609	4616	4622
2.0	3524	4082	4303	4419	4490	4538	4572	4597	4617	4633	4646	4657	4666	4674	4680	4686
2.1	3585	4147	4367	4482	4551	4598	4631	4655	4674	4690	4702	4712	4721	4728	4735	4740
2.2	3642	4206	4424	4537	4605	4649	4681	4705	4723	4738	4750	4759	4768	4774	4781	4786
2.3	3695	4259	4475	4585	4651	4694	4725	4748	4765	4779	4790	4799	4807	4813	4819	4824
2.4	3743	4308	4521	4628	4692	4734	4763	4784	4801	4813	4824	4832	4840	4846	4851	4855
2.5	3789	4352	4561	4666	4728	4767	4795	4815	4831	4843	4852	4860	4867	4873	4877	4882
2.6	3831	4392	4598	4700	4759	4797	4823	4842	4856	4868	4877	4884	4890	4895	4900	4903
2.7	3871	4429	4631	4730	4786	4822	4847	4865	4878	4888	4897	4903	4909	4914	4918	4921
2.8	3908	4463	4661	4756	4810	4844	4867	4884	4896	4906	4914	4920	4925	4929	4933	4936
2.9	3943	4494	4687	4779	4831	4863	4885	4901	4912	4921	4928	4933	4938	4942	4945	4948
3.0	3976	4523	4712	4800	4850	4880	4900	4915	4925	4933	4940	4945	4949	4952	4955	4958
3.1	4007	4549	4734	4819	4866	4894	4913	4927	4936	4944	4949	4954	4958			
3.2	4036	4573	4753	4835	4880	4907	4925	4937	4946	4953	4958					
3.3	4063	4596	4771	4850	4893	4918	4934	4946	4954							
3.4	4089	4617	4788	4864	4904	4928	4943	4953								
3.5	4114	4636	4803	4876	4914	4936	4950									
3.6	4138	4654	4816	4886	4922	4943										
3.7	4160	4670	4829	4896	4930	4950										
3.8	4181	4686	4840	4904	4937											
3.9	4201	4701	4850	4912	4943											
4.0	4220	4714	4860	4919	4948											
4.1	4239	4727	4869	4926	4953											
4.2	4256	4739	4877	4932												
4.3	4273	4750	4884	4937												
4.4	4289	4760	4891	4942												
4.5	4304	4770	4898	4946												
4.6	4319	4779	4903	4950												
4.7	4333	4788	4909													
4.8	4346	4796	4914													
4.9	4359	4804	4919													
5.0	4372	4811	4923													
5.8	4456	4858	4949													
6.4	4507	4882														
7.0	4548	4901														
10.0	4683	4951														
12.8	4752															
31.9	4900															
63.7	4950															

Table C *(continued)*

								Degrees of Freedom							
t	17	18	19	20	21	22	23	24	25	26	27	28	29	30	∞
0.1	0392	0393	0393	0393	0394	0394	0394	0394	0394	0394	0395	0395	0395	0395	0398
0.2	0781	0781	0782	0782	0783	0783	0784	0784	0785	0785	0785	0785	0786	0786	0793
0.3	0928	1162	1163	1164	1164	1165	1166	1166	1167	1167	1168	1168	1168	1169	1179
0.4	1529	1531	1532	1533	1534	1535	1536	1537	1537	1538	1538	1539	1540	1540	1554
0.5	1883	1884	1886	1887	1889	1890	1891	1892	1893	1894	1894	1895	1896	1896	1915
0.6	2218	2220	2222	2224	2225	2227	2228	2229	2230	2231	2232	2233	2234	2235	2257
0.7	2533	2536	2538	2540	2542	2544	2545	2547	2548	2549	2550	2551	2552	2553	2580
0.8	2826	2829	2832	2834	2837	2839	2841	2842	2844	2845	2847	2848	2849	2850	2881
0.9	3097	3100	3103	3106	3108	3111	3113	3115	3116	3118	3120	3121	3122	3124	3159
1.0	3343	3347	3351	3354	3357	3359	3361	3364	3366	3367	3369	3371	3372	3373	3413
1.1	3567	3571	3575	3578	3581	3584	3586	3589	3591	3593	3595	3597	3598	3600	3643
1.2	3767	3772	3776	3779	3782	3785	3788	3791	3793	3795	3797	3799	3801	3802	3849
1.3	3945	3950	3954	3958	3962	3965	3968	3970	3973	3975	3977	3979	3981	3982	4032
1.4	4103	4107	4112	4116	4119	4123	4126	4128	4131	4133	4136	4138	4139	4141	4192
1.5	4240	4245	4250	4254	4258	4261	4264	4267	4269	4272	4274	4276	4278	4280	4332
1.6	4360	4365	4370	4374	4377	4381	4384	4387	4389	4392	4394	4396	4398	4400	4452
1.7	4463	4468	4473	4477	4481	4484	4487	4490	4492	4495	4497	4499	4501	4503	4554
1.8	4552	4557	4561	4565	4569	4572	4575	4578	4580	4583	4585	4587	4589	4590	4641
1.9	4627	4632	4636	4640	4644	4647	4650	4652	4655	4657	4659	4661	4663	4665	4713
2.0	4691	4696	4700	4704	4707	4710	4713	4715	4718	4720	4722	4724	4725	4727	4772
2.1	4745	4750	4753	4757	4760	4763	4766	4768	4770	4772	4774	4776	4777	4779	4821
2.2	4790	4794	4798	4801	4804	4807	4809	4812	4814	4816	4817	4819	4820	4822	4861
2.3	4828	4832	4835	4838	4841	4843	4846	4848	4850	4851	4853	4854	4856	4857	4893
2.4	4859	4863	4866	4869	4871	4874	4876	4877	4879	4881	4882	4884	4885	4886	4918
2.5	4885	4888	4891	4894	4896	4898	4900	4902	4903	4905	4906	4907	4908	4909	4938
2.6	4906	4910	4912	4914	4916	4918	4920	4921	4923	4924	4925	4926	4927	4928	4953
2.7	4924	4927	4929	4931	4933	4935	4936	4937	4939	4940	4941	4942	4943	4944	
2.8	4938	4941	4943	4945	4946	4948	4949	4950	4951	4952	4953	4954	4955	4956	
2.9	4950	4952	4954	4956	4957	4958	4960								

Note: The column headed by ∞ is for use when $df > 30$. It coincides with column 1 of table A.

For $t > 2.9$ and $df > 16$, $A > 0.4950$.

All missing entries in the table have $A > 0.4950$. The reason this value of A was chosen is that for $A > 0.495$. p-value < 0.005. Therefore, if the value you want for A is missing, you may assume the significance level is 0.005 or higher.

For $t > 5.0$, selected values for t are given for $df = 1$, 2, or 3, to show when significance is reached at various significance levels.

For example, $t = 7.0$, $df = 2$ gives $A = 0.4901$, p-value $= 0.0099 (< 0.01)$. Therefore, there is significance at the .01 level. Any value of t less than 7.0 would not be significant at the .01 level.

From Leon F. Marzillier, *Elementary Statistics* ©1990 by Wm. C. Brown Publishers.

Table D   Values of *t* for Selected Confidence Levels

Level of Confidence

df	0.68	0.80	0.85	0.90	0.925	0.95	0.975	0.98	0.99	0.995
1	1.82	3.08	4.17	6.31	8.45	12.7	25.5	31.8	63.7	127
2	1.31	1.89	2.28	2.92	3.44	4.30	6.21	6.97	9.92	14.1
3	1.19	1.64	1.92	2.35	2.68	3.18	4.18	4.54	5.84	7.45
4	1.13	1.53	1.78	2.13	2.39	2.78	3.50	3.75	4.60	5.60
5	1.10	1.48	1.70	2.02	2.24	2.57	3.16	3.36	4.03	4.77
6	1.08	1.44	1.65	1.94	2.15	2.45	2.97	3.14	3.71	4.35
7	1.07	1.41	1.62	1.89	2.09	2.36	2.84	3.00	3.50	4.03
8	1.06	1.40	1.59	1.86	2.05	2.31	2.75	2.90	3.36	3.83
9	1.05	1.38	1.57	1.83	2.01	2.26	2.69	2.82	3.25	3.69
10	1.05	1.37	1.55	1.81	1.99	2.23	2.63	2.76	3.17	3.58
11	1.04	1.36	1.55	1.80	1.97	2.20	2.59	2.72	3.11	3.50
12	1.04	1.36	1.54	1.78	1.95	2.18	2.56	2.68	3.05	3.43
13	1.03	1.35	1.53	1.77	1.94	2.16	2.53	2.65	3.01	3.37
14	1.03	1.35	1.52	1.76	1.92	2.14	2.51	2.62	2.98	3.33
15	1.03	1.34	1.52	1.75	1.91	2.13	2.49	2.60	2.95	3.29
16	1.03	1.34	1.51	1.75	1.90	2.12	2.47	2.58	2.92	3.25
17	1.02	1.33	1.51	1.74	1.90	2.11	2.46	2.57	2.90	3.22
18	1.02	1.33	1.50	1.73	1.89	2.10	2.45	2.55	2.88	3.20
19	1.02	1.33	1.50	1.73	1.88	2.09	2.43	2.54	2.86	3.17
20	1.02	1.33	1.50	1.72	1.88	2.09	2.42	2.53	2.85	3.15
21	1.02	1.32	1.50	1.72	1.87	2.08	2.41	2.52	2.83	3.14
22	1.02	1.32	1.49	1.72	1.87	2.07	2.41	2.51	2.82	3.12
23	1.02	1.32	1.49	1.71	1.86	2.07	2.40	2.50	2.81	3.10
24	1.02	1.32	1.49	1.71	1.86	2.06	2.39	2.49	2.80	3.09
25	1.01	1.32	1.49	1.71	1.86	2.06	2.38	2.49	2.79	3.08
26	1.01	1.31	1.48	1.71	1.85	2.06	2.38	2.48	2.78	3.07
27	1.01	1.31	1.48	1.70	1.85	2.05	2.37	2.47	2.77	3.06
28	1.01	1.31	1.48	1.70	1.85	2.05	2.37	2.47	2.76	3.05
29	1.01	1.31	1.48	1.70	1.85	2.05	2.36	2.46	2.76	3.04
30	1.01	1.31	1.48	1.70	1.84	2.04	2.36	2.46	2.75	3.03
∞	1.00	1.28	1.44	1.64	1.78	1.96	2.24	2.33	2.58	2.81

If you want to construct a 0.95 confidence interval estimate with *df* =15, use *t*=2.13. Use the last line when *df* > 30. The values in it coincide with the normal table values.

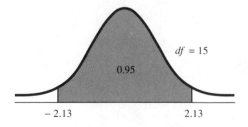

From Leon F. Marzillier, *Elementary Statistics* ©1990 by Wm. C. Brown Publishers.

Table E   Critical Values of the *F* Distribution

$\alpha = .05$

df_1

df_2	1	2	3	4	5	6	7	8	9	10	11	12
1	161	200	216	225	230	234	237	239	241	242	243	244
2	18.5	19.0	19.2	19.2	19.3	19.3	19.4	19.4	19.4	19.4	19.4	19.4
3	10.1	9.6	9.3	9.1	9.0	8.9	8.9	8.8	8.8	8.8	8.8	8.7
4	7.7	6.9	6.6	6.4	6.3	6.2	6.1	6.0	6.0	6.0	5.9	5.9
5	6.6	5.8	5.4	5.2	5.1	5.0	4.9	4.8	4.8	4.7	4.7	4.7
6	6.0	5.1	4.8	4.5	4.4	4.3	4.2	4.2	4.1	4.1	4.0	4.0
7	5.6	4.7	4.4	4.1	4.0	3.9	3.8	3.7	3.7	3.6	3.6	3.6
8	5.3	4.5	4.1	3.8	3.7	3.6	3.5	3.4	3.4	3.4	3.3	3.3
9	5.1	4.3	3.9	3.6	3.5	3.4	3.3	3.2	3.2	3.1	3.1	3.1
10	5.0	4.1	3.7	3.5	3.3	3.2	3.1	3.1	3.0	3.0	2.9	2.9
11	4.8	4.0	3.6	3.4	3.2	3.1	3.0	3.0	2.9	2.8	2.8	2.8
12	4.8	3.9	3.5	3.3	3.1	3.0	2.9	2.8	2.8	2.8	2.7	2.7
13	4.7	3.8	3.4	3.2	3.0	2.9	2.8	2.8	2.7	2.7	2.6	2.6
14	4.6	3.7	3.3	3.1	3.0	2.8	2.8	2.7	2.6	2.6	2.6	2.5
15	4.5	3.7	3.3	3.1	2.9	2.8	2.7	2.6	2.6	2.6	2.5	2.5
16	4.5	3.6	3.2	3.0	2.8	2.7	2.7	2.6	2.5	2.5	2.4	2.4
17	4.4	3.6	3.2	3.0	2.8	2.7	2.6	2.6	2.5	2.4	2.4	2.4
18	4.4	3.6	3.2	2.9	2.8	2.7	2.6	2.5	2.5	2.4	2.4	2.3
19	4.4	3.5	3.1	2.9	2.7	2.6	2.5	2.5	2.4	2.4	2.3	2.3
20	4.4	3.5	3.1	2.9	2.7	2.6	2.5	2.4	2.4	2.4	2.3	2.3
21	4.3	3.5	3.1	2.8	2.7	2.6	2.5	2.4	2.4	2.3	2.3	2.2
22	4.3	3.4	3.0	2.8	2.7	2.6	2.5	2.4	2.3	2.3	2.3	2.2
23	4.3	3.4	3.0	2.8	2.6	2.5	2.4	2.4	2.3	2.3	2.2	2.2
24	4.3	3.4	3.0	2.8	2.6	2.5	2.4	2.4	2.3	2.3	2.2	2.2
25	4.2	3.4	3.0	2.8	2.6	2.5	2.4	2.3	2.3	2.2	2.2	2.2
26	4.2	3.4	3.0	2.7	2.6	2.5	2.4	2.3	2.3	2.2	2.2	2.2
27	4.2	3.4	3.0	2.7	2.6	2.5	2.4	2.3	2.2	2.2	2.2	2.1
28	4.2	3.3	3.0	2.7	2.6	2.4	2.4	2.3	2.2	2.2	2.2	2.1
29	4.2	3.3	2.9	2.7	2.6	2.4	2.4	2.3	2.2	2.2	2.1	2.1
30	4.2	3.3	2.9	2.7	2.5	2.4	2.3	2.3	2.2	2.2	2.1	2.1
40	4.1	3.3	2.9	2.7	2.5	2.4	2.3	2.3	2.2	2.1	2.0	2.0
50	4.0	3.2	2.8	2.6	2.4	2.3	2.2	2.1	2.1	2.0	2.0	2.0
60	4.0	3.2	2.8	2.5	2.4	2.2	2.2	2.1	2.0	2.0	2.0	1.9
70	4.0	3.1	2.7	2.5	2.4	2.2	2.1	2.1	2.0	2.0	1.9	1.9
80	4.0	3.1	2.7	2.5	2.3	2.2	2.1	2.0	2.0	2.0	1.9	1.9
100	3.9	3.1	2.7	2.5	2.3	2.2	2.1	2.0	2.0	1.9	1.9	1.8
120	3.9	3.1	2.7	2.4	2.3	2.2	2.1	2.0	2.0	1.9	1.9	1.8
∞	3.8	3.0	2.6	2.4	2.2	2.1	2.0	1.9	1.9	1.8	1.8	1.8

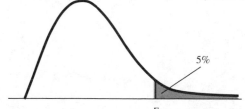

5%

F

From Leon F. Marzillier, *Elementary Statistics* ©1990 by Wm. C. Brown Publishers.

Table E   *(continued)*

α= .05

df_1

df_2	15	20	24	30	40	50	60	75	100	120	∞
1	246	248	249	250	251	252	253	253	253	253	254
2	19.4	19.4	19.4	19.4	19.4	19.5	19.5	19.5	19.5	19.5	19.5
3	8.7	8.7	8.6	8.6	8.6	8.6	8.6	8.6	8.6	8.6	8.5
4	5.9	5.8	5.8	5.8	5.7	5.7	5.7	5.7	5.7	5.7	5.6
5	4.6	4.6	4.5	4.5	4.5	4.4	4.4	4.4	4.4	4.4	4.4
6	3.9	3.9	3.8	3.8	3.8	3.8	3.7	3.7	3.7	3.7	3.7
7	3.5	3.4	3.4	3.4	3.3	3.3	3.3	3.3	3.3	3.3	3.2
8	3.2	3.2	3.1	3.1	3.0	3.0	3.0	3.0	3.0	3.0	2.9
9	3.0	2.9	2.9	2.9	2.8	2.8	2.8	2.8	2.8	2.8	2.7
10	2.8	2.8	2.7	2.7	2.7	2.6	2.6	2.6	2.6	2.6	2.5
11	2.7	2.6	2.6	2.6	2.5	2.5	2.5	2.5	2.4	2.4	2.4
12	2.6	2.5	2.5	2.5	2.4	2.4	2.4	2.4	2.4	2.3	2.3
13	2.5	2.5	2.4	2.4	2.3	2.3	2.3	2.3	2.3	2.2	2.2
14	2.5	2.4	2.4	2.3	2.3	2.2	2.2	2.2	2.2	2.2	2.1
15	2.4	2.3	2.3	2.2	2.2	2.2	2.2	2.2	2.1	2.1	2.1
16	2.4	2.3	2.2	2.2	2.2	2.1	2.1	2.1	2.1	2.1	2.0
17	2.3	2.2	2.2	2.2	2.1	2.1	2.1	2.0	2.0	2.0	2.0
18	2.3	2.2	2.2	2.1	2.1	2.0	2.0	2.0	2.0	2.0	1.9
19	2.2	2.2	2.1	2.1	2.0	2.0	2.0	2.0	1.9	1.9	1.9
20	2.2	2.1	2.1	2.0	2.0	2.0	2.0	1.9	1.9	1.9	1.8
21	2.2	2.1	2.1	2.0	2.0	1.9	1.9	1.9	1.9	1.9	1.8
22	2.2	2.1	2.1	2.0	2.0	1.9	1.9	1.9	1.8	1.8	1.8
23	2.1	2.0	2.0	2.0	1.9	1.9	1.9	1.8	1.8	1.8	1.8
24	2.1	2.0	2.0	1.9	1.9	1.9	1.8	1.8	1.8	1.8	1.7
25	2.1	2.0	2.0	1.9	1.9	1.8	1.8	1.8	1.8	1.8	1.7
26	2.1	2.0	2.0	1.9	1.8	1.8	1.8	1.8	1.8	1.8	1.7
27	2.1	2.0	1.9	1.9	1.8	1.8	1.8	1.8	1.7	1.7	1.7
28	2.0	2.0	1.9	1.9	1.8	1.8	1.8	1.8	1.7	1.7	1.6
29	2.0	1.9	1.9	1.8	1.8	1.8	1.8	1.7	1.7	1.7	1.6
30	2.0	1.9	1.9	1.8	1.8	1.8	1.7	1.7	1.7	1.7	1.6
40	1.9	1.8	1.8	1.7	1.7	1.7	1.6	1.6	1.6	1.6	1.5
50	1.9	1.8	1.7	1.7	1.6	1.6	1.6	1.6	1.5	1.5	1.4
60	1.8	1.8	1.7	1.6	1.6	1.6	1.5	1.5	1.5	1.5	1.4
70	1.8	1.7	1.7	1.6	1.6	1.5	1.5	1.5	1.4	1.4	1.4
80	1.8	1.7	1.6	1.6	1.5	1.5	1.5	1.4	1.4	1.4	1.3
100	1.8	1.7	1.6	1.6	1.5	1.5	1.4	1.4	1.4	1.4	1.3
120	1.8	1.7	1.6	1.6	1.5	1.5	1.4	1.4	1.4	1.3	1.2
∞	1.7	1.6	1.5	1.5	1.4	1.4	1.3	1.3	1.2	1.2	1.0

From Leon F. Marzillier, *Elementary Statistics* ©1990 by Wm. C. Brown Publishers.

Table E   *(continued)*

$\alpha = 0.01$

df_1

df_2	1	2	3	4	5	6	7	8	9	10	11	12
1	4052	4999	5403	5625	5764	5859	5928	5981	6022	6056	6082	6106
2	98.5	99.0	99.2	99.2	99.3	99.3	99.3	99.4	99.4	99.4	99.4	99.4
3	34.1	30.8	29.5	28.7	28.2	27.9	27.7	27.5	27.3	27.2	27.1	27.0
4	21.2	18.0	16.7	16.0	15.5	15.2	15.0	14.8	14.7	14.5	14.4	14.4
5	16.3	13.3	12.1	11.4	11.0	10.7	10.4	10.3	10.2	10.0	10.0	9.9
6	13.7	10.9	9.8	9.2	8.8	8.5	8.3	8.1	8.0	7.9	7.8	7.7
7	12.2	9.6	8.4	7.8	7.5	7.2	7.0	6.8	6.7	6.6	6.5	6.5
8	11.3	8.6	7.6	7.0	6.6	6.4	6.2	6.0	5.9	5.8	5.7	5.7
9	10.6	8.0	7.0	6.4	6.1	5.8	5.6	5.5	5.4	5.3	5.2	5.1
10	10.0	7.6	6.6	6.0	5.6	5.4	5.2	5.1	5.0	4.8	4.8	4.7
11	9.6	7.2	6.2	5.7	5.3	5.1	4.9	4.7	4.6	4.5	4.5	4.4
12	9.3	6.9	6.0	5.4	5.1	4.8	4.6	4.5	4.4	4.3	4.2	4.2
13	9.1	6.7	5.7	5.2	4.9	4.6	4.4	4.3	4.2	4.1	4.0	4.0
14	8.9	6.5	5.6	5.0	4.7	4.5	4.3	4.1	4.0	3.9	3.9	3.8
15	8.7	6.4	5.4	4.9	4.6	4.3	4.1	4.0	3.9	3.8	3.7	3.7
16	8.5	6.2	5.3	4.8	4.4	4.2	4.0	3.9	3.8	3.7	3.6	3.6
17	8.4	6.1	5.2	4.7	4.3	4.1	3.9	3.8	3.7	3.6	3.5	3.4
18	8.3	6.0	5.1	4.6	4.2	4.0	3.8	3.7	3.6	3.5	3.4	3.4
19	8.2	5.9	5.0	4.5	4.2	3.9	3.8	3.6	3.5	3.4	3.4	3.3
20	8.1	5.8	4.9	4.4	4.1	3.9	3.7	3.6	3.4	3.4	3.3	3.2
21	8.0	5.8	4.9	4.4	4.0	3.8	3.6	3.5	3.4	3.3	3.2	3.2
22	7.9	5.7	4.8	4.3	4.0	3.8	3.6	3.4	3.4	3.3	3.2	3.1
23	7.9	5.7	4.8	4.3	3.9	3.7	3.5	3.4	3.3	3.2	3.1	3.1
24	7.8	5.6	4.7	4.2	3.9	3.7	3.5	3.4	3.2	3.2	3.1	3.0
25	7.8	5.6	4.7	4.2	3.9	3.6	3.5	3.3	3.2	3.1	3.0	3.0
26	7.7	5.5	4.6	4.1	3.8	3.6	3.4	3.3	3.2	3.1	3.0	3.0
27	7.7	5.5	4.6	4.1	3.8	3.6	3.4	3.3	3.1	3.1	3.0	2.9
28	7.6	5.4	4.6	4.1	3.8	3.5	3.4	3.2	3.1	3.0	3.0	2.9
29	7.6	5.4	4.5	4.0	3.7	3.5	3.3	3.2	3.1	3.0	2.9	2.9
30	7.6	5.4	4.5	4.0	3.7	3.5	3.3	3.2	3.1	3.0	2.9	2.8
40	7.3	5.2	4.3	3.8	3.5	3.3	3.1	3.0	2.9	2.8	2.7	2.7
50	7.2	5.1	4.2	3.7	3.4	3.2	3.0	2.9	2.8	2.7	2.6	2.6
60	7.1	5.0	4.1	3.6	3.3	3.1	3.0	2.8	2.7	2.6	2.6	2.5
70	7.0	4.9	4.1	3.6	3.3	3.1	2.9	2.8	2.7	2.6	2.5	2.4
80	7.0	4.9	4.0	3.6	3.2	3.0	2.9	2.7	2.6	2.6	2.5	2.4
100	6.9	4.8	4.0	3.5	3.2	3.0	2.8	2.7	2.6	2.5	2.4	2.4
120	6.8	4.8	4.0	3.5	3.2	3.0	2.8	2.7	2.6	2.5	2.4	2.3
∞	6.6	4.6	3.8	3.3	3.0	2.8	2.6	2.5	2.4	2.3	2.2	2.2

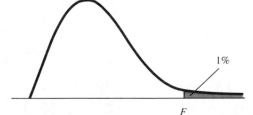

From Leon F. Marzillier, *Elementary Statistics* ©1990 by Wm. C. Brown Publishers.

Table E   *(continued)*

$$\alpha = 0.01$$
$$df_1$$

df_2	15	20	24	30	40	50	60	75	100	120	∞
1	6157	6209	6235	6261	6287	6302	6313	6323	6334	6339	6366
2	99.4	99.4	99.5	99.5	99.5	99.5	99.5	99.5	99.5	99.5	99.5
3	26.9	26.7	26.6	26.5	26.4	26.3	26.2	26.3	26.2	26.2	26.1
4	14.2	14.0	13.9	13.8	13.8	13.7	13.6	13.6	13.6	13.6	13.5
5	9.7	9.6	9.5	9.4	9.3	9.2	9.2	9.2	9.1	9.1	9.0
6	7.6	7.4	7.3	7.2	7.1	7.1	7.1	7.0	7.0	7.0	6.9
7	6.3	6.2	6.1	6.0	5.9	5.8	5.8	5.8	5.8	5.7	5.6
8	5.5	5.4	5.3	5.2	5.1	5.1	5.0	5.0	5.0	5.0	4.9
9	5.0	4.8	4.7	4.6	4.6	4.5	4.4	4.4	4.4	4.4	4.3
10	4.6	4.4	4.3	4.2	4.2	4.1	4.1	4.0	4.0	4.0	3.9
11	4.2	4.1	4.0	3.9	3.9	3.8	3.8	3.7	3.7	3.7	3.6
12	4.0	3.9	3.8	3.7	3.6	3.6	3.5	3.5	3.5	3.4	3.4
13	3.8	3.7	3.6	3.5	3.4	3.4	3.3	3.3	3.3	3.2	3.2
14	3.7	3.5	3.4	3.4	3.3	3.2	3.2	3.1	3.1	3.1	3.0
15	3.5	3.4	3.3	3.2	3.1	3.1	3.0	3.0	3.0	3.0	2.9
16	3.4	3.3	3.2	3.1	3.0	3.0	2.9	2.9	2.9	2.8	2.8
17	3.3	3.2	3.1	3.0	2.9	2.9	2.8	2.8	2.8	2.8	2.7
18	3.2	3.1	3.0	2.9	2.8	2.8	2.8	2.7	2.7	2.7	2.6
19	3.2	3.0	2.9	2.8	2.8	2.7	2.7	2.6	2.6	2.6	2.5
20	3.1	2.9	2.9	2.8	2.7	2.6	2.6	2.6	2.5	2.5	2.4
21	3.0	2.9	2.8	2.7	2.6	2.6	2.6	2.5	2.5	2.5	2.4
22	3.0	2.8	2.8	2.7	2.6	2.5	2.5	2.5	2.4	2.4	2.3
23	2.9	2.8	2.7	2.6	2.5	2.5	2.4	2.4	2.4	2.4	2.3
24	2.9	2.7	2.7	2.6	2.5	2.4	2.4	2.4	2.3	2.3	2.2
25	2.8	2.7	2.6	2.5	2.4	2.4	2.4	2.3	2.3	2.3	2.2
26	2.8	2.7	2.6	2.5	2.4	2.4	2.3	2.3	2.2	2.2	2.1
27	2.8	2.6	2.6	2.5	2.4	2.3	2.3	2.2	2.2	2.2	2.1
28	2.8	2.6	2.5	2.4	2.4	2.3	2.3	2.2	2.2	2.2	2.1
29	2.7	2.6	2.5	2.4	2.3	2.3	2.2	2.2	2.2	2.1	2.0
30	2.7	2.6	2.5	2.4	2.3	2.2	2.2	2.2	2.1	2.1	2.0
40	2.5	2.4	2.3	2.2	2.1	2.0	2.0	2.0	1.9	1.9	1.8
50	2.4	2.3	2.2	2.1	2.0	1.9	1.9	1.9	1.8	1.8	1.7
60	2.4	2.2	2.1	2.0	1.9	1.9	1.8	1.8	1.7	1.7	1.6
70	2.3	2.2	2.1	2.0	1.9	1.8	1.8	1.7	1.7	1.7	1.5
80	2.3	2.1	2.0	1.9	1.8	1.8	1.8	1.7	1.6	1.6	1.5
100	2.2	2.1	2.0	1.9	1.8	1.7	1.7	1.6	1.6	1.5	1.4
120	2.2	2.0	2.0	1.9	1.8	1.7	1.6	1.6	1.5	1.5	1.4
∞	2.0	1.9	1.8	1.7	1.6	1.5	1.5	1.4	1.4	1.3	1.0

From Leon F. Marzillier, *Elementary Statistics* ©1990 by Wm. C. Brown Publishers.

Table F   *p*-values for χ^2

χ^2	Degrees of Freedom 2	3	4
3.2	2019		
3.3	1920		
3.4	1827		
3.5	1738		
3.6	1653		
3.7	1572		
3.8	1496		
3.9	1423		
4.0	1353		
4.1	1287		
4.2	1225		
4.3	1165		
4.4	1108		
4.5	1054		
4.6	1003	2035	
4.7	0954	1951	
4.8	0907	1870	
4.9	0863	1793	
5.0	0821	1718	
5.1	0781	1646	
5.2	0743	1577	
5.3	0707	1511	
5.4	0672	1447	
5.5	0639	1386	
5.6	0608	1328	
5.7	0578	1272	
5.8	0550	1218	
5.9	0523	1166	2067
6.0	0498	1116	1991
6.1	0474	1068	1918
6.2	0450	1023	1847
6.3	0429	0979	1778
6.4	0408	0937	1712
6.5	0388	0897	1648
6.6	0369	0858	1586
6.7	0351	0821	1526
6.8	0334	0786	1468
6.9	0317	0752	1413
7.0	0302	0719	1359
7.1	0287	0688	1307

χ^2	Degrees of Freedom 2	3	4	5	6	7
7.2	0273	0658	1257	2062		
7.3	0260	0629	1209	1993		
7.4	0247	0602	1162	1926		
7.5	0235	0576	1117	1860		
7.6	0224	0550	1074	1797		
7.7	0213	0526	1032	1736		
7.8	0202	0503	0992	1676		
7.9	0193	0481	0953	1618		
8.0	0183	0460	0916	1562		
8.1	0174	0440	0880	1508		
8.2	0166	0421	0845	1456		
8.3	0158	0402	0812	1405		
8.4	0150	0384	0780	1355		
8.5	0143	0367	0749	1307	2037	
8.6	0136	0351	0719	1261	1974	
8.7	0129	0336	0691	1216	1912	
8.8	0123	0321	0663	1173	1851	
8.9	0117	0307	0636	1131	1793	
9.0	0111	0293	0611	1091	1736	
9.1	0106	0280	0586	1051	1680	
9.2	0101	0267	0563	1013	1626	
9.3	0096	0256	0540	0977	1574	
9.4	0091	0244	0518	0941	1523	
9.5	0087	0233	0497	0907	1473	
9.6	0082	0223	0477	0874	1425	
9.7	0078	0213	0456	0842	1379	
9.8	0074	0203	0439	0811	1333	2002
9.9	0071	0194	0421	0781	1289	1943
10	0067	0186	0404	0752	1247	1886

All missing entries on this page give *p*-values > 0.20.
See the next page for *p*-values corresponding to larger values of χ^2.

From Leon F. Marzillier, *Elementary Statistics* ©1990 by Wm. C. Brown Publishers.

Table F   *(continued)*

					Degrees of Freedom				
χ^2	2	3	4	5	6	7	8	9	10
11	0041	0117	0266	0514	0884	1386	2017		
12		0074	0174	0348	0620	1006	1512	2133	
13		0046	0113	0234	0430	0721	1118	1626	2237
14			0073	0156	0296	0512	0818	1223	1730
15			0047	0104	0203	0360	0591	0909	1321
16				0068	0138	0251	0424	0669	0996
17				0045	0093	0174	0301	0487	0744
18					0062	0120	0212	0352	0550
19					0042	0082	0149	0252	0403
20						0056	0103	0179	0293
21						0038	0071	0127	0211
22							0049	0089	0151
23								0062	0107
24								0043	0076
25									0053
26									0037

The missing entries in the top right-hand corner give p-value > 0.20. All other missing entries give p-value < 0.005.

From Leon F. Marzillier, *Elementary Statistics* ©1990 by Wm. C. Brown Publishers.

INDEX